MOLECULAR BIOLOGY
INTELLIGENCE
UNIT

Power Laws, Scale-Free Networks and Genome Biology

Eugene V. Koonin, Ph.D.
Yuri I. Wolf, Ph.D.
Georgy P. Karev, Ph.D., D.Sci.

National Center for Biotechnology Information
National Library of Medicine
National Institutes of Health
Bethesda, Maryland, U.S.A.

LANDES BIOSCIENCE / EUREKAH.COM
GEORGETOWN, TEXAS
U.S.A.

SPRINGER SCIENCE+BUSINESS MEDIA
NEW YORK, NEW YORK
U.S.A.

Power Laws, Scale-Free Networks and Genome Biology

Molecular Biology Intelligence Unit

Landes Bioscience / Eurekah.com
Springer Science+Business Media, Inc.

ISBN: 0-387-25883-3 Printed on acid-free paper.

Springer Science+Business Media, Inc., 233 Spring Street, New York, New York 10013, U.S.A.
http://www.springer.com

Please address all inquiries to the Publishers:
Landes Bioscience / Eurekah.com, 810 South Church Street, Georgetown, Texas 78626, U.S.A.
Phone: 512/ 863 7762; FAX: 512/ 863 0081
http://www.eurekah.com
http://www.landesbioscience.com

Printed in the United States of America.

9 8 7 6 5 4 3 2 1

Library of Congress Cataloging-in-Publication Data

Power laws, scale-free networks and genome biology / [edited by] Eugene V. Koonin, Yuri I. Wolf, Georgy P. Karev.
p. ; cm. -- (Molecular biology intelligence unit)
Includes bibliographical references and index.
ISBN 0-387-25883-3 (alk. paper)
1. Genomics. 2. Genomics--Mathematical models. 3. Computational biology. 4. Biological models. I. Koonin, Eugene V. II. Wolf, Yuri I. III. Karev, Georgy P. IV. Title. V. Series.
[DNLM: 1. Genomics. 2. Algorithms. 3. Computational Biology. 4. Models, Biological. QU 58.5 P887 2006]
QH447.P69 2006
572.8'6--dc22

2006001285

About the Editors...

EUGENE V. KOONIN is a Senior Investigator and Group Leader at the National Center for Biotechnology Information, within the National Institutes of Health in Bethesda, Maryland. His research interests include all aspects of comparative and evolutionary genomics, including, in particular, mathematical modeling of genome evolution and evolutionary systems biology. He received his PhD in Molecular Biology from the Department of Biology of Moscow State University (then USSR) and moved to Bethesda in 1991.

YURI I. WOLF is a Staff Scientist at the National Center for Biotechnology Information, within the National Institutes of Health in Bethesda, Maryland. His research is focused on quantitative aspects of evolutionary and comparative genomics. He received his MS in Molecular Biology from the Department of Biology of Moscow State University and his PhD in Genetics from the Institute of Cytology and Genetics in Novosibirsk, Russia.

GEORGY P. KAREV is a Research Scientist at the National Center for Biotechnology Information, within the National Institutes of Health in Bethesda, Maryland. His research is focused on mathematical modeling in biology. Main research interests include modeling of genome evolution, dynamics of heterogeneous populations and communities (demographic models, forest ecosystems), bifurcation approach to modeling complex biological system (epidemiological models, cancer modeling, neuron firing model), structural individual-based models, and stochastic theory of populations. He received his PhD in Mathematics from the Institute of Electronic Engineering in Moscow, Russia and Dr. Sci. degree in Biophysics from the Institute of Biophysics, Krasnoyarsk, Russia. He is a member of The Society for Mathematical Biology (SMB) and The European Society for Mathematical and Theoretical Biology (ESMTB).

CONTENTS

EDITORS

Eugene V. Koonin
Email: koonin@ncbi.nlm.nih.gov
Chapter 6

Yuri I. Wolf
Email: wolf@ncbi.nlm.nih.gov
Chapter 6

Georgy P. Karev
Email: karev@ncbi.nlm.nih.gov
Chapter 6

National Center for Biotechnology Information
National Library of Medicine
National Institutes of Health
Bethesda, Maryland, U.S.A.

CONTRIBUTORS

Eivind Almaas
Department of Physics
University of Notre Dame
Notre Dame, Indiana, U.S.A.
Email: almaas.1@nd.edu
Chapter 1

J.S. Bader
Department of Biomedical Engineering
Johns Hopkins University
Baltimore, Maryland, U.S.A.
Email: joel.bader@jhu.edu
Chapter 5

Albert-László Barabási
Department of Physics
University of Notre Dame
Notre Dame, Indiana, U.S.A.
Email: alb@nd.edu
Chapter 1

Sergey V. Buldyrev
Department of Physics
Yeshiva University
New York, New York, U.S.A.
Email: buldyrev@yu.edu
Chapter 9

T. Gregory Dewey
Keck Graduate Institute
of Applied Life Sciences
Claremont, California, U.S.A.
Email: greg_dewey@kgi.edu
Chapter 8

Nikolay V. Dokholyan
Department of Biochemistry
and Biophysics
The University of North Carolina
at Chapel Hill, School of Medicine
Chapel Hill, North Carolina, U.S.A.
Email: dokh@med.unc.edu
Chapter 7

Pau Fernández
ICREA-Complex Systems Lab
Universitat Pompeu Fabra (GRIB)
Barcelona, Spain
Email: pau.duran@gmail.com
Chapter 12

David J. Galas
Keck Graduate Institute
of Applied Life Sciences
Claremont, California, U.S.A.
Email: greg_dewey@kgi.edu
Chapter 8

Mark Gerstein
Department of Molecular Biophysics
and Biochemistry
and
Department of Computer Science
Yale University
New Haven Connecticut, U.S.A.
Email: mark.gerstein@yale.edu
Chapter 10

Kwang-Il Goh
School of Physics
Seoul National University
Seoul, Korea
Email: kwangil.goh@gmail.com
Chapter 2

Byungnam Kahng
School of Physics
Seoul National University
Seoul, Korea
Email: kahng@phya.snu.ac.kr
Chapter 2

Michael Kamal
Whitehead Institute
for Biomedical Research
Center for Genome Research
Cambridge, Massachusetts, U.S.A.
Email: kamal@broad.mit.edu
Chapter 10

Doochul Kim
School of Physics
Seoul National University
Seoul, Korea
Email: dkim@snu.ac.kr
Chapter 2

Nicholas M. Luscombe
Department of Molecular Biophysics
and Biochemistry
Yale University
New Haven Connecticut, U.S.A.
Email: nick@csb.yale.edu
Chapter 10

Sergei Maslov
Department of Physics
Brookhaven National Laboratory
Upton, New York, U.S.A.
Email: maslov@bnl.gov
Chapter 3

Jiang Qian
Department of Molecular Biophysics
and Biochemistry
Yale University
New Haven Connecticut, U.S.A.
Email: jiang.qian@jhmi.edu
Chapter 10

S. Rackovsky
Department of Pharmacology
and Biological Chemistry
and
Center for Biomathematics
Mount Sinai School of Medicine
of New York University
New York, New York, U.S.A.
Email: shelly@camelot.mssm.edu
Chapter 11

Eugene I. Shakhnovich
Department of Chemistry
and Chemical Biology
Harvard University
Cambridge, Massachusetts, U.S.A.
Email: eugene@belok.harvard.edu
Chapter 7

Kim Sneppen
Neils Bohr Institute
Blegdamsvej 17
Copenhagen, Denmark
Email: sneppen@nbi.dk
Chapter 3

Ricard V. Solé
ICREA-Complex Systems Lab
Universitat Pompeu Fabra (GRIB)
Barcelona, Spain
and
Santa Fe Institute
Santa Fe, New Mexico, U.S.A.
Email: ricard.sole@upf.edu
Chapter 12

Erik van Nimwegen
Division of Bioinformatics, Biozentrum
University of Basel
Basel, Switzerland
Email: erik.vannimwegen@unibas.ch
Chapter 14

Andreas Wagner
The Santa Fe Institute
and
Department of Biology
University of New Mexico
Albuquerque, New Mexico, U.S.A.
Email: wagnera@unm.edu
Chapter 4

Itai Yanai
Department of Molecular
and Cellular Biology
Harvard University
Cambridge, Massachusetts, U.S.A.
Email: yanai@mcb.harvard.edu
Chapter 13

PREFACE

The Genomic Revolution, Systems Biology, Power Laws, and Scale-Free Networks

The decade between 1995 and 2004 witnessed an ongoing revolution in biology. Certainly, sequencing of the human genome[1] serves as a legitimate symbol of this new revolution whereas its beginning is marked by the appearance of the first complete genome sequence of a cellular life form, the bacterium *Haemophilus influenzae*.[2] Indeed, comparative genomics is the core and foundation of the new biology. It brought about the appreciation of previously unimagined plasticity of genomes and the fundamental role of horizontal gene transfer and lineage-specific gene loss in evolution. However, equally importantly, genome comparisons corroborated and expanded the notion of fundamental conservation of the building blocks of life (genes and proteins) even as they are mixed and matched, and modified, and lost in the course of evolution.[3]

As the collection of sequenced genomes continues to expand at an ever increasing rate, gradually saturating the major branches of the tree of life (and changing this very concept in the process), the post-genomic phase of biology is taking shape. The advent of post-genomic biology has been made possible by the development of a new generation of experimental techniques which allow, at least in principle, an exhaustive analysis of various aspects of a cell or tissue, such as the complete repertoire of mRNAs, proteins or small molecules, or the complete set of protein-protein interactions or metabolite fluxes. These days, microarrays, proteomics methods, and large-scale protein-protein interaction measurements strive to study not just mRNAs, proteins or interactions but, respectively, the transcriptome, the proteome or the interactome of the given cell, tissue or whole organism. Even if the proliferation of various "omes" and "omics" irks many biologists brought up in the traditions of classical biochemistry and molecular biology,[4,5] the gist of the new biology is clear and defensible: only by knowing all components and all connections in an organism can we hope to "understand" it. The latest buzzword to denote this new direction is Systems Biology, an awkward phrase, perhaps, but rather appropriate as it captures the idea of understanding a cell or an organism as a system through a complete inventory of its parts and the interactions between them.[6,7]

This latest revolution in biology has been ushered in by new technologies, firstly, efficient whole-genome sequencing, and then, transcription microarrays, proteomics, and others. The corresponding conceptual developments have been quick to announce themselves. The key words for these concepts are **complexity**, **network**, and **power law**. Complexity with respect to biology has been defined in a variety of ways, and this is hardly the place to discuss these diverse definitions in any detail (e.g., see ref. 8). Intuitively, however, it is obvious that biological systems differ greatly in their organizational complexity which is a function of the number of distinct components and their interactions. Thus, the human proteome with ~20 thousand distinct gene products, many of which are represented by multiple alternative splice forms, is, arguably, much more complex than the proteome of the parasitic bacterium *Mycoplasma genitalium* with its 470 proteins, which is compatible with our intuition that humans are more complex than bacteria. However, we feel almost as strongly that humans are more complex than the tiny worm *Caenorhabditis elegans*, while the number of genes in the human and worm genomes is about the same.[9] Thus, beyond doubt, there are crucial aspects of biological complexity that we do not understand well at all.

A simple but potentially powerful insight into the nature of biological complexity is that, essentially, any complex system can be abstracted in the form of a network graph in which the vertices are the elements of the systems and the edges are interactions (connections) between them. The latter can be, in the most straightforward case, physical interactions between proteins, but also similarities between expression profiles of genes, relationships between regulators and the regulated genes, links between neurons or other cells, and a variety of other types of links between biological entities. These biological networks share with each other and with other types of networks, e.g., the World Wide Web, the networks of relationships (business, friendly or sexual) between members of human society, and others, certain simple but interesting mathematical properties. The distribution of the number of connections per node (node connectivity) in these networks more or less precisely follows a power law, i.e., described by the simple function $P(i) \cong ci^{-\gamma}$ where $P(i)$ is the frequency of nodes with exactly i connections or sets with exactly i members, γ is a parameter which typically assumes values between 1 and 3, and c is a normalization constant. Obviously, in double-logarithmic coordinates, the plot of P as a function of i is close to a straight line with a negative slope. An implication that has become widely known is that the networks with a power-law distribution of node connectivity are scale-free, i.e., show the same properties at different scales. This connects the study of networks with another famous and powerful concept, that of fractals.[10] Probably, the most remarkable feature of the scale-free networks is that, unlike random networks, they are resistant to error but vulnerable to attack.[11] In other words, if nodes are taken out randomly, the structure of the network will remain generally the same because most nodes have very few connections and are, in a sense, unimportant. However, if highly connected nodes, the so-called hubs, are specifically targeted, the network as a whole might not survive even the first hit.

The power law distributions transcend networks as such. The quantities that are so distributed include the number of genes in a family, the number of pseudogenes per gene, the number of people per city, the number of published papers per scientist, the number of citations per paper, and much, much more. In fact, the first distributions where power laws have been noticed are the distribution of people in a society by wealth (the Pareto law[12]) and the distributions of words in a text by frequency (Zipf law[13]).

A natural question is: why are power laws so common in so many widely different areas? Clearly, there are some general organizational principles behind these similar distributions, but are these principles just superficial or do they reflect profound commonalities between all these disparate systems? A simple but powerful insight has been offered by Barabasi and coworkers who noticed that power law distributions often appeared in **evolving** systems, be it the Internet or the biological networks.[14] One of the major modes of evolution in such systems is accretion of nodes under the so-called preferential attachment principle according to which the probability that a new node forms a connection with a preexisting one is proportional to the number of links the latter already had. In anthropomorphic terms, the rich get richer; using Darwinian terminology, which is likely to better reflect the situation, at least as far as biology is concerned, the fit get fitter.[15] Importantly, random networks described in the classic work of Erdos and Renyi[16] never show the power law distribution of node connectivity but instead have a distribution that is close to Poisson. Notably, power law distributions as asymptotic solutions are also readily produced by birth-and-death models which can be naturally applied to processes of genome evolution such as evolution of gene families.[17]

Evolution of networks via preferential attachment or a birth-and-death process leading to power laws is an important concept but is too general to be of much epistemological value in itself. The real question is: can we learn something new, preferably, something not readily discernible by other approaches, about life, through the analysis of power law distributions and scale-free networks in biological systems? Again, the first strong hint at a positive answer has been obtained by Barabasi and colleagues. They reasoned that, if scale-free networks indeed reflect biological reality, their hubs should be, in some meaningful sense, more important for the organism than the nodes with fewer connections. Indeed, the results of biological experiments on the effect of gene knockouts on the survival of yeast are compatible with this notion: the hubs of the yeast protein-protein interaction network (characterized in genome-scale two-hybrid experiments) are more likely to correspond to essential genes (those that cannot be knocked out without killing the organism) than weakly connected nodes.[18] Along the same lines, it has been observed that the hubs of the human gene coexpression network are, on average, genes that evolve slower than genes with low connectivity.[19] The connectivity of a gene (protein) in expression or interaction network seems to be a distinct property which is not readily reducible to anything else we can learn about that gene or protein. Therefore, the observations that this property correlates with empirically measurable quantities of clear biological significance, such as knockout effect or evolutionary rate, suggest that connectivity is, indeed, biologically important. Accordingly, these findings provide the rationale for deeper exploration of the biological counterparts of network organization. However, it also has been noticed that, while such correlations are often statistically significant, they are usually not overwhelmingly strong and explain but a small part of the variation of the respective quantity. Accordingly, debates abound in the literature as to which of the observed correlations are truly significant and which are secondary or even might arise from artifacts in the data.[20-27]

While the realization of the general importance of power law distributions is at least as old as the classic work of Pareto on the foundations of economics, the application to genome-wide analysis started in earnest only in the 21st century and so is still in its infancy. Even in this short time, excellent reviews on properties of scale-free and other networks and their role in biology have appeared[28-30] as well as several books, aimed either at lay readers (or, at least, "lay scientists")[15,31-33] or specialists.[34] Arguably, however, these works fall short of presenting a coherent, reasonably complete picture of the role, promise, and potential pitfalls of the analysis of power-law distributions and scale-free networks in its specific capacity as a major part of theoretical systems biology. Hence we replied with enthusiasm to the suggestion of Ron Landes to put together this book. Surely, most scientists today will agree that there are too many multi-author books around, while too few of them have any measurable impact. Understandably, every group of editors believes that their book is going to be different, and we are no exception. Our justification is twofold. From the beginning, we felt that the research field that we could define as "power laws and scale-free networks in genome biology" could gain from a multi-faceted overview in which different viewpoints, methodological approaches, and scientific cultures would be represented. What is more, we thought that compiling such an overview could be a relatively straightforward task because there were no existing comprehensive treatises to compete with. We approached with this idea a number of scientists known for pioneering contributions in this new field and were struck by their almost invariable willingness to contribute to the projected book; very few people declined, and then, for a good reason. Thus, we simply had to go ahead with the book, and here is the resulting collection.

A few words about each of the chapters, to give the reader an idea of the diversity of the contributions, and—we hope—the emerging synthesis. Almaas and Barabási discuss the occurrence of power laws in biological systems and the scale-free and hierarchical properties of biological networks. They emphasize the inhomogeneity and complexity of these networks, noting that "network biology" is still in its infancy. Goh, Kahng, and Kim describe graph-theoretic analysis of protein-protein interaction and metabolic networks and elucidate certain subtle structural properties of these networks, such as "dissortative mixing" whereby proteins with a small number of interaction partners tend to connect to the hubs of the network, and vice versa. The result is a distinctive network modularity. Maslov and Sneppen compare the large-scale organizations of two types of networks, protein-protein interaction and transcription-regulatory ones, and discover a remarkable, consistent effect of suppression of links between hubs in each of these networks, which results in distinct network modularity. Maslov and Sneppen observe that this property increases network robustness, and they suggest that, in the course of evolution, this could be a selected feature. Clearly, these results are very similar to those of Goh et al, even as the methodological approaches used by these authors are quite different. Bader investigates the protein-protein interaction network of the fruit fly and comes up with the unexpected observation that, when only reliable, biologically relevant interactions are considered, the network displays neither scale-free properties nor a power-law distribution of connectivity. When analyzed in this fashion, the number of connection per node decays faster than it would under power law and approaches an exponential distribution. This work emphasizes the caution that is due in interpreting the mathematical properties of networks.

Several chapters make the next key step of abstraction by analyzing models of evolution that lead to power law distributions. Dokholyan and Shakhnovich reveal the scale-free structure of the so-called protein domain universe graph (PDUG) and show that this organization could have evolved under a divergence but not under a convergence evolutionary model. Wagner examines the structure of protein-protein interaction networks and addresses the question whether their organization reflected in the power-law distribution of node connectivity was shaped and is maintained by natural selection. Wagner's answer is that the role of natural selection had been minor at best, with the structure of the network determined largely by physicochemical properties of proteins (but then, again, are these properties not a product of natural selection?) Karev, Wolf and Koonin describe Birth, Death and Innovation models of gene family evolution (BDIMs) and show that only nonlinear BDIMs, which include "interactions" between genes in a family, can produce evolutionary rates compatible with the observed distribution of family size. Again, as in Wagner's work, there is no explicit selection in BDIMs, but the "interactions" between genes in nonlinear models might actually represent a selective force. Yanai also addresses the interplay between neutral and selective forces in the evolution of genes families and concludes that selection does not need to be invoked to explain the general shape of the family size distribution. However, a more detailed analysis of the heavy tail of the distribution suggests that evolution of the largest families still might have evolved under positive selection. Dewey and Galas discuss expression networks derived from microarray data and transcription factor networks constructed from the data on gene regulation by specific transcription factors. An evolutionary model based on gene duplication seems to account nicely for the global properties of these networks. Kamal, Luscombe, Qian, and Gerstein present an evolutionary model that explains the observed distributions of protein fold frequencies on the basis of stochastic gene duplication, deletion, and acquisition of new fold.

Fernández and Solé consider regulatory networks at a higher level of abstraction by treating them as devices that perform computations. They note that the resistance of network to noise is achieved through redundant connections which are also the means for evolution to rewire a network without losing its function. Fernández and Solé posit that robustness of biological networks must be not maximum but optimal to ensure the necessary level of evolvability.

Van Nimwegen describes simple but truly remarkable observations on scaling of genes in different functional categories with genome size. The number of genes in each category increases as a power-law function of the total number of genes in the genome. It turns out, however, that the exponents are very different for different biological functions. In particular, regulatory and signal-transduction proteins tend to scale with the square of the total number of genes, which could be an important factor in limiting the complexity attainable by organisms. These trends await deeper explanations.

Rackovsky applies the network concepts to the analysis of the organization of the universes of proteins sequences and structures and describes substantial differences in the properties of these two virtual spaces. In Chapter 9, Buldyrev presents a review of the substantial body of work done on the power law properties of long-range correlations in DNA sequences and links these studies to more general physical models of critical phenomena. Somewhat paradoxically, although this avenue of research is at least 20 years older than systems biology and produced a number of elegant mathematical results, the biological implications are far from clear. Buldyrev makes the provocative but, we believe, plausible suggestion that long-range correlations in DNA sequences are, largely, a consequence of neutral evolution of junk DNA in complex genomes.

One way to conclude these introductory notes would be to quote the chapter by Fernández and Solé: "In summary, we are still very much puzzled by the question of how complex... networks are organized." On a more constructive note, however, we believe that the chapters collated in this book make it abundantly clear that theoretical systems biology has moved from the pure stamp collection phase (in this case, collection of examples of power law distributions and scale-free organization) to physics, i.e., search for models capable of explaining these observations. What is less clear although, probably, more important, is what new biology, if any, comes out of these analyses. Several examples outlined above may provide initial clues and more are to be found in the chapters comprising this book.

In the early days of computational molecular biology (bioinformatics), Gunnar von Heijne, one of the eminent practitioners in that field, provocatively entitled his book *Sequence Analysis in Molecular Biology: Treasure Trove or Trivial Pursuit.*[35] In the years since, the question had been answered definitively: blind alleys notwithstanding, bioinformatics is no trivial pursuit by any means. In the beginning of the 21st century, the same question lurks with regard to theoretical systems biology. It is our hope that this books suggests the same answer.

Hopefully, the book will be of interest and use to many biologists and physicists who already practice systems biology or think about venturing into this area, including graduate students. When soliciting contributions, we asked authors to follow Einstein's famous dictum and present their subjects as simply as possible but not simpler. We believe that everyone complied and, as a result, the book is not heavily mathematical although it does contain many equations; but such is the nature of the beast.

We are grateful to our publisher Ron Landes who came up with the idea of this book, to Cynthia Conomos and Celeste Carlton for expert help at all stages of the publication process and, certainly, to all the contributors for delivering their chapters without undue delays. Obviously, the coherence of the whole is the sole responsibility of the editors. The reader will judge whether or not we succeeded in reaching this objective.

Eugene V. Koonin, Ph.D.
Yuri I. Wolf, Ph.D.
Georgy P. Karev, Ph.D., D.Sci.

References

1. Lander ES, Linton LM, Birren B et al. Initial sequencing and analysis of the human genome. Nature 2001; 409:860-921.
2. Fleischmann RD, Adams MD, White O et al. Whole-genome random sequencing and assembly of Haemophilus influenzae Rd. Science 1995; 269:496-512.
3. Koonin EV, Galperin MY. Sequence - evolution - function. Computational Approaches in Comparative Genomics. New York: Kluwer Academic Publishers, 2002.
4. Petsko GA. Homologuephobia. Genome Biol 2001; 2:(COMMENT1002).
5. Koonin EV. An apology for orthologs - or brave new memes. Genome Biol 2001; 2:(COMMENT1005).
6. Ideker T, Galitski T, Hood L. A new approach to decoding life: Systems biology. Annu Rev Genomics Hum Genet 2001; 2:343-72.
7. Huang S. Back to the biology in systems biology: What can we learn from biomolecular networks? Brief Funct Genomic Proteomic 2004; 2:279-97.
8. Adami C. What is complexity? Bioessays 2002; 24:1085-94.
9. Finishing the euchromatic sequence of the human genome. Nature 2004; 431:931-45.
10. Mandelbrot BB. The fractal geometry of nature. San Francisco: W.H. Freeman,1982.
11. Albert R, Jeong H, Barabasi AL. Error and attack tolerance of complex networks. Nature 2000; 406:378-82.
12. Pareto V. Cours d'Economie politique. Paris: Rouge et Cie, 1897.
13. Zipf GK. Human Behaviour and the Principle of Least Effort. Boston: Addison-Wesley, 1949.
14. Barabasi AL, Albert R. Emergence of scaling in random networks. Science 1999; 286:509-12.
15. Barabasi AL. Linked: The New Science of Networks. New York: Perseus Press, 2002.
16. Erdos P, Renyi A. On the evolution of random graphs. Pupl Math Inst Hungar Acad Sci 1960; 7:17-61.
17. Karev GP, Wolf YI, Rzhetsky AY et al. Birth and death of protein domains: A simple model of evolution explains power law behavior. BMC Evol Biol 2002; 2:18.
18. Jeong H, Mason SP, Barabasi AL et al. Lethality and centrality in protein networks. Nature 2001; 411:41-2.
19. Jordan IK, Marino-Ramirez L, Wolf YI et al. Conservation and coevolution in the scale-free human gene coexpression network. Mol Biol Evol 2004; 21:2058-2070.
20. Fraser HB, Hirsh AE. Evolutionary rate depends on number of protein-protein interactions independently of gene expression level. BMC Evol Biol 2004; 4:13.
21. Fraser HB, Hirsh AE, Steinmetz LM et al. Evolutionary rate in the protein interaction network. Science 2002; 296:750-2.

22. Fraser HB, Wall DP, Hirsh AE. A simple dependence between protein evolution rate and the number of protein-protein interactions. BMC Evol Biol 2003; 3:11.
23. Jordan IK, Wolf YI, Koonin EV. No simple dependence between protein evolution rate and the number of protein-protein interactions: Only the most prolific interactors tend to evolve slowly. BMC Evol Biol 2003; 3:1.
24. Krylov DM, Wolf YI, Rogozin IB et al. Gene loss, protein sequence divergence, gene dispensability, expression level, and interactivity are correlated in eukaryotic evolution. Genome Res 2003; 13:2229-35.
25. Bloom JD, Adami C. Apparent dependence of protein evolutionary rate on number of interactions is linked to biases in protein-protein interactions data sets. BMC Evol Biol 2003; 3:21.
26. Bloom JD, Adami C. Evolutionary rate depends on number of protein-protein interactions independently of gene expression level: Response. BMC Evol Biol 2004; 4:14.
27. Rocha EP, Danchin A. An analysis of determinants of amino acids substitution rates in bacterial proteins. Mol Biol Evol 2004; 21:108-16.
28. Gisiger T. Scale invariance in biology: Coincidence or footprint of a universal mechanism? Biol Rev Camb Philos Soc 2001; 76:161-209.
29. Barabasi AL, Oltvai ZN. Network biology: Understanding the cell's functional organization. Nat Rev Genet 2004; 5:101-13.
30. Albert R, Barabasi AL. Statistical mechanics of complex networks. Reviews of Modern Physics 2002; 74:47-97.
31. Strogatz SH. Sync: The Emerging Science of Spontaneous Order. New York: Theia, 2003.
32. Watts DJ. Six Degrees: The Science of a Connected Age. New York: W.W. Norton & Company, 2003.
33. Buchanan, M. Nexus: Small Worlds and the Groundbreaking Science of Networks. New York; W.W. Norton & Company, 2002.
34. Mendes JF, Dorogovtsev SN. Evolution of Networks: From Biological Nets to the Internet and WWW. Oxford: Oxford University Press, 2003.
35. Von Heijne G. Sequence Analysis in Molecular Biology: Treasure Trove or Trivial Pursuit. San Diego: Academic Press, 1987.

CHAPTER 1

Power Laws in Biological Networks

Eivind Almaas and Albert-László Barabási*

Abstract

The rapidly developing theory of complex networks indicates that real networks are not random, but have a highly robust large-scale architecture, governed by strict organizational principles. Here, we focus on the properties of biological networks, discussing their scale-free and hierarchical features. We illustrate the major network characteristics using examples from the metabolic network of the bacterium *Escherichia coli*. We also discuss the principles of network utilization, acknowledging that the interactions in a real network have unequal strengths. We study the interplay between topology and reaction fluxes provided by flux-balance analysis. We find that the cellular utilization of the metabolic network is both globally and locally highly inhomogeneous, dominated by "hot-spots", representing connected high-flux pathways.

Introduction

The tremendous progress in the natural sciences we witnessed in the last century was based on the reductionist approach, allowing us to predict the behavior of a system from the understanding of its (often identical) elementary constituents and their individual interactions. However, our ability to understand simple fundamental laws governing individual "building blocks" is a far cry from being able to predict the overall behavior of a complex system.[5] Additionally, the building blocks of most complex systems, and hence the nature of their interactions, vary dramatically, rendering the traditional approaches obsolete. During the last few years, network approaches have shown great promise as a new tool to analyze and understand complex systems.[1,9,17,61] For example, technological information systems like the internet and the world-wide web are naturally modeled as networks, where the nodes are routers[23,63] or web-pages[2,10,47] and the links are physical wires or URL's respectively. The analysis of societies also lends itself naturally to a network description, with people as nodes and the connections between the nodes as friendships,[50] collaborations,[44,65] sexual contacts[48] or coauthorship of scientific papers[52,57a] to name a few possibilities. It seems that the closer we look at the world surrounding us, the more we realize that we are hopelessly entangled in myriads of interacting webs, and to describe them we need to understand the architecture of the various networks nature and technology offers us.

In biology, networks appear in many disparate systems, ranging from food webs in ecology to biochemical interactions in molecular biology. In particular in the cell the variety of interactions between genes, proteins and metabolites are well captured by networks. During

*Corresponding author: Albert-László Barabási—Department of Physics, University of Notre Dame, Notre Dame, Indiana 46556, U.S.A. Email: alb@nd.edu

Power Laws, Scale-Free Networks and Genome Biology, edited by Eugene V. Koonin, Yuri I. Wolf and Georgy P. Karev.

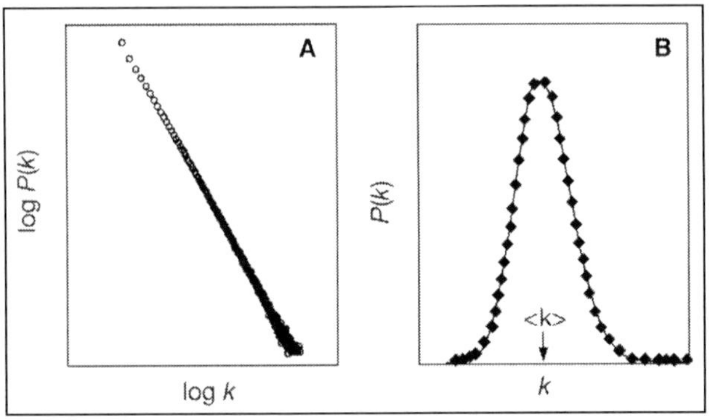

Figure 1. Characterizing degree distributions. For the power-law degree distribution (A), there exists no typical node, while for single peaked distributions (B), most nodes are well represented by the average (typical) node with degree $\langle k \rangle$.

the last decade, genomics has unleashed a downright flood of molecular interaction data. The nascent field of transcriptomics and proteomics have followed suit with analysis of protein levels under various conditions and genome wide analysis of gene expression at the mRNA level.[11,12,53] Thus, protein-protein interaction maps have been generated for a variety of organisms including viruses,[25] prokaryotes like *H. pylori*[54] and eukaryotes like *S. cerevisiae*[27,35,39,40,42,58,62] and *C. elegans*.[64] In this chapter we will discuss recent results and developments in the study and characterization of naturally occurring networks, with focus on cellular ones.

Power Laws in Network Topology

The complex network representation of different systems as networks has revealed surprising similarities, many of which are intimately tied to power laws. The simplest network measure is the average number of nearest neighbors of a node, or the average degree. However, this is a rather crude property, and to gain further insight into the topological organization of real networks, we need to determine the variation in the nearest neighbors, given by the degree distribution. For a surprisingly large number of networks, this degree distribution is best characterized by the power law functional form[6] (Fig. 1A),

$$P(k) \sim k^{-\alpha} \tag{1}$$

Important examples include the metabolic network of 43 organisms,[43] the protein interaction network of *S. cerevisiae*[42] and various food webs.[51] If the degree distribution instead was single-peaked (e.g., Poisson or Gaussian) as in Figure 1B, the majority of the nodes would be well described by the average degree, and hence the notion of a "typical" node. In contrast for networks with a power-law degree distribution, the majority of the nodes have only one or two neighbors while coexisting with many nodes with hundreds and some even with thousands of neighbors. For these networks there exists no typical node, and they are therefore often referred to as "scale-free".

The clustering of a node, the degree to which the neighborhood of a node resembles a complete subgraph, is another measure which sheds light on the structural organization of a network.[66] For a node i with degree k_i the clustering is defined as,

$$C_i = \frac{2n_i}{k_i(k_i - 1)} \tag{2}$$

representing the ratio of the number of actual connections between the neighbors of node i to the number of possible connections. For a node which is part of a fully interlinked cluster

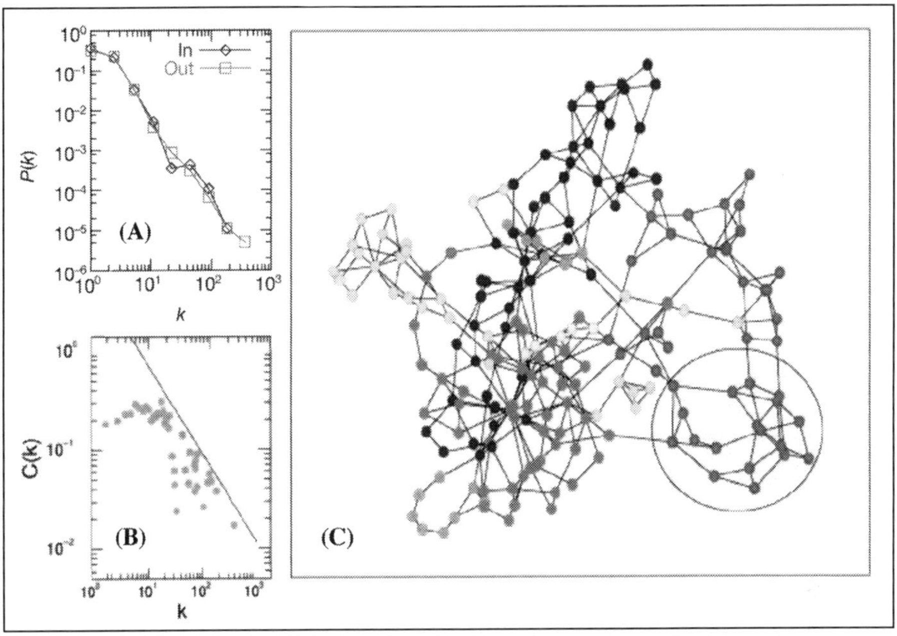

Figure 2. Properties of the metabolic network of *Escherichia coli*. A) The degree distribution displays a power law in both the in- and the out degrees.[43] B) The clustering coefficient varies with k as a power law. The solid line corresponds to k^{-1}. C) Three dimensional representation of the reduced metabolic network.[56]

$C_i = 1$, while $C_i = 0$ for a node which acts as a bridge between different clusters. Accordingly, the overall clustering coefficient of a network with N nodes is given by $\langle C \rangle = \Sigma\, C_i/N$, and represents a measure of a network's potential modularity. By studying the clustering of nodes with a given degree k, information about the actual modular organization of a network can be gleaned:[16,56,57,63] For all metabolic networks available, this behaves like the power law,

$$C(k) \sim k^{-\delta} \tag{3}$$

suggesting the existence of a hierarchy of nodes with different degrees of modularity (as measured by the clustering coefficient) overlapping in an iterative manner.[56] In Figure 2, we show the degree distribution (Fig. 2A) and the clustering as function of k (Fig. 2B) for the bacterium *Escherichia coli*. They both clearly adhere to a power-law behavior, suggesting that biological networks are both scale-free and hierarchical. Panel 2C is a three dimensional representation of a cleaned up version of the metabolic network,[56] demonstrating that modules are not clearly separated. Furthermore, the likelihood that a node appears in the shortest paths between other nodes on the network, the so-called betweenness-centrality g,[26,29] is also characterized by a power law distribution following $P(g) \sim g^{-\beta}$ for both biological and nonbiological networks,[31] suggesting that a few nodes act as bridges or linkers between the different parts of the network. In summary, we have seen strong evidence that biological networks are both scale-free[42,43] and hierarchical.[56]

Network Models

An important question now arises—we can characterize networks using the above mentioned quantities, but why is the power law behavior so pervasive? Several models building on very different principles are able to explain these observed features.

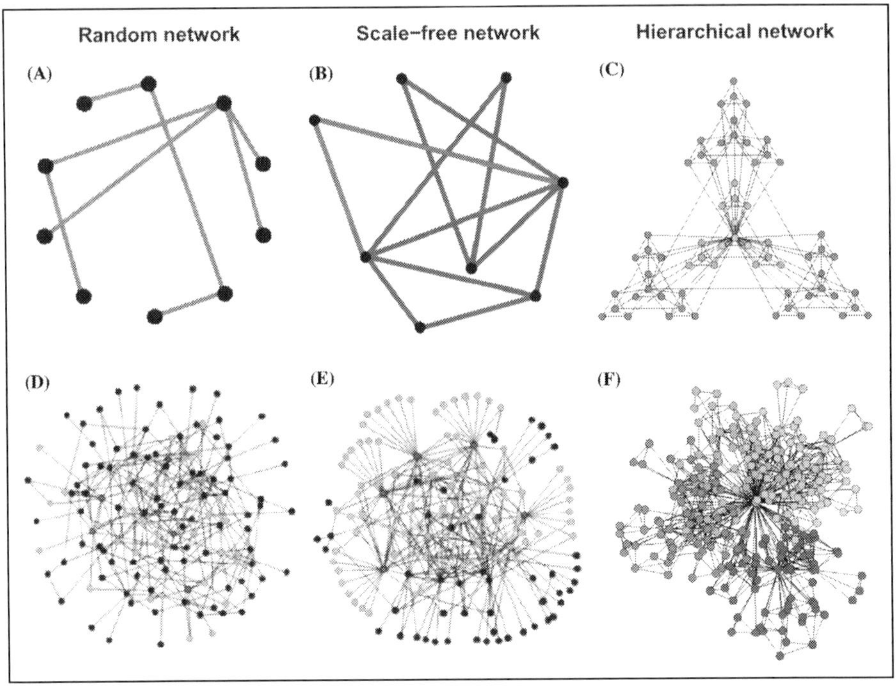

Figure 3. Graphical representation of three network models: A,D) The ER (random) model, B,E) the BA (scale-free) model and (C) and (F) the hierarchical model. The random network model is constructed by starting from N nodes before the possible node-pairs are connected with probability p. Panel (A) shows a particular realization of the ER model with 10 nodes and connection probability $p = 0.2$. In panel (B) we show the scale-free model at time t (green links) and at time $(t + 1)$ when we have added a new node (red links) using the preferential attachment probability (see Eq. (4)). Panel (C) demonstrates the iterative construction of a hierarchical network, starting from a fully connected cluster of four nodes (blue). This cluster is then copied three times (green) while connecting the peripheral nodes of the replicas to the central node of the starting cluster. By once more repeating this replication and connection process (red nodes), we end up with a 64-node scale-free hierarchical network. In panel (D) we display a larger version of the random network, and it is evident that most nodes have approximately the same number of links. For the scale-free model, (E) the network is clearly inhomogeneous: while the majority of nodes has one or two links, a few nodes have a large number of links. We emphasize this by coloring the five nodes with the highest number of links red and their first neighbors green. While in the random network only 27% of the nodes are reached by the five most connected nodes, we reach more than 60% of the nodes in the scale-free network, demonstrating the key role played by the hubs. Note that the networks in (D) and (E) consist of the same number of nodes and links. Panel (F) demonstrates that the standard clustering algorithms are not that successful in uncovering the modular structure of a scale-free hierarchical network. A color version of this figure is available online at http://www.Eurekah.com.

Random Network Models

While graph theory initially focused on regular graphs, since the 1950s large networks with no apparent design principles were described as random graphs,[8] proposed as the simplest and most straightforward realization of a complex network. According to the Erdos-Renyi (ER) model of random networks,[22] we start with N nodes and connect every pair of nodes with probability p, creating a graph with approximately $pN(N-1)/2$ randomly distributed edges (Fig. 3A,D). For this model the degrees follow a Poisson distribution (Fig. 4A), and as a consequence, the average degree $\langle k \rangle$ of the network describes the typical node. Furthermore, for this "democratic" network

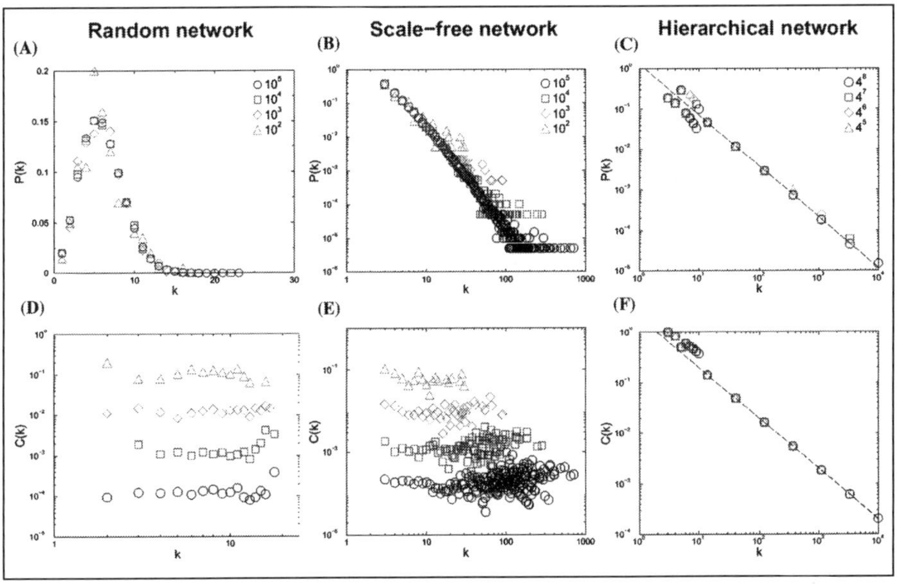

Figure 4. Properties of the three network models. A) The ER model sports a Poisson degree distribution $P(k)$ (the probability that a randomly selected node has exactly k links) which is strongly peaked at the average degree $\langle k \rangle$ and decays exponentially for large k. The degree distributions for the scale-free (B) and the hierarchical (C) network models do not have a peak, they instead decay according to the power-law $P(k) \sim k^{-\gamma}$. The average clustering coefficient for nodes with exactly k neighbors, $C(k)$, is independent of k for both the ER (D) and the scale-free (E) network model. F) In contrast, $C(k) \sim k^{-1}$ for the hierarchical network model (cf. Fig. 2).

model, the clustering is independent of the node degree k (Fig. 4D). As we have just seen in Figure 2, the ER model does not capture the properties of biological networks.

Scale-Free Network Model

In the network model of Barabási and Albert (BA), two crucial mechanisms, which both are absent from the classical random network model, are responsible for the emergence of a power-law degree distribution.[6] First, networks grow through the addition of new nodes linking to nodes already present in the system. Second, there is a higher probability to link to a node with a large number of connections in most real networks, a property called preferential attachment. These two principles are implemented as follows: starting from a small core graph consisting of m_0 nodes, a new node with m links is added at each time step and connected to the already existing nodes (Fig. 3B,E). Each of the m new links are then preferentially attached to a node i (with k_i neighbors) which is chosen according to the probability

$$\prod_i = k_i / \sum_j k_j \tag{4}$$

The simultaneous combination of these two network growth rules gives rise to the observed power-law degree distribution (Fig. 4B). In panel 3B, we illustrate the growth process of the scale-free model by displaying a network at time t (green links) and then at time $(t + 1)$, when we have added a new node (red links) using the preferential attachment probability. Compared to random networks, the probability that a node is highly connected is statistically significant in scale-free networks. Consequently, many network properties are determined by a relatively small number of highly connected nodes, often called "hubs". To make the effect of the hubs on the network structure visible, we have colored the five nodes with largest degrees red in Figure 3D,E and their nearest neighbors green. While in the ER network only 27% of the nodes are reached by the five most connected ones, we reach more than 60% of the nodes in the scale-free network, demonstrating the key role played by the hubs. Another consequence of the hub's dominance of the network topology is that scale-free networks are highly tolerant of random failures (perturbations) while being extremely sensitive to targeted attacks.[3] Comparing the properties of the BA network model with those of the ER model, we note that the clustering of the BA network is larger, however $C(k)$is approximately constant (Fig. 4E), indicating the absence of a hierarchical structure.

Hierarchical Network Model

Many real networks are expected to be fundamentally modular, meaning that the network can be seamlessly partitioned into a collection of modules where each module performs an identifiable task, separable from the function(s) of other modules.[33,34,37,46,55,60] Therefore, we must reconcile the scale-free property with potential modularity. In order to account for the modularity as reflected in the power-law behavior of Figure 2B and a simultaneous scale-free degree distribution Figure 2A, we have to assume that clusters combine in an iterative manner, generating a hierarchical network.[56,63] Such a network emerges from a repeated duplication and integration process of clustered nodes,[56] which in principle can be repeated indefinitely. This process is depicted in panel 3c, where we start from a small cluster of four densely linked nodes (blue). We next generate three replicas of this hypothetical initial module (green) and connect the three external nodes of the replicated clusters to the central node of the old cluster, thus obtaining a large 16-node module. Subsequently, we again generate three replicas of this 16-node module (red), and connect the 16 peripheral nodes to the central node of the old module, obtaining a new module of 64 nodes. This hierarchical network model seamlessly integrates a scale-free topology with an inherent modular structure by generating a network that has a power law degree distribution (Fig. 4C) with degree exponent $\gamma = 1 + \ln 4/\ln 3 \approx 2.26$ and a clustering coefficient $C(k)$ which proves to be dependent on k^{-1} (Fig. 4F). However, note that modularity does not imply clear-cut sub-networks linked in well-defined ways.[36,56] In fact, the boundaries of modules are often blurred (see Fig. 3F), bridged by highly connected nodes which interconnect modules.

Power Laws in Network Utilization

Despite their successes, purely topologic approaches have important intrinsic limitations. For example, the activity of the various metabolic reactions or regulatory interactions differs widely, some being highly active under most growth conditions while others are switched on only for some rare environmental circumstances. Therefore, an ultimate description of cellular networks requires us to consider the intensity (i.e., strength), the direction (when applicable) and the temporal aspects of the interactions. While so far we know little about the temporal aspects of the various cellular interactions, recent results have shed light on how the strength of the interactions is organized in metabolic and genetic-regulatory networks.[4]

In metabolic networks the flux of a given metabolic reaction, representing the amount of substrate being converted to a product within unit time, offers the best measure of interaction strength. Recent metabolic flux-balance approaches (FBA)[18-20,38,59] that allow us to calculate the flux for each reaction, have significantly improved our ability to generate quantitative predictions on the relative importance of the various reactions, leading to experimentally testable hypotheses. Starting from a stoichiometric matrix of the K12 MG1655 strain of *E. coli*, containing 537 metabolites and 739 reactions,[18-20,38] the steady state concentrations of all metabolites satisfy,

$$\frac{d}{dt}[A_i] = \sum_j S_{ij} v_j = 0 \tag{5}$$

where S_{ij} is the stoichiometric coefficient of metabolite A_i in reaction j and v_j is the flux of reaction j. We use the convention that if metabolite A_i is a substrate (product) in reaction j, $S_{ij} < 0$ ($S_{ij} > 0$) and we constrain all fluxes to be positive by dividing each reversible reaction into two "forward" reactions with positive fluxes. Any vector of positive fluxes $\{v_j\}$ which satisfies Eq. (5) corresponds to a state of the metabolic network, and hence, a potential state of operation of the cell.

Assuming that cellular metabolism is in a steady state and optimized for the maximal growth rate,[18,38] FBA allows us to calculate the flux for each reaction using linear optimization, providing a measure of each reaction's relative activity.[4] A striking feature of the flux distribution of *E. coli* is its overall inhomogeneity: reactions with fluxes spanning several orders of magnitude coexist under the same conditions (Fig. 5A). This is captured by the flux distribution for *E. coli,* which follows (the by now familiar) power law where the probability that a reaction has flux v is given by $P(v) \sim (v + v_0)^{-\alpha}$. The flux exponent is predicted to be $\alpha = 1.5$ by FBA methods.[4] In a recent experiment[21] the strength of the various fluxes of the central metabolism was measured, revealing[4] the power-law flux dependence $P(v) \sim v^{-\alpha}$ with $\alpha \cong 1$ (Fig. 5B). This power law behavior indicates that the vast majority of reactions have quite small fluxes, while coexisting with a few reactions with extremely large flux values.

The observed flux distribution is compatible with two quite different potential **local** flux structures.[4] A homogeneous local organization would imply that all reactions producing (consuming) a given metabolite have comparable fluxes. On the other hand, a more delocalized "hot backbone" is expected if the local flux organization is heterogeneous, such that each metabolite has a dominant source (consuming) reaction. To distinguish between these two scenarios for each metabolite i produced (consumed) by k reactions, we define the measure[7,14]

$$Y(k,i) = \sum_{j=1}^{k} \left(\frac{\hat{v}_{ij}}{\sum_{l=1}^{k} \hat{v}_{il}} \right)^2 \tag{6}$$

where $\hat{v}_{ij}$ is the mass carried by reaction j which produces (consumes) metabolite i. If all reactions producing (consuming) metabolite i have comparable $\hat{v}_{ij}$ values, $Y(k,i)$ scales as $1/k$. If, however, a single reaction's activity dominates Eq. (6), we expect $Y(k,i) \sim 1$, i.e. $Y(k,i)$, is independent of k. For the *E. coli* metabolism optimized for succinate and glutamate uptake (Fig. 5) we find that both the in and out degrees follow the power law $Y(k,i) \sim k^{-0.27}$, representing an intermediate behavior between the two extreme cases.[4] This indicates that the large-scale inhomogeneity observed in the overall flux distribution is increasingly valid at the level of the individual metabolites as well: the more reactions consume (produce) a given metabolite, the more likely it is that a single reaction carries the majority of the flux. This implies that the majority of the metabolic flux is carried along linear pathways—the metabolic high flux backbone (HFB).[4]

A power law pattern is also observed when one investigates the strength of the various genetic regulatory interactions provided by microarray datasets. Assigning each pair of genes a correlation coefficient which captures the degree to which they are coexpressed, one finds that the distribution of these pair-wise correlation coefficients follows a power law.[24,45] That is, while the majority of gene pairs have only weak correlations, a few gene pairs display a significant correlation coefficient. These highly correlated pairs likely correspond to direct regulatory and protein interactions. This hypothesis is supported by the finding that the correlations are

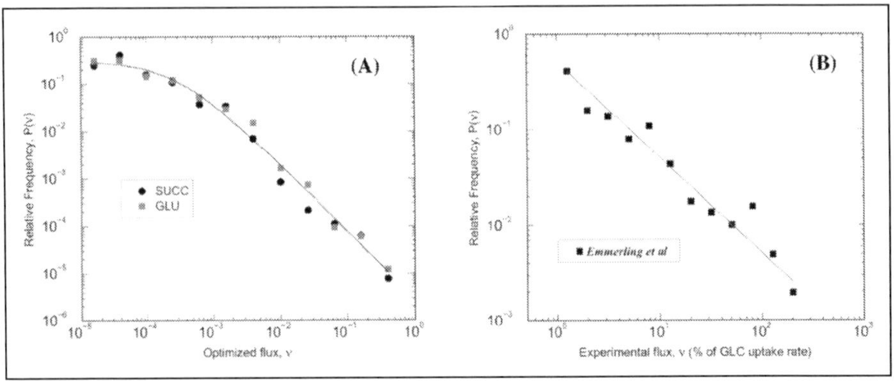

Figure 5. Flux distribution for the metabolism of *E. coli.* A) Flux distribution for optimized biomass production on succinate (black) and glutamate (red) rich uptake substrates. The solid line corresponds to the power law fit $P(v) \sim (v + v_0)^{-\alpha}$ with $v_0 = 0.00003$ and $\alpha = 1.5$. B) The distribution of experimentally determined fluxes (see ref 21) from the central metabolism of *E. coli* also displays power-law behavior with a best fit to $P(v) \sim v^{-\alpha}$ with $\alpha = 1$. A color version of this figure is available online at http://www.Eurekah.com.

larger along the links of the protein interaction network and between proteins occurring in the same complex than for pairs of proteins that are not known to interact directly.[15,28,32,41]

Taken together, these results indicate that the biochemical activity in both the metabolic and genetic networks is dominated by several 'hot links' that represent a few high activity interactions embedded into a web of less active interactions. This attribute does not seem to be a unique feature of biological systems: hot links appear in a wide range of nonbiological networks where the activity of the links follows a wide distribution.[13,30] The origin of this seemingly universal property is, again, likely rooted in the network topology. Indeed, it seems that the metabolic fluxes and the weights of the links in some nonbiological system[13,30] are uniquely determined by the scale-free nature of the network. A more general principle that could explain the correlation distribution data as well is currently lacking.

Conclusions

Power laws are abundant in nature, affecting both the construction and the utilization of real networks. The power-law degree distribution has become the trademark of scale-free networks and can be explained by invoking the principles of network growth and preferential attachment. However, many biological networks are inherently modular, a fact which at first seems to be at odds with the properties of scale-free networks. However, these two concepts can coexist in hierarchical scale-free networks. In the utilization of complex networks, most links represent disparate connection strengths or transportation thresholds. For the metabolic network of *E. coli* we can implement a flux-balance approach and calculate the distribution of link weights (fluxes), which (reflecting the scale-free network topology) displays a robust power-law, independent of exocellular perturbations. Furthermore, this global inhomogeneity in the link strengths is also present at the local level, resulting in a connected "hot-spot" backbone of the metabolism. Similar features are also observed in the strength of various genetic regulatory interactions. Despite the significant advances witnessed the last few years, network biology is still in its infancy, with future advances most notably expected from the development of theoretical tools, development of new interactive databases and increased insights into the interplay between biological function and topology.

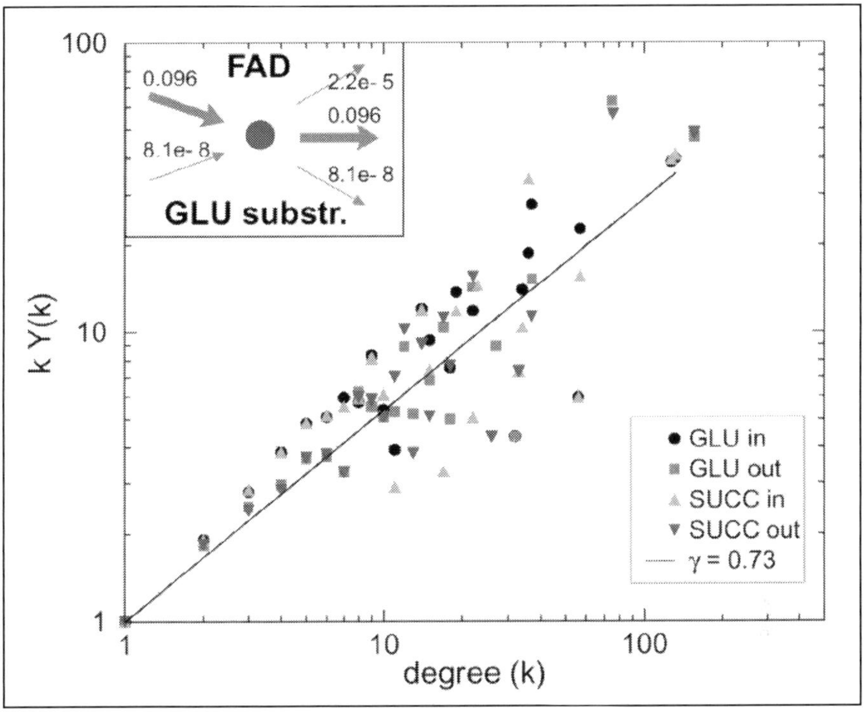

Figure 6. Characterizing the local inhomogeneity of the metabolic flux distribution. The measured $kY(k)$ (see Eq. (6)) shown as function of k for incoming and outgoing reactions for fluxes calculated on both succinate and glutamate rich substrates, averaged over all metabolites, indicating, $Y(k) \sim k^{-0.27}$ as the straight line in the figure has slope $\gamma = 0.73$. Inset: The nonzero mass flows $\hat{v}_{ij}$ producing (consuming) flavin adenine dinucleotide (FAD) on a glutamate rich substrate.

References

1. Albert R, Barabási A-L. Statistical mechanics of complex networks. Rev Mod Phys 2002; 74:47-97.
2. Albert R, Jeong H, Barabási A-L. Diameter of the world-wide web. Nature 1999; 401:130-1.
3. Albert R, Jeong H, Barabási A-L. Attack and error tolerance of complex networks. Nature 2000; 406:378-82.
4. Almaas E, Kovacs B, Vicsek T et al. Global organization of metabolic fluxes in the bacterium Escherichia coli. Nature 2004; 427:839-843.
5. Anderson PW. More is different. Science 1972; 177:393-6.
6. Barabási A-L, Albert R. Emergence of scaling in random networks. Science 1999; 286:509-12.
7. Barthelemy M, Gondran B, Guichard E. Spatial structure of the Internet traffic. Physica A 2003; 319:633-42.
8. Bollobas B. Random Graphs. London: Academic Press, 1985.
9. Bornholdt S, Schuster HG. Handbook of graphs and networks: From the genome to the Internet. Berlin, Germany: Wiley-VCH, 2003.
10. Broder A, Kumar R, Maghoul F et al. Graph structure in the web. Comput Netw 2000; 33:309-20.
11. Burge C. Chipping away at the transcriptome. Nature Genet 2001; 27:232-4.
12. Caron H, van Schaik B, van der Mee M et al. The human transcriptome map: Clustering of highly expressed genes in chromosomal domains. Science 2001; 291:1289-92.
13. deMenezes MA, Barabási A-L. Fluctuations in network dynamics. Phys Rev Lett 2004; 92:article no. 028701.

14. Derrida B, Flyvbjerg H. Statistical properties of randomly broken objects and of multivalley structures in disordered-systems. J Phys A: Math Gen 1987; 20:5273-88.
15. Dezso Z, Oltvai ZN, Barabási A-L. Bioinformatics analysis of experimentally determined protein complexes in the yeast, Saccharomyces cerevisiae. Genome Res 2003; 13:2450-4.
16. Dorogovtsev SN, Goltsev AV, Mendes JFF. Pseudofractal scale-free web. Phys Rev E 2002; 65:066122.
17. Dorogovtsev SN, Mendes JFF. Evolution of networks: From biological nets to the Internet and WWW. Oxford: Oxford University Press, 2003.
18. Edwards JS, Ibarra RU, Palsson BO. In silico predictions of Escherichia coli metabolic capabilities are consistent with experimental data. Nat Biotechnol 2001; 19:125-30.
19. Edwards JS, Palsson BO. The Escherichia coli MG1655 in silico metabolic genotype: Its definition, characteristics, and capabilities. Proc Natl Acad Sci USA 2000; 97:5528-33.
20. Edwards JS, Ramakrishna R, Palsson BO. Characterizing the metabolic phenotype: A phenotype phase plane analysis. Biotechn Bioeng 2002; 77:27-36.
21. Emmerling M, Dauner M, Ponti A et al. Metabolic flux responses to pyruvate kinase knockout in Escherichia coli. J Bacteriol 2002; 184:152-64.
22. Erdos P, Renyi A. On the evolution of random graphs. Publ Math Inst Hung Acad Sci 1960; 5:17-61.
23. Faloutsos M, Faloutsos P, Faloutsos C. On power-law relationships of the Internet topology. Comput Commun Rev 1999; 29:251-62.
24. Farkas IJ, Jeong H, Vicsek T et al. The topology of the transcription regulatory network in the yeast, Saccharomyces cerevisiae. Physica A 2003; 318:601-12.
25. Flajolet M, Rotondo G, Daviet L et al. A genomic approach to the hepatitis C virus. Gene 2000; 242:369-79.
26. Freeman L. A set of measures of centrality based upon betweenness. Sociometry 1977; 40:35-41.
27. Gavin AC, Bosche M, Krause R et al. Functional organization of the yeast proteome by systematic analysis of protein complexes. Nature 2002; 415:141-7.
28. Ge H, Liu Z, Church GM et al. Correlation between transcriptome and interactome mapping data from Saccharomyces cerevisiae. Nature Genet 2001; 29:482-6.
29. Girvan M, Newman MEJ. Community structure in social and biological networks. Proc Natl Acad Sci 2002; 99:7821-26.
30. Goh K-I, Kahng B, Kim D. Fluctuation-driven dynamics of the internet topology. Phys Rev Lett 2002a; 88:108701.
31. Goh K-I, Oh E, Jeong H et al. Classification of scale-free networks. Proc Natl Acad Sci 2002b; 99:12583-88.
32. Grogoriev A. A relationship between gene expression and protein interactions on the proteome scale: Analysis of the bacteriophage T7 and yeast Saccharomyces cerevisiae. Nucleic Acids Res 2001; 29:3513-9.
33. Hartwell LH, Hopfield JJ, Leibler S et al. From molecular to modular cell biology. Nature 1999; 402:C47-52.
34. Hasty J, McMillen D, Isaacs F et al. Computational studies of gene regulatory networks: In numero molecular biology. Nature Rev Genet 2001; 2:268-79.
35. Ho Y, Gruhler A, Heilbut A et al. Systematic identification of protein complexes in Saccharomyces cerevisiae by mass spectrometry. Nature 2002; 415:180-3.
36. Holme P, Huss M, Jeong H. Subnetwork hierarchies of biochemical pathways. Bioinformatics 2003; 19:532-9.
37. Holter NS, Maritan A, Cieplak M et al. Dynamic modeling of gene expression data. Proc Natl Acad Sci 2001; 98:1693-8.
38. Ibarra RU, Edwards JS, Palsson BO. Escherichia coli K-12 undergoes adaptive evolution to achieve in silico predicted optimal growth. Nature 2002; 420:186-9.
39. Ito T, Chiba T, Ozawa R et al. A comprehensive two-hybrid analysis to explore the yeast protein interactome. Proc Natl Acad Sci 2001; 98:4569-74.

40. Ito T, Tashiro K, Muta S et al. Towards a protein-protein interaction map of the budding yeast: A comprehensive system to examine two-hybrid interactions in all possible combinations between the yeast proteins. Proc Natl Acad Sci 2000; 97:1143-47.
41. Jansen R, Greenbaum D, Gerstein M. Relating whole-genome expression data with protein-protein interactions. Genome Res 2002; 12:37-46.
42. Jeong H, Mason SP, Barabási A-L et al. Lethality and centrality in protein networks. Nature 2001; 411:41-2.
43. Jeong H, Tombor B, Albert R et al. The large-scale organization of metabolic networks. Nature 2000; 407:651-4.
44. Kochen M, ed. The small-world. Norwood: Ablex, 1989.
45. Kutznetsov VA, Knott GD, Bonner RF. General statistics of stochastic processes of gene expression in eukaryotic cells. Genetics 2002; 161:1321-32.
46. Lauffenburger D. Cell signaling pathways as control modules: Complexity for simplicity. Proc Natl Acad Sci 2000; 97:5031-33.
47. Lawrence S, Giles CL. Accessibility of information on the web. Nature 1999; 400:107-9.
48. Liljeros F, Edling CR, Amaral LAN et al. The web of human sexual contacts. Nature 2001; 411:907-8.
49. McGraith S, Holtzman T, Moss B et al. Genome-wide analysis of vaccinia virus protein-protein interactions. Proc Natl Acad Sci 2000; 97:4879-84.
50. Milgram S. The small-world problem. Psychology Today 1967; 2:60-7.
51. Montoya JM, Sole RV. Small-world patterns in food webs. J Theor Biol 2002; 214:405-12.
52. Newman MEJ. The structure of scientific collaboration networks. Proc Natl Acad Sci 2001; 98:404-9.
53. Pandey A, Mann M. Proteomics to study genes and genomes. Nature 2000; 405:837-46.
54. Rain J-C, Selig L, DeReuse H et al. The protein-protein interaction map of Helicobacter pylori. Nature 2001; 409:211-15.
55. Rao CV, Arkin AP. Control motifs for intracellular regulatory networks. Annu Rev Biomed Eng 2001; 3:391.
56. Ravasz E, Somera AL, Mongru DA et al. Hierarchical organization of modularity in metabolic networks. Science 2002; 297:1551-5.
57. Ravasz E, Barabási A-L. Hierarchical organization in complex networks. Phys Rev E 2003; 67:026112.
57a. Redner S. How popular is your paper? An empirical study of the citation distribution. European Physical Journal B 1998; 4:131-134.
58. Schwikowski B, Uetz P, Fields S. A network of protein-protein interactions in yeast. Nature Biotechn 2000; 18:1257-61.
59. Segre D, Vitkup D, Church GM. Analysis of optimality in natural and perturbed metabolic networks. Proc Natl Acad Sci 2002; 99:15112-7.
60. Shen-Orr SS, Milo R, Mangan S et al. Network motifs in the transcriptional regulation network of Escherichia coli. Nature Genet 2001; 31:64-8.
61. Strogatz SH. Exploring complex networks. Nature 2001; 410:268-76.
62. Uetz P, Giot L, Cagney G et al. A comprehensive analysis of protein-protein interactions in Saccharomyces cerevisiae. Nature 2000; 403:623-27.
63. Vázquez A, Pastor-Satorras R, Vespignani A. Large-scale topological and dynamical properties of the Internet. Phys Rev E 2002; 65:066130.
64. Walhout A, Sordella R, Lu X et al. Protein interaction mapping in C. elegans using proteins involved in vulva development. Science 2000; 287:116-22.
65. Wasserman S, Faust K. Social Network Analysis: Methods and Applications. Cambridge: Cambridge University Press, 1994.
66. Watts DJ, Strogatz SH. Collective dynamics of small-world networks. Nature 1998; 393:440-2.

CHAPTER 2

Graphical Analysis of Biocomplex Networks and Transport Phenomena

Kwang-Il Goh, Byungnam Kahng* and Doochul Kim

Many biocomplex networks such as the protein interaction networks and the metabolic networks exhibit an emerging pattern that the distribution of the number of connections of a protein or substrate follows a power law. As the network theory is developed recently, several quantities describing network structure such as modularity and degree-degree correlation have been introduced. Here we investigate and compare the structural properties of the yeast protein networks for different datasets with those quantities. Moreover, we introduce a new quantity, called the load, characterizing the amount of signal passing through a vertex. It is shown that the load distribution also follows a power law, and its characteristics are related to the structure of the core part of the biocomplex networks.

Introduction

Recently biocomplex systems have drawn considerable attentions since their emergent behaviors, arising from diverse interactions and adaptations, are more than the sum of individual components.[1,2] Such complex systems may be described in terms of graphs, consisting of vertices and edges, where vertices and edges represent substrate or proteins, and their mutual reactions or interactions in metabolic networks or protein interaction networks, respectively.[3-6] In the last century, biologists mainly focused their interests on the identification of individual molecules and their functions in relation to macroscopic biological phenomena. However, it is recently believed that thousands of genes and their products such as proteins, RNA and small molecules, function in a complete and concerted way.[7] Thus it is natural to invoke the graph theory which helps us to visualize how molecules in a given organism function together in concerted ways.

The cellular components such as genes, proteins, and other molecules, connected by all physiologically relevant interactions, form a full weblike molecular architecture in a cell.[8] In such an architecture, genes are known to play a structural role, determining the scope and passing the information in a hereditary manner to subsequent generations. The functional role of gene is expressed through protein. At the biological level, proteins rarely act alone; rather they interact with other proteins to perform particular cellular functions. Thus protein-protein interactions play pivotal roles in various aspects of the structural and functional organization of the cell and their complete description is indispensible to thorough understanding of the cell. Proteins can be viewed as vertices of a protein-protein interaction network in which two proteins are connected if they can physically attach to each other, forming a complex network called the protein interaction network (PIN). Recently, high-throughput data-collection methods such as protein chips or

*Corresponding author: Byungnam Kahng—School of Physics, Seoul National University NS50, Seoul 151-747, Korea. Email: kahng@phya.snu.ac.kr

Power Laws, Scale-Free Networks and Genome Biology, edited by Eugene V. Koonin, Yuri I. Wolf and Georgy P. Karev.

semi-automated yeast two-hybrid screens have been introduced, that help to determine which proteins interact with each other in large scale. In particular, organisms with sequenced genomes such as the yeast *Saccharomyces cerevisiae* provide important test beds for analyzing such a PIN.[9]

In this manuscript, we investigate the structural property of the PIN in graph theoretic aspect and also the transport phenomena on such complex networks. We first introduce several quantities describing network structure in the following section. Then we specifically analyze the structural properties of the *S. cerevisiae* PIN and we consider a transport problem on complex networks. The final section is devoted to the conclusions and discussions.

The Degree Distribution, the Degree Correlation Function and the Clustering Coefficient

Retrospectively, the graphical approach was initiated by Erdôs and Rényi (ER)[10] in 1960, who were the first to study the statistical aspect of random graphs using the probabilistic method. Thus, modeling random networks has a long history, and has been particularly active as a branch of combinatorial graph theory. In graph theory, one of interesting quantities is the degree, defined as the number of edges connecting to a given vertex. The degree distribution of the ER network follows a Poisson distribution. Recently, however, there were findings that the degree distribution of the PIN follows a power law,

$$P_D(k) \sim k^{-\gamma}, \tag{1}$$

where k means degree and γ is the degree exponent. The network displaying a power-law degree distribution is called scale-free (SF) network. Besides the PIN, SF networks[11] are ubiquitous in real-world networks such as the world-wide web (WWW),[12-14] the Internet,[15-17] the citation network[18] and the author collaboration network of scientific papers,[19-20] and the metabolic networks in biological organisms.[21] The SF behavior of the degree distribution can be generalized into the Pareto form,

$$P_D(k) \sim (k + k_0)^{-\gamma}, \tag{2}$$

with a constant k_0.

In fact, the degree distribution of the yeast PIN fits better to this Pareto form, which will be discussed later.

It is known that the degrees of the two vertices located at the ends of an edge are correlated to each other. As the first step, such degree-degree correlation can be quantified in terms of the average of the degrees over neighbors of proteins with degree k as a function of k, denoted by $\langle k_{nn}\rangle(k)$. In most biological networks, the function $\langle k_{nn}\rangle(k)$ exhibits a decreasing behavior with increasing k. The decaying behavior is expressed roughly by another power law as

$$\langle k_{nn}\rangle(k) \sim k^{-\nu}. \tag{3}$$

On the other hand, the degree-degree correlation can also be described in terms of the assortativity coefficient introduced by Newman, which is defined as

$$r = \frac{\langle k_1 k_2\rangle - \langle (k_1+k_2)/2\rangle^2}{\langle (k_1^2+k_2^2)/2\rangle - \langle (k_1+k_2)/2\rangle^2}, \tag{4}$$

where k_1 and k_2 are the degree of two end vertices, respectively, of an edge, and $\langle ... \rangle$ denotes the average over all edges. It is nothing but the Pearson correlation coefficient for the degrees of two end vertices over all edges, normalized so that $-1 \le r \le 1$. r is negative when the function $\langle k_{nn}\rangle(k)$ exhibits decreasing behavior like the case of the PIN. In fact, the assortativity coefficient was introduced to characterize social networks, which have positive values of r in general. Thus vertices with higher degree tend to connect to those with lesser (similar) degrees in PIN (social networks).

Many real-world biocomplex networks have modular structures within them. Such modular structures are characterized in terms of the clustering coefficient. Let C_i be the local clustering coefficient of a vertex i, defined as $C_i = 2e_i/k_i(k_i - 1)$, where e_i is the number of edges present among the neighbors of vertex i, out of its maximum possible number $k_i(k_i - 1)/2$. The clustering coefficient of a network, C, is the average of C_i over all vertices. $C(k)$ means the mean clustering coefficient over the vertices with degree k. When a network is modular and hierarchical, the clustering function follows a power law,

$$C(k) \sim k^{-\beta}, \quad (5)$$

for large k, and C is independent of system size N.[22,23]

Graph Theoretic Analysis of the Yeast Protein Interaction Network

There are a number of existing databases[24-26] or large-scale data sets[7,9,27-30] that store the information on the protein interactions in yeast. As all biological data are subject to some errors and incompleteness, which database to use is not a trivial problem. Without having a unified one only, we have tried to access as many data as we can, including those from the four major large-scale datasets, (i) the large-scale yeast two-hybrid data by Uetz et al[9,28] and (ii) by Ito et al[27] as well as the curated databases, (iii) the Munich Information Center for Protein Sequences (MIPS)[24] and (iv) the database of the interacting proteins (DIP)[25] as of March 2003. We also collected data from following additional sources: (a) Two-hybrid data by Tong et al.[29] (b) Mass spectrometry protein complexes analysis data (filtered one) by Ho et al.[30] After trimming the synonyms and other redundant entries manually, the resulting network consists of 16174 interactions (excluding self-interactions) between 5002 vertices in terms of distinct open reading frames. We denote this data as "integrated" one. Topological features of the resulting integrated network are summarized in Table 1 and Figure 1, which contain the comparison with topological features from individual databases. We measure various quantities describing the structural properties of the PIN based on our dataset as follows:

i. **Giant cluster**—Among 5002 proteins, as many as 4927 (98%) forms a giant cluster.
ii. **Mean degree**—The mean degree $\langle k \rangle$, i.e., the average number of interaction partners per protein, is $\langle k \rangle \approx 6.44$ excluding self-interactions, which is larger than previous estimates, $\langle k \rangle \approx 2\sim3$ based on references 9,27,31.
iii. **Degree distribution**—It has been reported that $P_D(k)$ follows a power law, Eq. (1), with $\gamma \approx 2.4\sim2.7$[32] or a power law with exponential cutoff in the form of $P_D(k) \sim (k + k_0)^{-\gamma}\exp(-k/k_c)$ with $\gamma \approx 2.45$, $k_0 = 1$, and $k_c \approx 20$.[31] Based on our dataset, we found, however, that the connectivity distribution fits better to the generalized Pareto function, Eq. (2) with $\gamma \approx 3.5$ and $k_0 \approx 8.4$. That is, the PIN is scale-free. Note that the exponent $\gamma \approx 3.5$ is rather larger than previous measured values, $\gamma \approx 2.4\sim2.7$.
iv. **Assortativity**—The assortativity coefficient r[33] is negative as $r = -0.137$, i.e., the PIN is dissortatively mixed, meaning that proteins with a small number of interaction partner are likely to connect to those with a large number of interaction partner, and vice versa, compared with its random counterpart whose r value is typically null.
v. **Average of neighbor's degree**—The function $\langle k_{nn} \rangle(k)$ exhibits a decreasing behavior with increasing k, a common behavior to dissortatively mixed networks. The decaying behavior is expressed roughly by another power law, Eq. (3), with $\nu \approx 0.2 \sim 0.3$, where the value ν is smaller than a previous estimated value $0.5 \sim 0.6$[34] based on the dataset by Ito et al.
vi. **Clustering**—The clustering coefficient, C, is obtained to be $C = 0.131$, larger than the values based on the data by Uetz et al and by Ito et al.
vii. **Hierarchical modularity**—The average clustering function $C(k)$ is likely to be constant for small k, while it decreases with increasing k for large k. Such a behavior is comparable to the ones measured from other databases as shown in Figure 1.

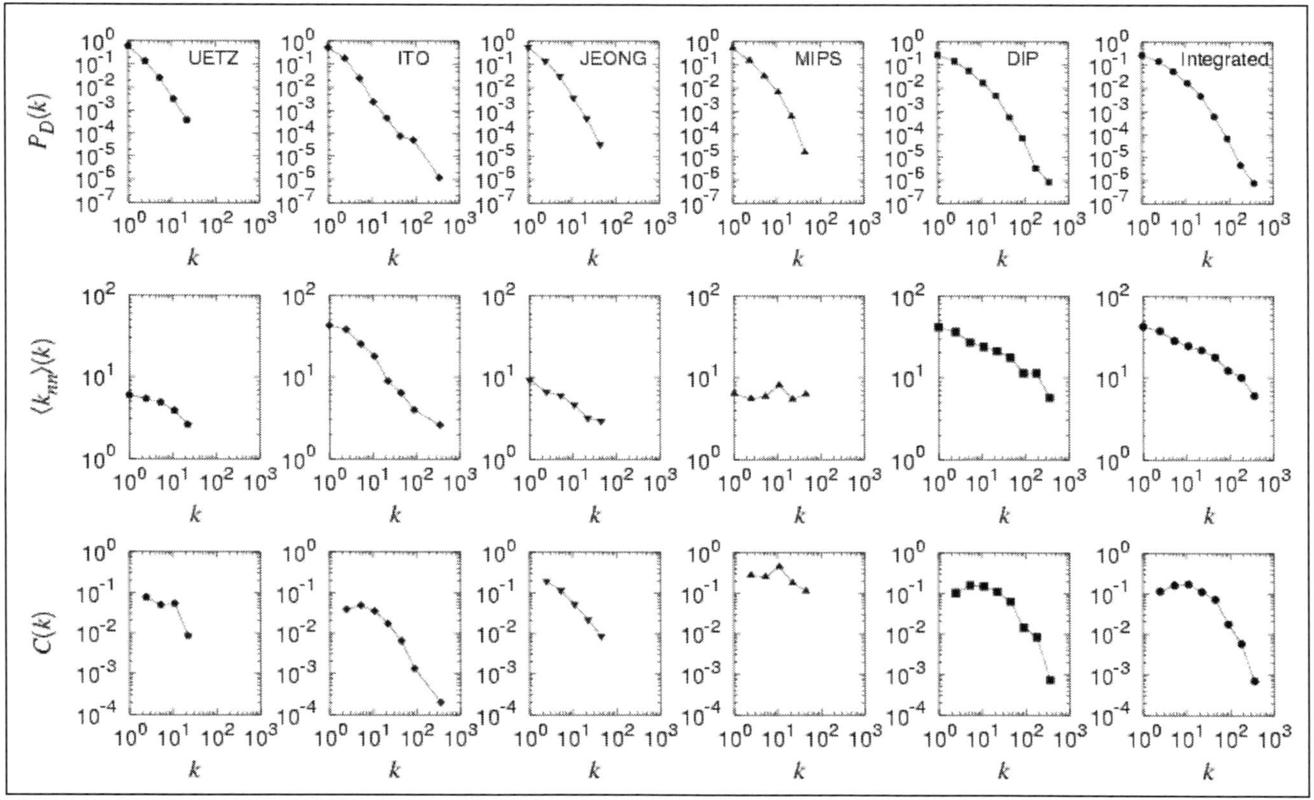

Figure 1. Topological characteristics of the Yeast PIN for various datasets: That of Uetz et al[9] Ito et al[27] Jeong et al[31] MIPS,[24] DIP,[25] and the integrated one. Shown are the degree distribution $P_D(k)$, the average of the neighbor degree $\langle k_{\text{nn}} \rangle (k)$, and the local clustering function $C(k)$. All data points are logarithmically binned. The ranges of abscissae and ordinates are fixed for easy comparison.

Table 1. Topological characteristics of the Yeast PIN for various datasets

	Uetz	ITO	JEONG	MIPS	DIP	Integrated
N	1331	3279	1846	1991	4713	5002
$\langle k \rangle$	2.10	2.68	2.39	2.66	6.30	6.44
r	-0.145	-0.176	-0.162	0.055	-0.136	-0.137
C	0.071	0.037	0.153	0.271	0.122	0.131
N_1	924	2839	1458	1439	4626	4927
N_2	8	6	7	11	3	3

N is the number of proteins with at least one interacting partner, $\langle k \rangle$ the mean degree, r the assortativity coefficient, C the clustering coefficient, N_1 the size of the giant cluster, and N_2 the size of the second giant cluster. Self-interactions are eliminated throughout the analysis.

Putting all these together, the yeast protein interaction network is scale-free, dissortatively mixed, highly clustered, and organized in a highly modular manner. The topological characteristics from our dataset and its comparison to other ones are summarized in Table 1 and in Figure 1. Such structural properties are universal for different species, so that they could be used as a test bed to find incomplete protein interactions.

Classification of Scale-Free Networks

While the emergence of the scale-free behavior in complex networks is intriguing and has a number of important consequences in its own right, there may exist other hidden orders in the scale-free networks. In this section, we introduce a candidate for this, the load distribution, and show that we can classify a range of real-world and model-generated scale-free networks into two distinct classes. We argue that such classification is rooted from the distinct topological features of the *shortest pathways* in the network.

Load Distribution

Let us suppose that a signal is sent from a vertex i to j ($i \rightarrow j$), along the shortest pathway between them.[35] In the information network such as the Internet, data packet is normally transmitted along the shortest pathways, however, for biological networks, it is not, even though the shortest pathways are the major flux canal. Nevertheless, here we consider the signal transport along the shortest pathways for simplicity. If there exist more than one shortest pathways, the signal would encounter one or more branching points. In this case, the signal is presumed to take one of them with equal probability, and the signal is effectively divided evenly over the branches at each branching point as it travels. Then the load $\ell_k^{i\rightarrow j}$ at a vertex k is defined as the amount of signals passing through that vertex k. Note that $\ell_k^{i\rightarrow j} = 0$ for vertices which do not fall on the shortest pathway ($i \rightarrow j$). Also note that the contribution from the pathway ($i \rightarrow j$), may be different from that of ($j \rightarrow i$), $\ell_k^{j\rightarrow i}$, even for undirected networks. Then we define the load ℓ_k of a vertex k as the accumulated sum of $\ell_k^{i\rightarrow j}$ over all pairs of senders and receivers:

$$\ell_k = \sum_{i,j} \ell_k^{i\rightarrow j}.$$

Here, we do not take into account the time delay of signal transfer at each vertex or edge, so that all signals are delivered in a unit time, regardless of the distance between any two vertices. So the load is a static variable for a given number of vertices N. The definition of the load is illustrated in Figure 2. Since the packets are conserved, the total load contributed by one pair is simply related to the shortest pathway length d_{ij} between them, by

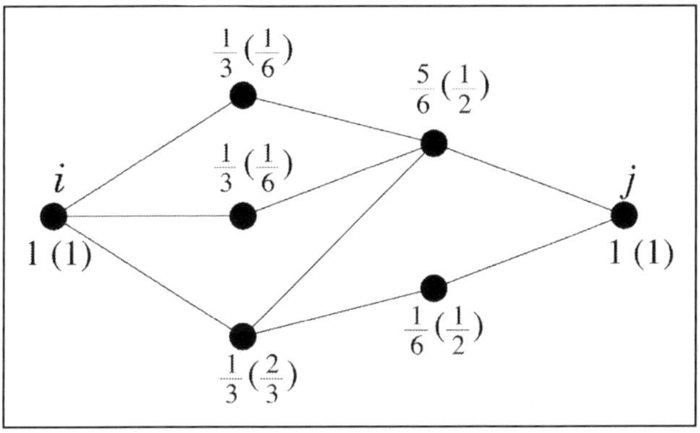

Figure 2. Illustration of the definition of load: The load at each vertex due to a unit packet transfer from the vertex i to the vertex j. In this diagram, only the vertices along the shortest paths between (i, j) are shown. The quantity in parenthesis is the load due to the one from j to i.

$$\sum_k \ell_k^{i\to j} = d_{ij} + 1.$$

Thus we have the sum rule for :

$$\sum_k \ell_k = \sum_{i,j} \left(d_{ij} + 1\right) \equiv N(N-1)(D+1) \sim N^2 D, \tag{6}$$

where D is called the diameter. The quantity we defined as load is closely related to the one used in sociology called "betweenness centrality" (BC) which quantifies how much power is centralized to a person in social networks.[36,37]

We focus our interest on the manner how ℓ_k are distributed. Once a SF network is generated artificially or adopted from the real world, we select an ordered pair of vertices (i, j) on the network, and identify the shortest pathway(s) between them and measure the load on each vertex along the shortest pathway using the modified version of the breath-first search algorithm introduced by Newman[37] and independently by Brandes.[38]

We have measured load ℓ_k of each vertex k for SF networks with various γ. It is found numerically that the load distribution $P_L(\ell)$ follows the power law,[35]

$$P_L(\ell) \sim \ell^{-\delta}. \tag{7}$$

When the indices of the vertices are ordered according to the rank of the load, we have $\ell_1 \geq \ldots \geq \ell_N$. Then, the power-law behavior of the load distribution implies that

$$\frac{\ell_i}{\sum_j \ell_j} \sim \frac{1}{N^{1-\alpha}} \frac{1}{i^{\alpha}}, \tag{8}$$

with

$$\delta = 1 + 1/\alpha. \tag{9}$$

The relation, Eq. (8), is valid in the region,[39] $\ell_{\min} < \ell < \ell_{\max}$, where

$$\ell_{\min} \sim \ell_{\max} / N^{\alpha} \sim \begin{cases} ND & \text{if } \alpha < 1 \\ ND/\ln N & \text{if } \alpha = 1 \\ N^{2-\alpha} D & \text{if } \alpha > 1. \end{cases} \tag{10}$$

Based on numerical measurements of load exponents for a variety of SF networks, we find that the load exponent is likely to be robust, independent of the details of network structure such as the degree exponent γ as long as γ is in the range $2 < \gamma < 3$ and other details such as the mean degree, the directionality of edge, and so on.[35] Thus we may categorize the SF networks according to the load distributions of them. We found two classes, say, class I and II.[40] For the class I, the load exponent is $\delta \approx 2.2(1)$ and for the class II, it is $\delta \approx 2.0(1)$. We conjecture the load exponent for the class II to be exactly $\delta = 2$ since it can be derived analytically for simple models. We will show that such different universal behaviors in the load distribution originate from different generic topological features of networks.

Real-World and Artificial Networks Investigated

A few network examples that we find to belong to the class I with $\delta \approx 2.2(1)$ include:

i. The protein interaction network of the yeast *S. cerevisiae* compiled by Jeong et al[31] (PIN_1), where vertices represent proteins and the two proteins are connected if they interact.
ii. The core of protein interaction network of the yeast *S. cerevisiae* obtained by Ito et al (PIN_2).[27]
iii. The metabolic networks for 5 species of eukaryotes and 32 species of bacteria in reference 21, where vertices represent substrates and they are connected if a reaction occurs between two substrates via enzymes. The reaction normally occurs in one direction, so that the network is directed.
iv. The Barabási-Albert (BA) model[41] when the number of incident edges of an incoming vertex $m \geq 2$.
v. The stochastic model for the protein interaction networks introduced by Solé et al.[42]

For both (i) and (v), the degree distribution is likely to follow a generalized power-law with a cut-off. Despite this abnormal behavior in the degree distribution for finite system, the load distribution follows a pure power law with the exponent $\delta \approx 2.2(1)$. The representative load distributions for real world networks (ii) and (iii) are shown in Figure 3A.

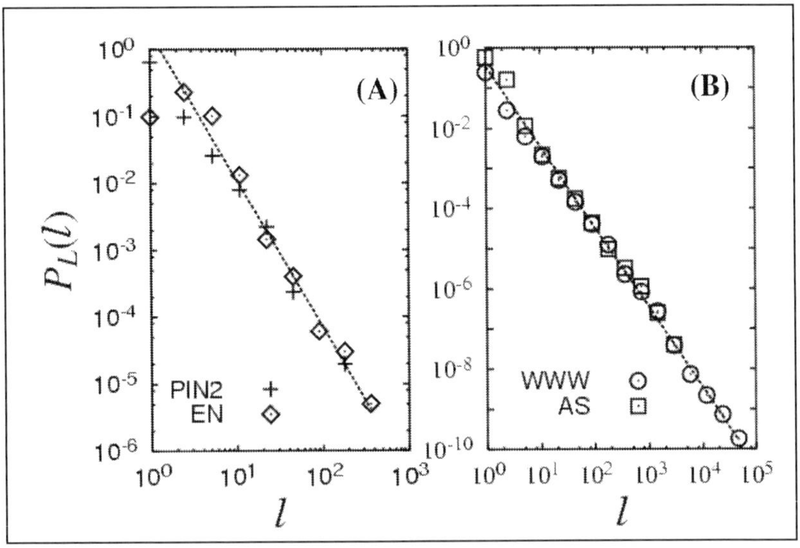

Figure 3. Load distributions for the two classes: A) The PIN of the yeast (ii) and the metabolic network of a eukaryote *Emericella nidulans* (iii), belonging to the class I. B) WWW within www.nd.edu domain (xi) and the Internet ASes (xiii) which belong to the class II. From Goh KI et al, Proc Natl Acad Sci USA 99:12583-8, ©2002 National Academy of Sciences, USA, with permission.[40]

The networks that we find to belong to the class II with $\delta = 2.0$ include:

vi. The Internet at the autonomous systems (AS) level as of October, 2001.[43]

vii. The metabolic networks for 6 species of archaea in reference 21.

viii. The WWW within www.nd.edu domain.[12]

ix. The BA model with $m = 1$.[41]

x. The deterministic model by Jung et al.[44]

In particular, the networks (ix) and (x) are of tree structure, where the edge load distribution can be solved analytically. The load distributions for real-world networks (vi) and (viii) are shown in Figure 3B.

Topology of the Shortest Pathways

To understand the generic topological features of the networks in each class, we particularly focus on the topology of the shortest pathways between two vertices separated by a distance d. We define the *mass-distance relation* $M(d)$ as the mean number of vertices on the shortest pathways between a given pair of vertices, averaged over all pairs separated by the same distance d. If the shortest pathway topology is simple and resembles a fractal with the fractal dimension D_F, $M(d)$ would behave like $\sim d^{D_F}$ for large d, while if is tree-like, one would expect $M(d) \sim d$. We find that the mass-distance relation behaves differently for each class; for the class I, $M(d)$ behaves nonlinearly (Fig. 4A-B), while for the class II, it is roughly linear (Fig. 4C-D).

For the networks belonging to the class I such as the PIN2 (iii) and the metabolic network for eukaryotes (iv), $M(d)$ exhibits a nonmonotonic behavior (Fig. 4A,B), *viz.*, it exhibits a hump at $d_h \approx 10$ for (iii) or $d_h \approx 14$ for (iv). To understand why such a hump arises, we visualize the topology of the shortest pathways between a pair of vertices, taken from the metabolic network of a eukaryote organism, *Emericella nidulans* (*EN*), as a prototypical example for the class I. Figure 5A shows such a graph with linear size 26 edges ($d = 26$), where an edge between

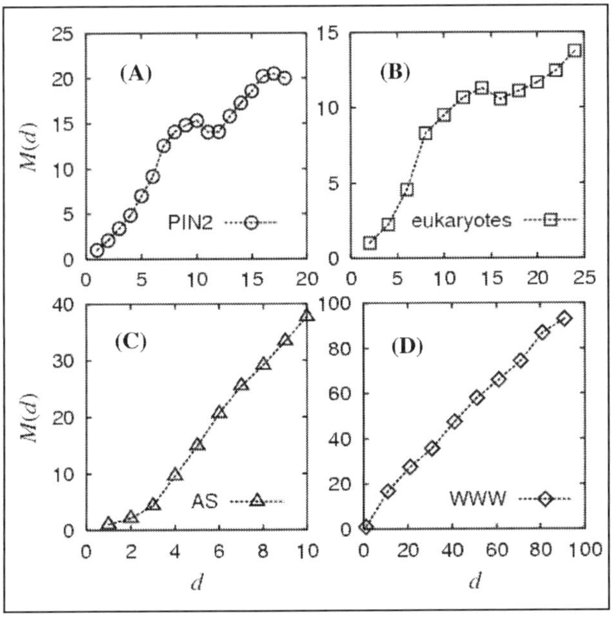

Figure 4. Mass-distance relation for prototypical SF networks: The yeast PIN (A), the metabolic networks of eukaryotes (B), the Internet at the AS level (C), and the WWW within nd.edu domain (D). From Goh KI et al, Proc Natl Acad Sci USA 99:12583-8, ©2002 National Academy of Sciences, USA, with permission.[40]

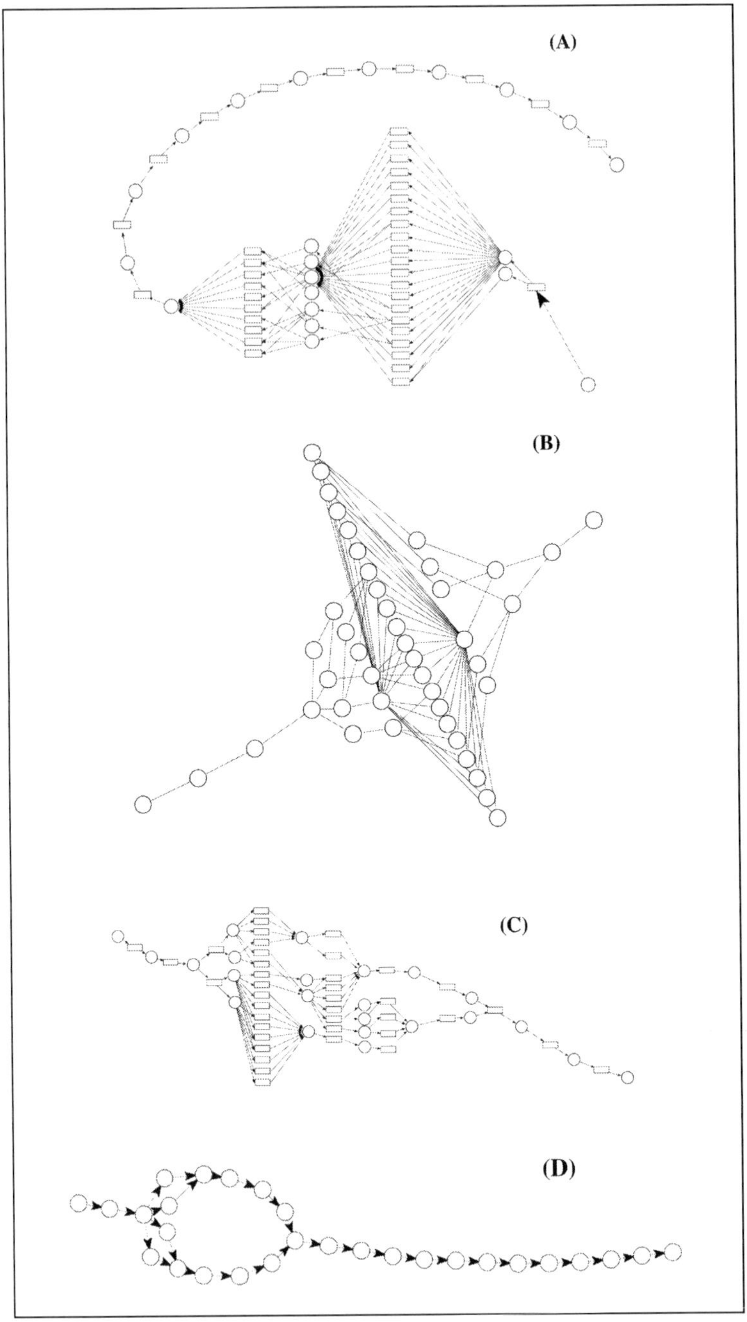

Figure 5. Topology of the shortest pathways: A) The metabolic network of a eukaryote *E. nidulans* of length 26. B) The Internet at AS level of length 10. C) The metabolic network of an archae *Methanococcus jannaschii* of length 20. D) WWW of www.nd.edu with length 20. In (A) and (C), circles denote substrates and rectangles denote intermediate states. From Goh KI et al, Proc Natl Acad Sci USA 99:12583-8, ©2002 National Academy of Sciences, USA, with permission.[40]

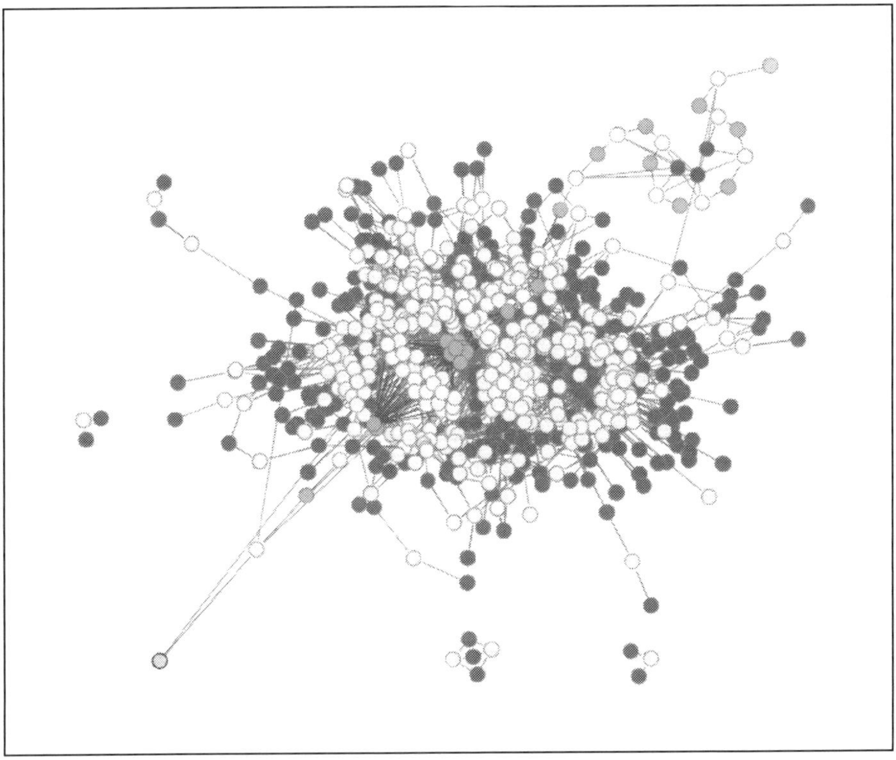

Figure 6. Global snapshot of the metabolic network of *E. nidulans*. The metabolites are shown in blue and the enzymes in light blue. Highlighted in orange (metabolites) and yellow (enzymes) are the shortest pathways of longest length, $d = 26$, whose starting and end points are indicated in green. A color version of this figure is available online at www.Eurekah.com.

a substrate and an enzyme is taken as the unit of length. From Figure 5A, one can see that there exists a blob structure inside which vertices are multiply connected, while vertices outside are singly connected. The characteristic of the class I is that the blob is localized in a small region. To give a visual image of the existence of the localized blob, we show the global snapshot of the shortest pathways in the *EN* metabolic network in Figure 6.

For the class II, the mass depends on distance linearly, $M(d) \sim Ad$ for large d (Fig. 4C,D). Despite the linear dependence, the shortest pathway topology for the case of $A > 1$ is more complicated than that of the simple tree structure where $A \cong 1$. Therefore, the SF networks in the class II are subdivided into two types, called the class IIa and IIb, respectively. For the class IIa, $A > 1$ and the topology of the shortest pathways includes multiply connected vertices (Fig. 5B and C), while for the class IIb, $A \cong 1$ and the shortest pathway is almost singly connected (Fig. 5D). Examples in real world networks in the class IIa are the Internet at the AS level ($A \sim 4.5$) and the metabolic network for archaea (A~2.0), while that in the class IIb is the WWW ($A \sim 1.0$).

The WWW is an example belonging to the class IIb. For this network, the mass-distance relation exhibits $M(d) \sim 1.0d$, suggesting that the topology of the shortest pathway is almost singly connected, which is confirmed in Figure 5D. When a SF network is of tree structure, one can solve the distribution of load running through each edge analytically, and obtain the load exponent to be $\delta = 2$.

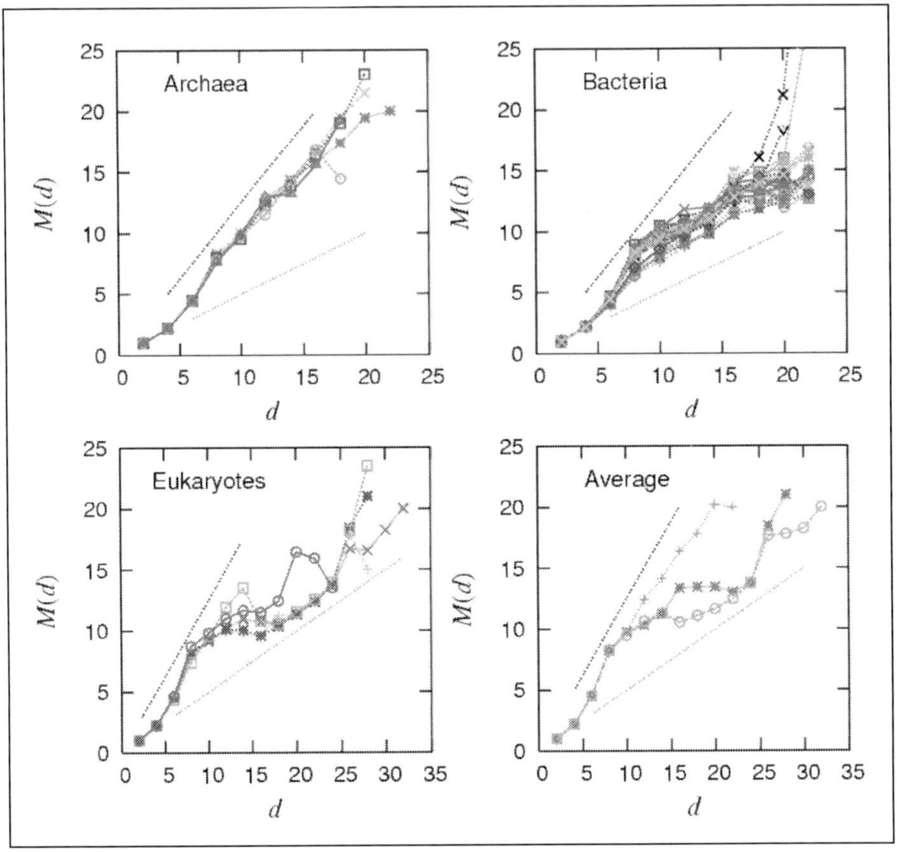

Figure 7. Mass-distance relations for the metabolic networks of the three domains of life: 6 archaea, 32 bacteria, and 5 eukaryotes, respectively, are plotted. In the bottom-right panel, $M(d)$ averaged over all species in each domain are compared. + stands for the data for archaea, ≠ for bacteria, and ° for eukayotes. The straight lines have slopes 1.25 (black) and 0.5 (orange), respectively, drawn for the eye. Note that since we count only the metabolites in $M(d)$, $M(d) = 0.5d$ for singly-connected shortest pathways. From Goh KI et al, Proc Natl Acad Sci USA 99:12583-8, ©2002 National Academy of Sciences, USA, with permission.[40] A color version of this figure is available online at www.Eurekah.com.

Application to the Metabolic Networks

In biological perspectives, the power of the shortest pathway analysis and the resulting classification is exemplified by the success in the categorizing the domains of life. In Figure 7, we show the mass-distance relations of the metabolic networks of all 43 species that we considered, grouped by the domains. Evidently, $M(d)$ for archaea behave differently from that for bacteria and eukaryotes. The eukaryotes have the class I-type metabolic networks and the archaea have the class II-type ones. The existence of the blob in eukaryotes and lack thereof in archaea implies the formation of such architecture might be driven by evolutionary pressure. One advantage of having the class I-type topology is that it is more resilient to the targeted attack on highly connected vertices.[40] It would be interesting to extend such idea to a more realistic situation for the metabolic stability.

Conclusion and Discussion

We have studied the structural properties of the yeast protein interaction networks and the transport phenomena along the shortest pathways on biocomplex networks from the graph theoretic viewpoint. Thanks to recent development of data collection and graph analysis methods, the structural properties of the yeast protein interaction networks have been unveiled rapidly. Here we analyzed the degree distribution, the degree-degree correlation, and the clustering coefficient of the yeast interaction networks for several different datasets available[9,24,25,27,31] and also for an integrated data we constructed. The yeast PIN is found to be strongly dissortative and highly modular. We believe that such analysis could be helpful for understanding the evolution of the protein interaction networks and finding protein interactions yet undiscovered. Moreover, we investigate the transport problem along the shortest pathways on biocomplex networks such as metabolic networks. We found that the load distribution follows a power law, and its exponent is robust, insensitive to detailed structural properties. We could classify real-world networks into two classes based on this property and also on the topological features of the shortest pathways. In particular, we find the metabolic networks for archaea belongs to the different class from that for bacteria and eukaryotes. The shortest pathway structure is simple for archaea. While further theoretical understandings are needed in relation to the robustness of the load distribution, at the moment, it would be interesting to notice that the load distribution is closely related to the structure of the core part of biocomplex networks.

Acknowledgement

This work is supported by the KOSEF Grant No. R14-2002-059-01000-0 in the ABRL program.

References

1. Ziemelis K, Allen L. Complex systems, and following review articles on complex systems. Nature 2001; 410:241.
2. Gallagher R, Appenzeller T. Complex systems, and following viewpoint articles on complex systems. Science 1999; 284:87.
3. Strogatz SH. Exploring complex networks. Nature 2001; 410:268-276.
4. Albert R, Barabási A-L. Statistical mechanics of complex networks. Rev Mod Phys 2002; 74:47.
5. Dorogovtsev SN, Mendes JFF. Evolution of networks. Oxford: Oxford University Press, 2003.
6. Newman MEJ. The structure and function of complex networks. SIAM Rev 2003; 45:167.
7. Gavin AC et al. Functional organization of the yeast proteome by systematic analysis of protein complexes. Nature 2002; 415:141-147.
8. Marcotte EM et al. A combined algorithm for genome-wide prediction of protein function. Nature 1999; 402:83-86.
9. Uetz P et al. A comprehensive analysis of protein-protein interactions in Saccharomyces cerevisiae. Nature 2000; 403:623-627.
10. Erdos P, Rényi A. On the evolution of random graph. Publ Math Inst Hung Acad Sci Ser A 1960; 5:17.
11. Barabási A-L, Albert R, Jeong H. Mean-field theory of scale-free networks. Physica A 1999; 272:173.
12. Albert R, Jeong H, Barabási A-L. Diameter of the world wide web. Nature 1999; 401:130-131.
13. Huberman BA, Adamic LA. Growth dynamics of the world wide web. Nature 1999; 401:131.
14. Broder A et al. Graph structure of the world wide web. Computer Networks 2000; 33:309.
15. Faloutsos M. Faloutsos P, Faloutsos C. On power-law relationship in the internet topology. Comput Commun Rev 1999; 29:251.
16. Pastor-Satorras R, Vázquez A, Vespignani A. Dynamical and correlation properties of the Internet. Phys Rev Lett 2001; 87:258701.
17. Goh K-I, Kahng B, Kim D. Fluctuation-driven dynamics of the Internet topology. Phys Rev Lett 2002; 88:108701.

18. Redner S. How popular is your paper? Eur Phys J B 1998; 4:131.
19. Newman MEJ. The structure of scientific collaboration. Proc Natl Acad Sci USA 2001; 98:404.
20. Barabási A-L, Jeong H, Ravasz R et al. On the topology of the scientific collaboration networks. Physica A 2002; 311:590-614.
21. Jeong H, Tombor B, Albert R et al. Large-scale organization of metabolic networks. Nature 2000; 407:651.
22. Ravasz E et al. Hierachical organization of modularity in metabolic networks. Science 2002; 297:1551-1555.
23. Ravasz E, Barabási A-L. Hierachical organization in complex networks. Phys Rev E 2003; 67:026112.
24. Mews HW et al. MIPS: Analysis and annotation of proteins from whole genomes. Nucl Acids Res 2004; 32:D41-D44.
25. Salwinski L et al. The database of interacting proteins: 2004 update. Nucl Acids Res 2004; 32:D449-D451.
26. Bader GD, Betel D, Hogue CW. BIND: The biomolecular interaction network database. Nucl Acids Res 2003; 31:248-250.
27. Ito T et al. A comprehensive two-hybrid analysis to explore the yeast protein interactome. Proc Natl Acad Sci USA 2001; 98:4569-4574.
28. Schwikowski B, Uetz P, Fields S. A network of protein-protein interactions in yeast. Nat Biotechnol 2000; 18:1257-1261.
29. Tong AH et al. A combined experimental and computational strategy to define protein interaction networks for peptide recognition modules. Science 2002; 295:321-324.
30. Ho Y et al. Systematic identification of protein complexes in Saccharomyces cerevisiae by mass spectrometry. Nature 2002; 415:180-183.
31. Jeong H et al. Lethality and centrality in protein networks. Nature 2001; 411:41-42.
32. Wagner A. How the global structure of protein interaction networks evolves. Proc R Soc Lond B 2003; 270:457-466.
33. Newman MEJ. Assortative mixing in networks. Phys Rev Lett 2002; 89:208701.
34. Maslov S, Sneppen K. Specificity and stability in topology of protein networks. Science 2002; 296:910-913.
35. Goh K-I, Kahng B, Kim D. Universal behavior of load distribution in scale-free networks. Phys Rev Lett 2001; 87:278701.
36. Freeman LC. A set of measure of centrality based on betweenness. Sociometry 1977; 40:35.
37. Newman MEJ. Scientific collaboration networks II: Shortest paths, weighted networks, and centrality. Phys Rev E 2001; 64:016132.
38. Brandes U. A faster algorithm for betweenness centrality. J Math Sociol 2001; 25:163.
39. Goh K-I, Kahng B, Kim D. Packet transport and load distribution in scale-free network models. Physica A 2003; 318:72.
40. Goh K-I, Oh E, Jeong H et al. Classification of scale-free networks. Proc Natl Acad Sci USA 2002; 99:12583.
41. Barabási A-L, Albert R. Emergence of scaling in random networks. Science 1999; 286:509.
42. Solé R, Pastor-Satorras R, Smith E et al. A model of large-scale proteome evolution. Adv Complex Syst 2002; 5:43.
43. Meyer D. University of Oregon Route Views Archive Project 2001 (http://archive.routeviews.org).
44. Jung S, Kim S, Kahng B. Geometric fractal growth model for scale-free networks. Phys Rev E 2002; 65:056101.

Chapter 3

Large-Scale Topological Properties of Molecular Networks

Sergei Maslov* and Kim Sneppen

Abstract

Bio-molecular networks lack the top-down design. Instead, selective forces of biological evolution shape them from raw material provided by random events such as gene duplications and single gene mutations. As a result individual connections in these networks are characterized by a large degree of randomness. One may wonder which connectivity patterns are indeed random, while which arose due to the network growth, evolution, and/or its fundamental design principles and limitations?

Here we introduce a general method allowing one to construct a random null-model version of a given network while preserving the desired set of its low-level topological features, such as, e.g., the number of neighbors of individual nodes, the average level of modularity, preferential connections between particular groups of nodes, etc. Such a null-model network can then be used to detect and quantify the nonrandom topological patterns present in large networks.

In particular, we measured correlations between degrees of interacting nodes in protein interaction and regulatory networks in yeast. It was found that in both these networks, links between highly connected proteins are systematically suppressed. This effect decreases the likelihood of cross-talk between different functional modules of the cell, and increases the overall robustness of a network by localizing effects of deleterious perturbations. It also teaches us about the overall computational architecture of such networks and points at the origin of large differences in the number of neighbors of individual nodes.

Introduction

Complex networks appear in biology on many different levels:

- All biochemical reactions taking place in a single cell constitute its metabolic network, where nodes are individual metabolites, and edges are metabolic reactions converting them to each other.
- Virtually every one of these reactions is catalyzed by an enzyme and the specificity of this catalytic function is ensured by the key and lock principle of the physical interaction with its substrate. Often the functional enzyme is formed by several mutually interacting proteins. Thus the structure of the metabolic network is shaped by the network of physical interactions of cell's proteins with their substrates and each other.

*Corresponding author: Sergei Maslov—Department of Physics, Brookhaven National Laboratory, Upton, New York 11973, U.S.A. Email: maslov@bnl.gov

Power Laws, Scale-Free Networks and Genome Biology, edited by Eugene V. Koonin, Yuri I. Wolf and Georgy P. Karev. ©2006 Eurekah.com and Springer Science+Business Media.

- The abundance and the level of activity of each of the proteins in the physical interaction network in turn is controlled by the regulatory network of the cell. Such regulatory network includes all of the multiple mechanisms in which proteins in the cell exert control on each other including transcriptional and translational regulation, regulation of mRNA editing and its transport out of the nucleus, specific targeting of individual proteins for degradation, modification of their activity e.g., by phosphorylation/dephosphorylation or allosteric regulation, etc.
- On yet higher level individual cells in a multicellular organism exchange signals with each other. This gives rise to several new networks such as e.g., nervous, hormonal, and immune systems of animals. The inter-cellular signaling network stages the development of a multicellular organism from the fertilized egg.
- Finally, on the grandest scale, the interactions between individual species in ecosystems determine their food webs.

In this review we concentrate on large-scale topological properties of complex biological networks operating on the levels of physical protein-protein interactions and transcriptional regulation.

Topological Properties of Protein Networks

Single-Node Topological Properties

An interesting property of many biological networks that was recently brought to attention of the scientific community[1-3] is an extremely broad distribution of nodes' degrees (often called connectivities in the network literature) defined as the number of immediate neighbors of a given node in the network. While the majority of nodes have just a few edges connecting them to other nodes in the network, there exist some nodes, that we will refer to as "hubs", with an unusually large number of neighbors. The degree of the most connected hub in such a network is typically several orders of magnitude larger than the average degree in the network. Often the number of nodes $N(K)$ with a given degree K can be approximated by a scale-free power law form $N(K) \cong K^{-\gamma}$ in which case the network is referred to as scale-free.[1]

In this review we concentrate on large-scale properties of physical interaction and regulatory protein networks. In Figure 1 we show the presently known[4] set of transcriptional regulations in a procaryotic bacterium *Escherichia coli*. For comparison, Figure 2 shows the presently known[5] transcriptional regulations in a simple single-cell eucaryote, *Saccharomyces cerevisiae* (baker's yeast).

Both yeast and *E. coli* regulatory networks are characterized by the above mentioned broad distribution of out-degrees K_{out} of its protein-nodes defined as the number of directed arrows emanating from individual regulatory proteins. Clearly visible in Figures 1 and 2 are the hub regulatory proteins that control the expression level of an unusually large number other proteins. For example, in the *E. coli* network one can see an extremely highly connected node in the lower half of Figure 1. It is the CAP protein that senses the glucose level, and in response to it orchestrates a cooperative action of a large battery of other proteins related to its utilization.

By comparing Figures 1 and 2 one gets an impression that the apparent growth in complexity of the transcription regulatory network from procaryotes to eucaryotes is achieved mostly by the virtue of an increase in the typical number of regulatory inputs of a protein (in-degree) K_{in}.

To quantify this further in Figure 3A we compare distributions of nodes' in-degrees in transcriptional regulatory networks of yeast (diamonds, dashed-line) and *E. coli* (circles, solid-line). This figure also includes the set of currently known transcriptional regulations in human (*Homo sapiens*) as extracted by Ariadne Genomics from abstracts of publications cited in MEDLINE. One can clearly see that the distribution of in-degrees in human is broader than that in yeast, which in its turn is significantly broader than that in the *E. coli*. Indeed, while in

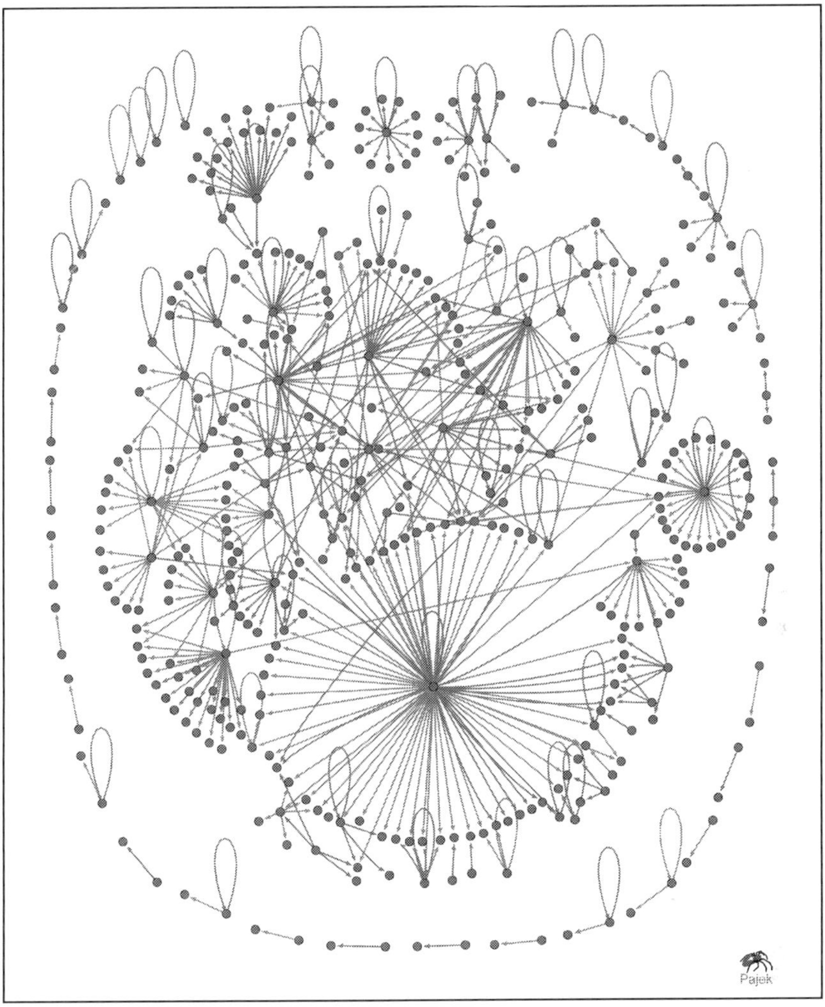

Figure 1. Presently known[4] transcriptional regulations in *E. coli*. Green and red arrows denote positive and negative regulations correspondingly. Nodes in this network represent operons (groups of genes transcribed onto a single mRNA) and arrows (edges) – direct transcriptional regulation of a downstream operon by a transcription factor encoded in the upstream operon. This network consists of 606 regulations of 424 operons by transcription factors contained in 116 different operons. A color version of this figure is available online at http://www.Eurekah.com.

the *E. coli* K_{in} has an exponential distribution ranging only between 0 and 6, in yeast its range is already between 0 and 15 and in human—between 0 and 18 and the tails of the K_{in} distribution in both eucaryots start to significantly deviate from the exponential functional form.

The above observations are in agreement with two recent empirical studies: C.K. Stover et al[6] found that the number of transcription factors (N_{tr}) in procaryotic organisms grows as a square of the number of genes (N): $N_{tr} \propto N^2$. Very recently E. van Nimwegen[7] has extended this result to eucaryotes where he also observed a superlinear scaling $N_{tr} \propto N^{1.26}$. The exact equation

$$N_{tr} / N = \langle K_{in} \rangle / \langle K_{out} \rangle \quad (1)$$

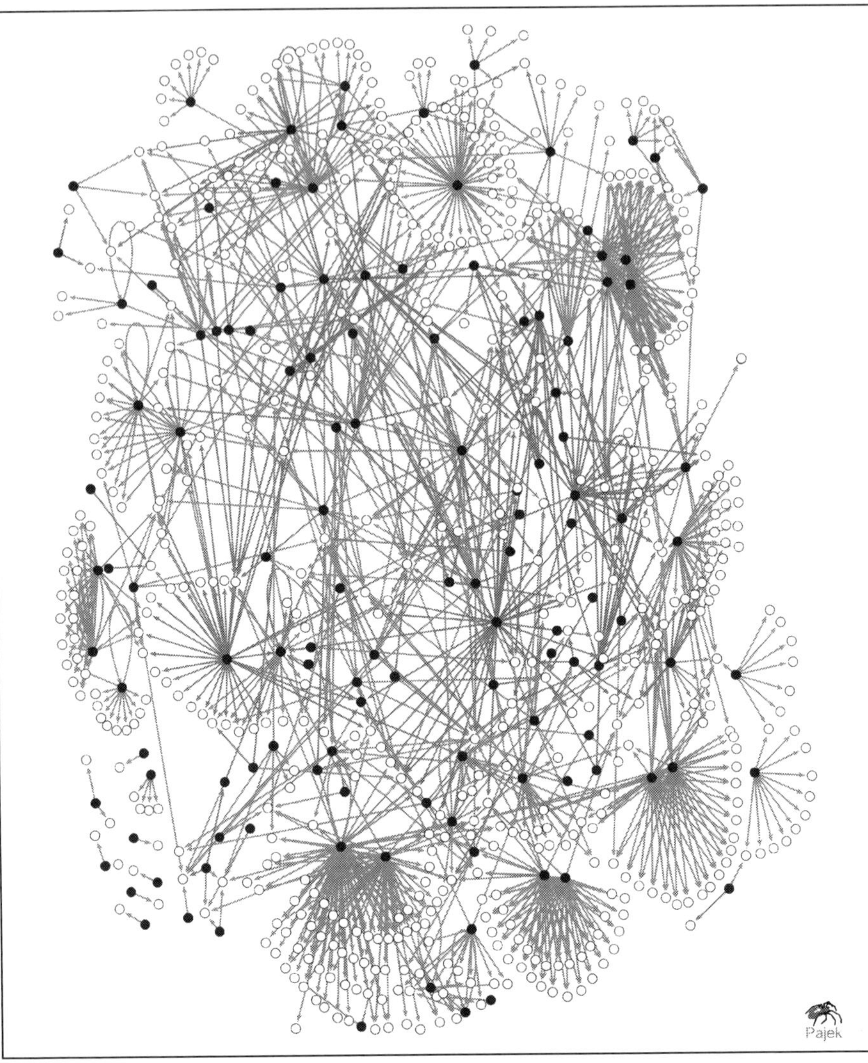

Figure 2. Presently known[5] transcriptional regulations in baker's yeast *S. cerevisiae*. This network consists of 1289 regulations of 682 proteins by 125 transcription factors. Green and red arrows denote positive and negative regulations correspondingly. Vertices corresponding to transcription factors are filled while those of remaining proteins are left empty. Apart from the absence of clear signs of modularity (the network has a unique giant connected component or module and only a few small small disconnected modules), one notices several striking features related to hub proteins that each regulate many other proteins: (1) They tend to regulate genes with just a few regulatory inputs. As a result of this they are well separated form each other, and positioned on a periphery of the network. This will be later quantified in the correlation profile of this network (Figs. 7, 9). (2) It is much more frequent for a protein to regulate many other proteins, than to be regulated by many.

relates the fraction of transcription factors in the genome of an organism to the average in- and out-degrees of its transcription regulatory network. Thus a direct consequence of the growth of the ratio N_{tr}/N with N is the increase in complexity of regulation of individual genes: $\langle K_{in} \rangle$.

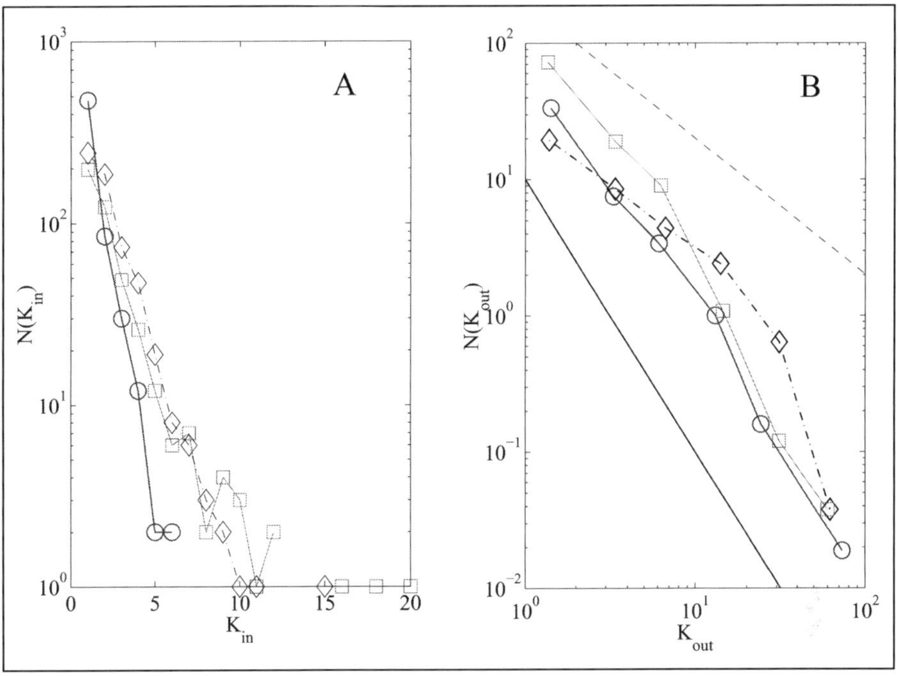

Figure 3. A) The histogram $N(K_{in})$ of nodes' in-degrees K_{in} in transcription regulatory networks of human (squares, solid line), yeast (diamonds, dot-dashed line), and *E. coli* (circles, solid line). This histogram in human is noticeably broader than in yeast, which in its term broader than in the *E. coli*. B) The histogram $N(K_{out})$ of nodes' out-degrees K_{out} in transcription regulatory network in human (squares, solid line), yeast (diamonds, dot-dashed line), and *E. coli* (circles, solid line). Overall, these three histograms are rather similar to each other. Straight lines are power law fits with the slope -2 (solid) and -1 (dashed). To improve the statistics all histograms in this panel were logarithmically binned into 3 bins per decade.

The distribution of K_{out} shown in Figure 3B appears to be about equally broad in *E. coli*, yeast and human. It ranges between 1 and about 70 regulations in all three networks. The power-law fit $N(K_{out}) \sim K_{out}^{-\gamma}$ gives $\gamma \approx 2$ in *E-coli* and human, while in yeast the distribution seems to have an initial slope characterized by $\gamma \approx 1$ followed by a sharper decay for $K_{out} > 30$. However, due to a limited range and an incomplete and possibly anthropogenically biased nature of the data (databases of research articles) one should not take these fits too seriously: at the very least they all indicate an unusually broad distribution of out-degrees in transcriptional regulatory networks.

Comparison of the Figure 3A and B also shows that in all organisms the in-degree distribution is much more narrow than that of the out-degree. That is a simple consequence of the fact that regulatory proteins (those with a nonzero K_{out}) constitute just a small fraction of all proteins in the cell.

Apart from transcriptional regulatory networks, metabolic networks,[2] and protein-protein physical interaction networks[3] are characterized by a very broad distribution in the number of neighbors of their individual nodes. A small part of such physical interaction network in baker's yeast is visualized in Figure 4.

One aspect of a broad distribution of node degrees in protein interaction and regulatory networks, is the possibility of amplification and exponential spread of signals propagating in the network. The upper bound of the one step amplification of some biochemical signal propagating in a directed network is given by

$$A^{(dir)} = \frac{\langle K_{in} K_{out} \rangle}{\langle K_{in} \rangle}. \tag{2}$$

This amplification factor $A^{(dir)}$ measures the average number of neighbors to which the signal can be potentially broadcasted in one propagation step. The above formula, derived by Newman in reference 9 follows from the observation that a signal enters a given node with a probability proportional to its in-degree K_{in}, and leaves along any of its K_{out} outgoing links. For $A^{(dir)} \leq 1$ any signal eventually dies out and hence affects only a small fraction of nodes in the network. On the other hand, for $A^{(dir)} > 1$ signals propagating in the network might be exponentially amplified, and thus each of them could influence (and possibly interfere with) other signals over the entire network.

The degree K in undirected networks cannot be decomposed into in- and out- components. Hence the upper bound on amplification of signals is given by the amplification factor $A^{(undir)}$:[9]

$$A^{(undir)} = \frac{\langle K(K-1) \rangle}{\langle K \rangle}. \tag{3}$$

In the above equation we take into account the fact that the signal cannot reach new nodes along the edge by which it came to a given node. Hence the use of $K - 1$ in the enumerator. The amplification factor $A^{(undir)}$ in scale-free networks with $\gamma < 3$ is very large and sensitive to the degrees of the highest connected hub-nodes. Here the borderline case $A^{(undir)} = 1$ also separates two different regimes. For $A^{(undir)} \leq 1$ the network breaks into many components isolated from each other, while for $A^{(undir)} > 1$ it consists of a unique "giant" component, containing the majority of all nodes, and a few small disconnected components.

The direct calculation of the directed amplification ratio $A^{(dir)}$ in the transcription regulatory network gives $A_{ec}^{(dir)} = 1.08$ in the *E. coli* and $A_{yeast}^{(dir)} = 0.58$. Hence as directed networks they are both below or approximately at (in *E. coli*) the critical point $A_c = 1$. Therefore, signals propagating in these networks cannot exponentially amplify, which limits the extent of cross-talk between them. However if both these regulatory networks are treated as undirected (i.e., one temporarily forgets about the arrows on their edges) one gets significantly overcritical amplification ratios $A^{(undir)} >> 1$: $A_{ec}^{(undir)} = 10.5$ in the *E. coli* and $A_{yeast}^{(undir)} = 13.4$ in yeast. This explains why the majority of nodes in Figures 1 and 2 belong to the largest connected component, and why the size of disconnected components is so small. Apparently the cross-talk presents much bigger potential problem in the network of physical interactions between yeast proteins (Fig. 4), where $A_{PPI}^{(undir)} = 26.3$. In the last chapter of this review we would return to the question of cross-talk and demonstrate how higher-level topological properties detected in both physical and regulatory networks in yeast[10] help to reduce such undesirable interference between signals.

Local Rewiring Algorithm: Constructing a Randomized Null-Model Network

The set of degrees of individual nodes is an example of a low-level (single-node) topological property of a network. While it answers the question about how many neighbors a given node has, it gives no information about the identity of those neighbors. It is clear that most functional properties of networks are defined at a higher topological level in the exact pattern of connections of nodes to each other. However, such multi-node connectivity patterns are rather difficult to quantify and compare between networks.

In this chapter we concentrate on multi-node topological properties of protein networks. These networks (as any other biological networks) lack the top-down design. Instead, selective

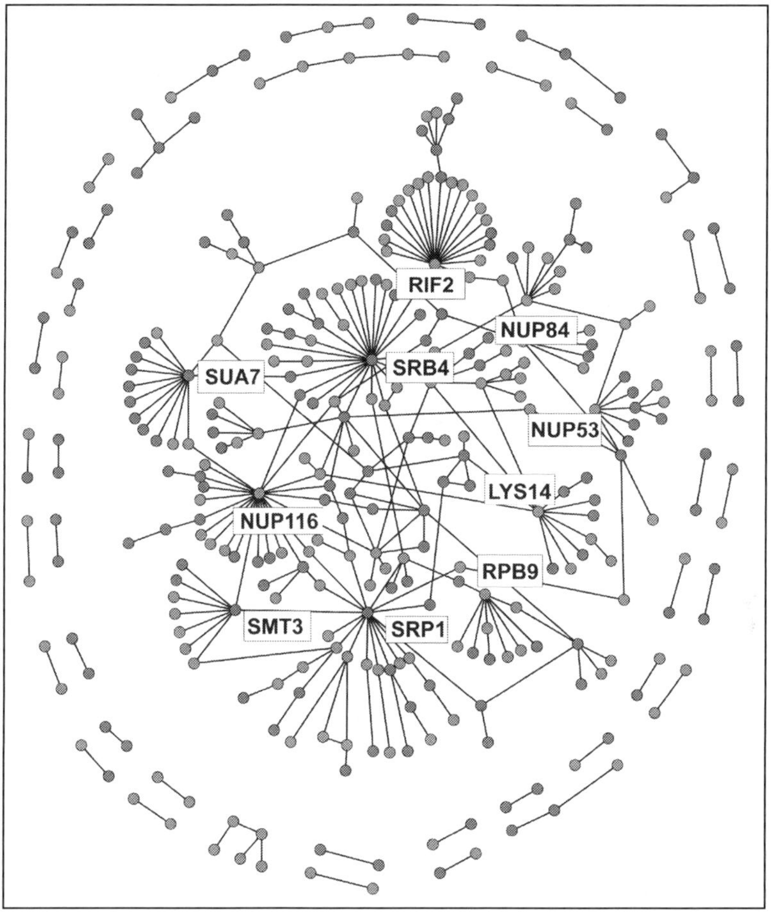

Figure 4. Network of physical interactions between nuclear proteins in yeast. Here we show the subset of protein-protein physical interactions reported in the full set of reference 8 consisting of 318 interactions between proteins that are known to be localized in the yeast nucleus.[5] The resulting network involves 329 proteins. Note that most neighbors of highly connected proteins have rather low connectivity. This feature will be later quantified in the correlation profile of this network (Figs. 6, 8).

forces of biological evolution shape them from raw material provided by random events such as mutations within individual genes, and gene duplications. As a result their connections are characterized by a large degree of randomness. One may wonder which connectivity patterns are indeed random, and which arose due to the network growth, evolution, ans/or its fundamental design principles and limitations?

To this end we first construct a proper randomized version (null model) of a given network. As was pointed out in the general context of complex scale-free networks,[9] a broad distribution of degrees indicates that the degree itself is an important individual characteristic of a node and as such it should be preserved in the randomized null-model network.[10] In addition to degrees one may choose to preserve some other low-level topological properties of the network in question.[11] Any measurable topological quantity, such as e.g., the total number of edges connecting pairs of nodes with given degrees, the number of loops of a certain type, the number and sizes of

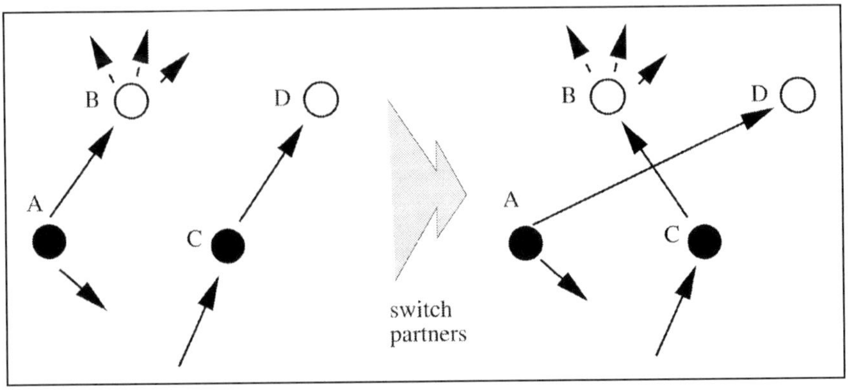

Figure 5. One step of the random local rewiring algorithm. A pair of edges A→B and C→D is randomly selected. The two edges are then rewired in such a way that A becomes connected to D, while C to B, provided that none of these new edges already exist in the network, in which case the rewiring step is aborted and a new pair of edges is selected. An independent random network is obtained when the above local switch move is performed a large number of times, say several times in excess of the total number of edges in the system. Note that for directed networks this rewiring algorithm separately conserves both the in- and out-degrees of each individual node.

components, the diameter of the network, can then be measured in the real complex network and separately in its randomized version. One then concentrates only on those topological properties of the real network that significantly deviate from its null model counterpart.[10-13]

An algorithm giving rise to a random network with the same set of individual node degrees as in a given complex network was proposed in references 10, 14 and 15. It consists of multiple repetitions of the following simple switch move (elementary rewiring step) illustrated in Figure 5:

Randomly select a pair of edges A→B and C→D and rewire them in such a way that A becomes connected to D, while C to B.

To prevents the appearance of multiple edges connecting the same pair of nodes, the rewiring step is aborted and a new pair of edges is selected if one or two of the new edges already exist in the network. A repeated application of the above rewiring step leads to a randomized version of the original network. The set of MATLAB programs generating such a randomized version of any complex network can be downloaded from.[16]

Sometimes it is desirable that the null-model random network in addition to nodes' degrees conserves some other topological quantity of the real network. In this case one could supplement[11] the random rewiring algorithm described above with the Metropolis acceptance/rejection criterion[17] of a switch move.

For the sake of concreteness let's assume that one wants to generate a random network with the same set of nodes' degrees and the same number N of triangles as the real undirected network.[11] Indeed, the number of triangles in a network is related to its "clustering coefficient" routinely used as a measure of its modularity.[18] Hence, by conserving N one generates a null-model with the same average level of modularity as the original complex network.

The Metropolis version[11] of the random rewiring algorithm uses an artificial energy function H that favors the number of triangles in a random network $N^{(r)}$ to be as close as possible to its value N in the real network:

$$H = \frac{(N^{(r)} - N)^2}{N} \qquad (4)$$

The Metropolis rules in this case allow for any local rewiring step that lowers the energy H or leaves it unchanged. However, those steps that lead to a ΔH increase in the "energy" H are accepted only with a probability $\exp(-\Delta H/T)$. Here the exact rules of the algorithm depend on (typically very small) "temperature" T introduced to prevent the sequence of rewiring steps from getting stuck in a local (often suboptimal or nonrepresentative) energy minimum. In order to get a random network with $N^{(r)}$ sufficiently close to N the temperature should be selected to be as small as possible without sacrificing the ergodicity of the problem. In the end one could always "prune" the resulting ensemble of random networks by leaving only networks with $N^{(r)} = N$.

Multi-Node Properties: Correlation Profile

The correlation profile of any large complex network quantifies correlations between degrees of its neighboring nodes. We have calculated correlation profiles of:

1. The protein interaction network consisting of 4475 physical interactions between 3279 yeast proteins as measured in the most comprehensive high-throughput yeast two-hybrid screen.[8] A subset of this network is shown in Figure 4.
2. The transcriptional regulatory network in yeast (Fig. 2), consists of 1289 (1047 positive and 242 negative) regulations by 125 transcription factors[5] within the set of 682 proteins.
3. While the regulatory network is naturally directed, the network of physical interactions among proteins in principle lacks directionality. Randomized versions of these two molecular networks were constructed by randomly rewiring their edges, while preventing "unphysical" multiple connections between a given pair of nodes, as described in the previous chapter. By construction this algorithm separately conserves the in- and out-degrees of each node. Therefore, in a randomized version of the regulatory network each protein has the same numbers of regulators and regulated proteins as in the original network. Taking in consideration the bait-prey asymmetry mentioned in,[10] when generating random counterpart of the interaction network we chose to separately conserve numbers of interaction partners of the bait-hybrid and the prey-hybrid of every protein.

The topological property of the network giving rise to its correlation profile is the number edges $N(K_0,K_1)$ connecting pairs of nodes with degrees K_0 and K_1. To find out if in a given complex network the degrees of interacting nodes are correlated, $N(K_0,K_1)$ should be compared to its value $N_r(K_0,K_1) \pm \Delta N_r(K_0,K_1)$ in a randomized network, generated by the edge rewiring algorithm. When normalized by the total number of edges E, $N(K_0,K_1)$ defines the joint probability distribution $P(K_0,K_1) = N(K_0,K_1)/E$ of degrees of interacting nodes. Any correlations would manifest themselves as systematic deviations of the ratio

$$R(K_0,K_1) = P(K_0,K_1)/P_r(K_0,K_1) \qquad (5)$$

away from 1. Statistical significance of such deviations is quantified by their Z-score

$$Z(K_0,K_1) = (P(K_0,K_1) - P_r(K_0,K_1))/\sigma_r(K_0,K_1), \qquad (6)$$

where $\sigma_r(K_0,K_1) = \Delta N_r(K_0,K_1)/N$ is the standard deviation of $P_r(K_0,K_1)$ in an ensemble of randomized networks.

Figures 6 and 7 show the ratio $R(K_0,K_1)$ as measured in yeast interaction and transcription regulatory networks, respectively. In the interaction network K_0 and K_1 are numbers of neighbors of the two interacting proteins, while in the regulatory network K_0 is the out-degree of the regulatory protein and K_1—the in-degree of its regulated partner. Thus by its very construction $P(K_0,K_1)$ is symmetric for the physical interaction network but not for the regulatory network. Figures 8 and 9 plot the statistical significance $Z(K_0,K_1)$ of deviations visible in Figures 6 and 7 correspondingly. To arrive at these Z-scores 1000 randomized networks were sampled and degrees were logarithmically binned into two bins per decade.

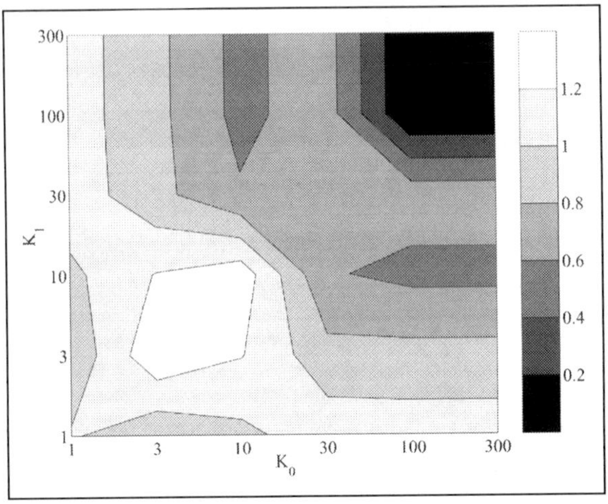

Figure 6. Correlation profile of the protein interaction network in yeast. The ratio $R(K_0,K_1)=P(K_0,K_1)/P_r(K_0,K_1)$, where $P(K_0,K_1)$ is the probability that a pair of proteins with K_0 and K_1 interaction partners correspondingly, directly interact with each other in the full set of reference 8 while $P_r(K_0,K_1)$ is the same probability in a randomized version of the same network, generated by the random rewiring algorithm described in the text. Note the logarithmic scale of both axes.

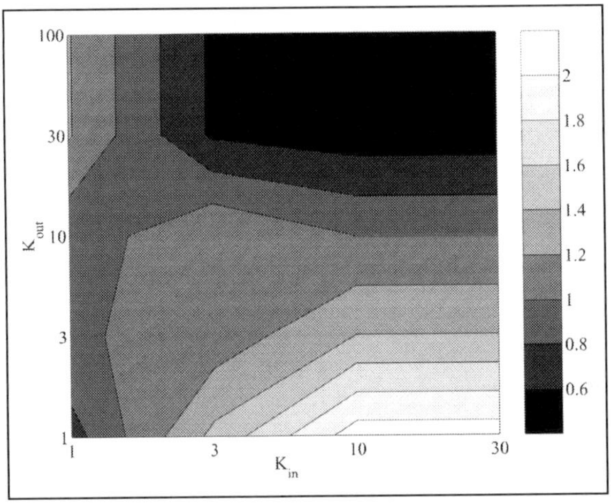

Figure 7. Correlation profile of the transcription regulatory network in yeast. The ratio $R(K_{out},K_{in}) = P(K_{out},K_{in})/P_r(K_{out},K_{in})$, where $P(K_{out},K_{in})$ is the probability that a protein node with the out-degree K_{out} transcriptionally regulates the protein node with the in-degree K_{in} in the transcription regulatory network obtained from the YPD database[5] (Fig. 2), while $P_r(K_{out},K_{in})$ is the same probability in a randomized version of the same network, generated by the random rewiring algorithm described in the text. Note the logarithmic scale of both axes.

The combination of R- and Z-profiles reveals the regions on the K_0 - K_1 plane, where connections between proteins in the real network are significantly enhanced or suppressed, compared to the null model. In particular, the blue/green region in the upper right corner of Figures 6-9 reflects the reduced likelihood that two hubs are directly linked to each other, while

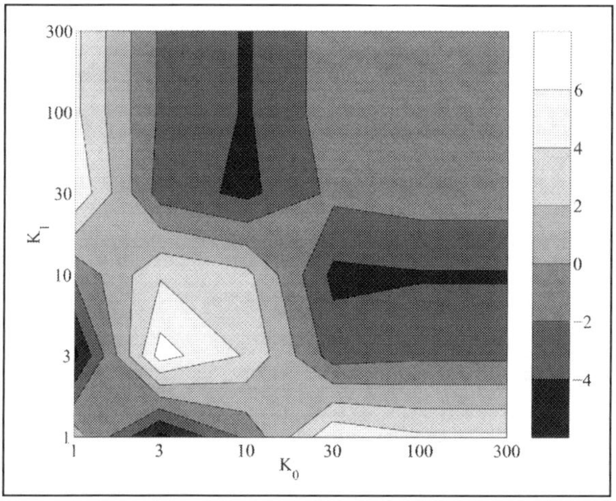

Figure 8. Statistical significance of correlations present in the protein interaction network in yeast. The Z-score of correlations $Z(K_0,K_1) = (P(K_0,K_1) - P_r(K_0,K_1))/\sigma_r(K_0,K_1)$, where $P(K_0,K_1)$ is the probability that a pair of proteins with K_0 and K_1 interaction partners correspondingly, directly interact with each other in the full set of reference 8 while $P_r(K_0,K_1)$ is the same probability in a randomized version of the same network, generated by the random rewiring algorithm described in the text, and $\sigma_r(K_0,K_1)$ is the standard deviation of $P_r(K_0,K_1)$ measured in 1000 realizations of a randomized network. Note the logarithmic scale of both axes.

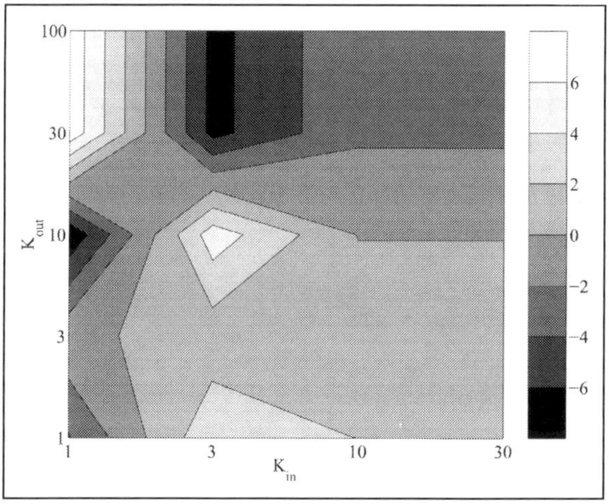

Figure 9. Statistical significance of correlations present in the transcription regulatory network in yeast. The ratio $Z(K_{out},K_{in}) = (P(K_{out},K_{in}) - P_r(K_{out},K_{in}))/\sigma_r(K_{out},K_{in})$, where $P(K_{out},K_{in})$ is the probability that a protein node with the out-degree K_{out} transcriptionally regulates the protein node with the in-degree K_{in} in the network from the YPD database,[5] while $P_r(K_{out},K_{in})$ is the same probability in a randomized version of the same network, generated by the random rewiring algorithm described in the text, and $\sigma_r(K_{out},K_{in})$ is the standard deviation of $P_r(K_{out},K_{in})$ measured in 1000 realizations of a randomized network. Note the logarithmic scale of both axes.

red regions in the upper left and the lower right corners of these figures reflect the tendency of hubs to associate with nodes of low degree. One should also note a prominent feature on the diagonal of the Figure 6 and 8 corresponding to an enhanced affinity of proteins with between 4 and 9 physical interaction partners towards each other. This feature can be tentatively attributed to members of multi-protein complexes interacting with other proteins from the same complex. The above range of degrees thus correspond to a typical number of direct interaction partners of a protein in a multi-protein complex. When we studied pairs of interacting proteins in this range of degrees we found 39 of such pairs to belong to the same complex in the recent high-throughput study of yeast protein complexes.[19] This is about 4 times more than one would expect to find by pure chance alone.

Robustness of the Correlation Profile with Respect to Potential Errors in the Data

When analyzing molecular networks one should consider possible sources of errors in the underlying data. Two-hybrid experiments in particular are known to contain a significant number of false positives and probably even more of false negatives.

The evidence of a significant number of false negatives lies in the fact that only a small fraction of functionally plausible interactions were detected in both directions (the bait-hybrid of a protein A interacting the prey-hybrid of a protein B as well as the prey-hybrid of a protein A interacting the bait-hybrid of a protein B). It is also attested by a relatively small overlap in interactions detected in the two independent high-throughput two hybrid experiments.[8,20] There exist a number of plausible explanations of these false negatives. First of all, binding may not be observed if the conformation of the bait or prey chimeric protein blocks relevant interaction sites or if it altogether fails to fold properly. Secondly, it is not entirely clear if the number of cells in batches used in high-throughput two hybrid experiments is sufficient for any given bait-prey pair to meet in at least one cell. Finally, 391 out of potential 5671 baits in reference 8 were not experimentally tested because they were found to activate the transcription of the reporter gene in the absence of any prey proteins.

Several sources of false positives are also commonly mentioned in the literature:

- In one scenario spurious interactions of highly connected baits are thought to arise due to a low-frequency indiscriminate activation of the reporter gene in the absence of any prey proteins. Such false positives (if they exist) are easy to eliminate by using curated high-throughput datasets which contain only protein pairs that were observed, say, at least 3 times in the course of the experiment. We have shown that all qualitative features of the correlation profile of the protein interaction network reported above remain unchanged when one uses such curated datasets.[21]
- In another scenario the interaction between proteins is real but it never happens in the course of the normal life cycle of the cell due to spatial or temporal separation of participating proteins. However, it is hard to believe that such nonfunctional interactions would be preserved for a long time in the course of evolution. Hence, it is dubious that such false-positives would be ubiquitous.
- In yet another scenario an indirect physical interaction is mediated by one or more unknown proteins localized in the yeast nucleus. However, since in two-hybrid experiments bait and prey proteins are typically highly overexpressed, it is only very abundant intermediate proteins that can give rise to an indirect binding. The relative insignificance of indirect bindings is attested by a relatively small number of triangles (178 vs ~ 100 in a randomized version) in the protein interaction network. Indeed, an indirect interaction of a protein A with a protein B effectively closes the triangle of direct interactions A - C and C - B with an intermediate protein C.

Discussion: What It May All Mean?

The large-scale organization of molecular networks deduced from correlation profiles of protein interaction and transcription regulatory networks in yeast is consistent with compartmentalization and modularity characteristic of many cellular processes.[22] Indeed, the suppression of connections between highly-connected proteins (hubs) suggests the picture of semi-independent modules centered around or regulated by individual hubs. On the other hand, the very fact that these molecular networks do not separate into many isolated components but are dominated by one "giant component" suggests that this tendency towards modularity is not taken to its logical end. The observed patterns can in fact be characterized as "soft modularity", where interactions between individual modules are suppressed but not completely eliminated. Thus on sufficiently large scale molecular networks exhibit system-wide properties making their behavior different from that of a set of mutually independent modules.

A further implication of the deficit of connections between highly connected proteins (Figs. 6, 7) is in the suppression of propagation of deleterious perturbations over the network. It is reasonable to assume that certain perturbations such as e.g., a significant change in the concentration of a given protein (including it vanishing altogether in a null-mutant cell) with a ceratin probability can affect its first, second, and sometimes even more distant neighbors in the corresponding network. While the number of immediate neighbors of a node is by definition equal to its own degree K_0, the average number of its second neighbors is bound from above by $K_0 \langle (K_1 - 1) \rangle_{K_0}$ and thus depends on the correlation profile of the network. Since highly connected nodes serve as powerful amplifiers for the propagation of deleterious perturbations it is especially important to suppress this propagation beyond their immediate neighbors. It was argued that scale-free networks in general are very vulnerable to cascading failures started at individual hubs.[23,24] The deficit of edges directly connecting hubs to each other reduces the branching ratio around these nodes and thus provides a certain degree of protection against such accidents.

Finally, we would like to mention that the tendency of highly connected proteins to be positioned at the periphery of signaling and regulatory networks teaches us something about the overall computational architecture of such networks and origins of their broad degree distributions. Indeed, the peripheral position of hubs indicates that they presumably execute collective orders of other more "computationally-involved" regulators, rather than performing computations and making decision on their own. This principle is nicely illustrated in the lambda-phage regulatory network (see Fig. 10), where the decision making/computation is done by CI , CII, and Cro proteins, which (with the exception of CI) are characterized by low-to-intermediate out-degrees and high in-degrees. Their orders on the other hand are executed through the N and LexA hub-proteins which have high out-degree and low in-degree.

Broad degree distributions observed in molecular networks presumably reflect the widely different needs associated with different functions that a living cell needs to cope with changes in its environment. Thus highly connected regulatory proteins usually correspond to rather complicated tasks such as e.g., the heat shock response, where about 40 chaperones are controlled by a single sigma factor, or the chemotaxis where a few regulatory proteins switch on a large number of proteins associated with flagella, flagellar motor, and sensing of the environment.

To summarize the above discussion, it is feasible that molecular networks operating in living cells have organized themselves in a particular computational architecture that makes their dynamical behavior both robust and specific. Topologically the specificity of different functional modules is enhanced by limiting interactions between hubs and suppressing the average degree of their neighbors. On a larger scale there is evidence for interconnections

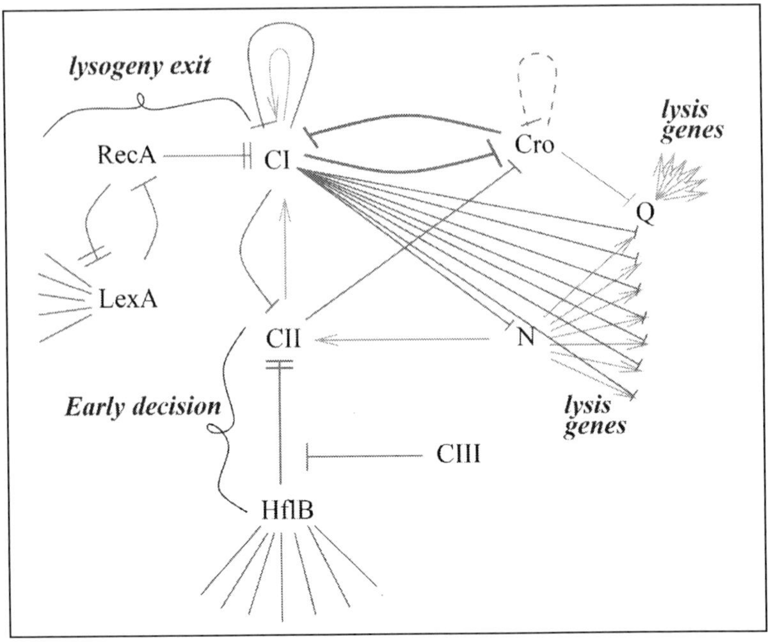

Figure 10. Lambda-phage regulatory network. The actual computation is done by centrally positioned Cro and CII that have low-to-intermediate out-degree and relatively large in-degree. Their decision is transmitted to peripherally positioned, highly connected hub-proteins such as N and LexA, which in their turn broadcast it to the whole battery of response genes. As a curiosity, note that the HflB protease from *E. coli*'s heat-shock response network interacts with the lambda-phage regulatory network. Another curiosity: the HflB directly regulates DnaK, which at least indirectly has substantial influence on the overall transcription of ribosomal RNAs of the *E. coli*. Thus the lambda network integrates as a small subnetwork in the overall bacterial regulatory network of *E. coli*. The notation used in this figure: $\uparrow$ indicates positive regulation, $\perp$ indicates passive negative regulation; $\pm$ indicates active degradation through the protease activity.

between these modules, although the principles of such global organization of living cells remain unclear from the present day data and analysis tools.

References

1. Barabasi A-L, Albert R. Emergence of scaling in random networks. Science 1999; 286:509-512.
2. Jeong H, Tombor B, Albert R et al. The large-scale organization of methabolic networks. Nature 2000; 407:651-654.
3. Jeong H, Mason S, Barabasi A-L et al. Centrality and lethality of protein networks. Nature 2001; 411:41-42.
4. Huerta AM, Salgado H, Thieffry D et al. As reported in the Regulon database. RegulonDB: A database on transcriptional regulation in Escherichia coli. Nucleic Acid Res 1998; 26:55.
5. Costanzo MC, Crawford ME, Hirschman JE et al. As reported in the YPD database. YPD, PombePD, and WormPD: Model organism volumes of the BioKnowledge library, an integrated resource for protein information. Nucleic Acids Res 2001; 29:75-79.
6. Stover CK, Pham XQ, Erwin AL et al. Complete genome sequence of Pseudomonas aerugiosa PA01, an opportunistic pathogen. Nature 2000; 406:959-398.
7. van Nimwegen E. Scaling laws in the functional content of genomes. Trends Genet 2003; 19(9):479-484.

8. Ito T, Chiba T, Ozawa R et al. A comprehensive two-hybrid analysis to explore the yeast protein interactome. Proc Natl Acad Sci USA 2001; 98:4569-4574.
9. Newman MEJ, Strogatz SH, Watts DJ. Random graphs with arbitrary degree distributions and their applications. Phys Rev E 2001; 64(026118):1-17.
10. Maslov S, Sneppen K. Specificity and stability in topology of protein networks. Science 2002; 296:910-913.
11. Maslov S, Sneppen K, Zaliznyak A. Pattern detection in complex networks: Correlation profile of the internet. Physica A 2004; 333:529-540. (Preprint at arXiv.org e-Print archive available at http://arxiv.org/abs/cond-mat//0205379. 2002)
12. Shen-Orr S, Milo R, Mangan S et al. Network motifs in the transcriptional regulation of Escherichia coli. Nat Genet 2002; 31(1):64-68.
13. Milo R, Shen-Orr S, Itzkovitz S et al. Network motifs: Simple building blocks of complex networks. Science 2002; 298:824-827.
14. Gale D, ed. Early Studies of These Algorithms in the Context of Matrices Were Reported. A Theorem of Flows in Networks. Pacific J Math 1957; 7:1073-1082.
15. Ryser H.J. Matrices of zeros and ones in combinatorial mathematics. Recent Advances in Matrix Theory. Madison: Univ of Wisconsin Press, 1964:103-124. (For more recent references including applications to graphs see e.g.: Kannan R., Tetali P., Vempala S. Simple Markov-chain algorithms for generating bipartite graphs and tournaments. Random Structures and Algorithms 1999; 14:293-308).
16. The set of MATLAB programs can be downloaded at http://www.cmth.bnl.gov/~maslov/matlab.htm
17. Metropolis N et al. J Chem Phys 1953; 21:1087.
18. Watts D, Strogatz S. Nature 1998; 400-403.
19. Gavin A-C, Bosche M, Krause R et al. Functional organization of the yeast proteome by systematic analysis of protein complexes. Nature 2002; 415:141-147.
20. Uetz P, Giot L, Cagney G et al. A comprehensive analysis of protein-protein interactions in Saccharomyces cerevisiae. Nature 2000; 403:623-627.
21. Maslov S, Sneppen K. Protein interaction networks beyond artifacts. FEBS Letters 2002; 530:255-256.
22. Hartwell LH, Hopfield JJ, Leibler S et al. From molecular to modular cell biology. Nature 1999; 402(6761 Suppl):C47-C52.
23. Albert R, Jeong H, Barabasi A-L. Error and attack tolerance of complex networks. Nature 2000; 406:378-382.
24. Vogelstein B, Lane D, Levine AJ. Surfing the p53 network. Nature 2000; 408:307-310.

CHAPTER 4

The Connectivity of Large Genetic Networks:

Design, History, or Mere Chemistry?

Andreas Wagner*

Abstract

I review evolutionary explanations of broad-tailed connectivity or degree distributions observed in metabolic networks and protein interaction networks. Self-assembled chemical reaction networks show degree distributions similar to those observed for metabolic networks, which argues against the postulated role of natural selection in maintaining this degree distribution. In addition, metabolic networks contain traces of their ancient history in the form of highly connected metabolites. Similarly to the degree distribution of metabolic networks, that of protein interaction networks can be explained without resorting to natural selection on the network level. I present data suggesting that highly connected proteins are not distinguishably older than other proteins, and explain this finding with a simple model of how a protein's degree changes in evolutionary time.

Introduction

Graph representations of biological networks have become popular with the recent accumulation of functional genomic data on such networks. Graphs are mathematical objects consisting of nodes and edges connecting these nodes. The *degree* or *connectivity d* of a node is the number of edges emanating from it, or, equivalently, the number of its neighbors in the graph. Multiple biological networks show a connectivity or degree distribution that is broad-tailed and often consistent with a power-law. That is, when choosing a node from such a network at random, the probability $P(d)$ that it has d interaction partners is proportional to $d^{-\gamma}$, γ being some constant that is characteristic of the network. Most prominently, this holds for metabolic networks, whose nodes can be substrates, reactions, or both, depending on the network representation one chooses, and protein interaction networks, where two nodes (proteins) are connected if they interact physically inside the cell. Broad-tailed degree distributions have also been demonstrated for other cellular networks.[1-3]

The degree distribution of a genetic network can be viewed as a feature of an organism like any other feature. It raises the same basic question: Why this and not some other degree distribution? There are three possible answers. First, a network's degree distribution could be a mere consequence of chemistry, the chemistry of DNA, RNA, and proteins, and the patterns of molecular interactions this chemistry allows. This possibility may seem far-fetched, given that

*Andreas Wagner—University of New Mexico and The Santa Fe Institute, Department of Biology, 167A Castetter Hall, Albuquerque, New Mexico 87131-1091, U.S.A. Email: wagnera@unm.edu

Power Laws, Scale-Free Networks and Genome Biology, edited by Eugene V. Koonin, Yuri I. Wolf and Georgy P. Karev.

molecular networks have many biological functions which may constrain their structure. However, this possibility is not without precedent. An illustrative example exists on a lower level of biological organization, protein structure. The thousands of currently known protein structures have a highly skewed distribution. There is a small number of 'frequent' tertiary structures, such as the TIM-barrel or the Rossman fold, found in nucleotide-binding proteins. While these folds are small in number, many proteins adopt them. Conversely, the majority of tertiary structures are 'unifolds' that may have originated only once in evolution, and are adopted by few proteins.[4-6] Does this skewed distribution of protein structures contain important information about design principles of proteins? For instance, do frequent structures have superior properties that lead to their frequent occurrence in proteins? The likely answer is no. Similarly skewed distributions of structures—a small number of excessively frequent structures and a vast majority of rare structures—occur in simple models of protein folding, models where polymers composed of parts with properties similar to amino acids fold into three-dimensional structures.[7,8] The distribution of protein structures may be a mere consequence of polymer chemistry.

The second possibility is that the degree distribution of genetic networks might somehow reflect their history, much like the jumble of streets in a medieval city reflects the city's growth over centuries. An important class of mathematical models, originally devised to explain power-law degree distributions in growing networks like the internet, do indeed link a network's history to its degree distribution. In their original and simplest incarnation, such models involve only two simple rules that change the structure of a network.[9] First, the network grows through addition of nodes. Second, newly added nodes connect to previously existing nodes, such that already highly connected nodes are more likely to receive a new connection than nodes of lesser connectivity. Over many cycles of node addition and linking to existing nodes, a power law degree distribution emerges. A great variety of variations to this model have been proposed (reviewed in ref. 10). They differ greatly in detail but retain in some way or another the rule that new connections preferably involve highly connected nodes. Importantly, most such models make a key prediction: Highly connected nodes are old nodes, nodes having been added very early in a network's history. In this sense, they link a network' degree distribution to its history.

The third possibility is that molecular networks have their degree distribution, because this structure is somehow best suited to the network's biological function. From an 'organismal design' perspective, this is the most interesting possibility. It means that natural selection has shaped the global connectivity pattern of a network, and that network structure reveals something about the design principles of biological networks.

A recent hypothesis postulates that the observed broad-tailed degree distribution of biological networks is indeed a product of natural selection.[11-13] This 'selectionist' hypothesis is based on the following observation. In networks with a broad-tailed degree distribution, the mean distance between network nodes that can be reached from each other (via a path of edges) is very small and it increases only very little upon random removal of nodes.[11] (In contrast, this mean distance or mean path length increases drastically when highly connected nodes are removed.) A network's mean path length can be thought of as a measure of how 'compact' the network is. In graphs with other degree distributions, mean path length increases more substantially upon random node removal, and the network becomes more easily fragmented into disconnected components. These observations have led to the proposition that robustly compact networks confer some advantages on cells, and that a broad-tailed degree distribution reflects the action of natural selection on the degree distribution itself. The nature of this advantage is unknown, except in the case of metabolic networks, where one can venture an informed guess.[14,15] A possible advantage of small mean path lengths in metabolic networks stems from the importance of minimizing transition times between metabolic states in response to environmental changes.[16-18] Networks with robustly small diameter may adjust more rapidly to environmental perturbations.

Metabolic Networks and Planetary Atmospheres

While the above speculation makes a weak case for a selectionist explanation of broad-tailed degree distributions in metabolic networks, another line of evidence makes a more solid case against it. One can ask whether power-law degree distributions might not be features of many or all large chemical reaction networks, whether or not part of an organism, whether or not they have a biological function which benefits from a robust network diameter. If so, then metabolic network degree distributions would join the club of other power-laws (such as Zipf's law of word frequency distributions in natural languages) whose existence does not owe credit to a benefit they provide. There is indeed evidence supporting this possibility.

Gleiss and collaborators[19] have compiled publicly available information on a class of large chemical reaction networks that exist not only outside the living, but on spatial scales many orders of magnitude larger than organisms. These are the chemical reaction networks of planetary atmospheres, networks whose structure is largely determined by the photochemistry of their component substrates. The available data stems not only from earth's atmosphere, but also from other solar planets including Venus and Jupiter, planets with chemically diverse atmospheres. These planets' atmospheres have been explored through remote spectroscopic sensing methods and by planetary probes. The chemical reaction networks in these atmospheres, despite being vastly different in chemistry, have a degree distribution consistent with a power law.[19] This suggests that power-law distributions may be very general features of chemical reaction networks. The reasons why we observe them in cellular reaction networks may have nothing to do with the robustness they may provide.

Although such comparisons to 'self-assembled' networks suggest an important influence of chemistry on metabolic network structure, another aspect of metabolic networks should not be overlooked. Metabolic networks have a history. They have not been assembled in their present state at once. They have grown, perhaps over a billion years, as organisms increased their metabolic and biosynthetic abilities. In understanding their structure, we have to take this history of biological networks into account.

We may never know enough about the history of life and metabolism to distinguish between different ways in which metabolism might have grown. However, we can address the key prediction of many network growth models I discussed above. *Are highly connected metabolites old metabolites?* The answer will contain a speculative element, because the oldest metabolites are those that arose in the earliest days of the living, close to life's origins. In addition, life forms as different as bacteria and humans have core metabolisms with a very similar structure. This suggests that the growth of metabolism has essentially been completed at the time the common ancestor of extant life emerged. Because this common ancestor does no longer exist, the detailed structure of its metabolism will remain in the dark forever. However, various hypotheses about life's origin make predictions on the chemical compounds expected to have been part of early organisms. There are several of these hypotheses, and they are complementary in the respect most important here: They emphasize the origins of different aspects of life's chemistry. Some emphasize the origins of the earliest genetic material, RNA. Others make postulates about the composition of the earliest proteins. Yet others ask about the earliest metabolites in energy metabolism. Each of them makes a statement about a different aspect of early life's chemistry.

Figure 1 shows the twelve most highly connected metabolites of the *E. coli* metabolic network graph.[14] Every single one of them has been part of early organisms according to at least one origin-of-life hypothesis. Colored in green are compounds such as coenzyme A thought to have been a part of early RNA-based organisms.[20] The RNA moieties such compounds

Twelve key metabolites in *E. coli* ranked by degree ("connectivity")

glutamate (51)
pyruvate (29)
coenzyme A (29)
α-ketoglutarate (27)
glutamine (22)
aspartate (20)
acetyl-CoA (17)
phosphoribosyl pyrophosphate (16)
tetrahydrofolate (15)
succinate (14)
3-phosphoglycerate (13)
serine (13)

Figure 1. Highly connected metabolites in *Escherichia coli* are evolutionarily old. The list shows the 12 most highly connected metabolites in the *E. coli* core intermediary metabolic network. The numbers in parentheses show the degree (number of neighbors) of a metabolite in the substrate network as defined by Wagner and Fell.[14] Green indicates proposed remnants of a surface metabolism or an RNA world. Red indicates proposed early amino acids. Blue indicates proposed early metabolites (in the tricarboxylic acid cycle or glycolysis). The network was generated after the elimination of the compounds NAD, ATP, and their derivatives. These are even more highly connected than the compounds shown here. They are also evolutionarily ancient. A color version of this figure is available online at http://www.Eurekah.com.

contain are present in all organismal lineages. Some compounds in this group, such as tetrahydrofolate and coenzyme A, are thought to have played a role in precellular life that may have taken place on polykationic surfaces. These compounds are elongate molecules with one anionic terminus. They are therefore able to flexibly tether other molecules to the substrate, thus localizing them while simultaneously increasing their potential to react with other compounds.[21] Colored in red in Figure 1 are amino acids that were part of early proteins, based on likely scenarios for the early evolution of the genetic code.[22] Shown in blue are compounds likely to have been a part of early energy and biosynthetic metabolism. Glycolysis and the TCA cycle are perhaps the most ancient metabolic pathways, and various of their intermediates (α-ketoglutarate, succinate, pyruvate, 3-phosphoglycerate) occur in Figure 1.[20,22-26] The potential relation between evolutionary history and connectivity of metabolites corroborates a postulate put forth by Morowitz,[23] namely that intermediary metabolism recapitulates the evolution of biochemistry.

In sum, the observation that power law degree distributions occur in self-assembled chemical reaction networks that were never under the influence of natural selection suggests that such distributions are a rather common feature of such networks. Natural selection on the level of this degree distribution is thus unnecessary to understand their origin. Metabolic networks have grown by addition of new metabolites, and their degree distribution is in tentative agreement with a general prediction of many network growth models: Highly connected metabolites tend to be phylogenetically old metabolites, metabolites that have been added very early in the evolution of metabolism.

Protein Interaction Networks

In contrast to chemical reaction networks, large and self-assembled protein interaction networks do not exist outside living cells. Thus, we can not hope to use arguments from self-assembled networks to argue for or against the role of natural selection in explaining a protein network's degree distribution. However, two different lines of evidence speak to this question for protein networks. The first class of evidence regards a corollary of the hypothesis that the degree distribution observed in genetic networks is a by-product of selection for 'robust compactness'. In networks with a broad-tailed degree distribution, mean path length increases drastically upon removal of highly connected nodes, as opposed to the removal of lowly connected nodes, which does not change dramatically. If it is network compactness that matters to the organism, then removal of highly connected nodes should have more severe effects on the fitness of the organism than removal of less highly connected nodes. This prediction of the selectionist hypothesis can be tested with a publicly available collection of yeast gene-knockout (synthetic-null) mutant strains.[27] Each strain of this collection lacks one gene, and the resulting change in growth rate has been measured under a variety of environmental conditions.[27-29] Jeong and collaborators[13] first showed that a correlation between the effect of a gene-knockout mutation and the encoded protein's degree exists. Figure 2 illustrates this correlation with more recent data.[28]

The interpretation of data like that shown in Figure 2 faces multiple problems, aside from the fact that the association between protein degree and mutational effect is weak. The first problem is conceptual. While removal of highly connected proteins may have more severe effects on a cell, the reasons might have nothing to do with an altered network topology. For example, high connectedness may simple be an indicator that a protein acts in a variety of different cellular processes, hence the more severe defect when the protein is eliminated from a cell. Other problems in interpreting associations like that shown in Figure 2 are technical. First, the resolution at which the effect of a gene knock-out mutation on growth rate can be measured is very low. Much smaller fitness differences between wild-type and mutant cells than one can observe in the laboratory may lead to elimination of a mutant in the wild. Second, gene knock-out effects are usually measured only in one or a few laboratory environments, not in the myriad of conditions in which they could manifest themselves in the wild. Third, laboratory assays of gene knock-out effects usually measure only one or a few components of fitness—most prominently growth rate—and leave others, such as cell survival under starvation untouched. Because of these problems, it is not clear whether laboratory gene knock-out experiments measure quantities that reliably indicate the effects of such mutations on an organism's ability to survive and reproduce in the wild.

These technical problems—but not the previous, conceptual one—could be overcome with an evolutionary approach. Here, one assesses not gene knockout effects but the rate at which different proteins in a protein interaction network evolve. Specifically, one asks whether highly connected proteins have evolved more slowly than lowly connected proteins. If this is the case, then one can argue that their evolution is more severely constrained. Several pertinent studies are available.[30-33] Their results differ in details, partly because they are sensitive to which of several available protein interaction data sets one uses.[30] However, their main conclusion is the same. If there are differences in the evolutionary rates of proteins in a network, they are not due to the differential effects these proteins have on a network's compactness. Thus, evolutionary studies do not support the notion that natural selection for robust compactness is responsible for the broad-tailed degree distribution of protein interaction networks.

A completely different approach to testing the selectionist hypothesis is encapsulated in the following question. *Can we explain the structure of protein interaction networks from processes of molecular evolution whose rates we can estimate, without resorting to natural selection acting on*

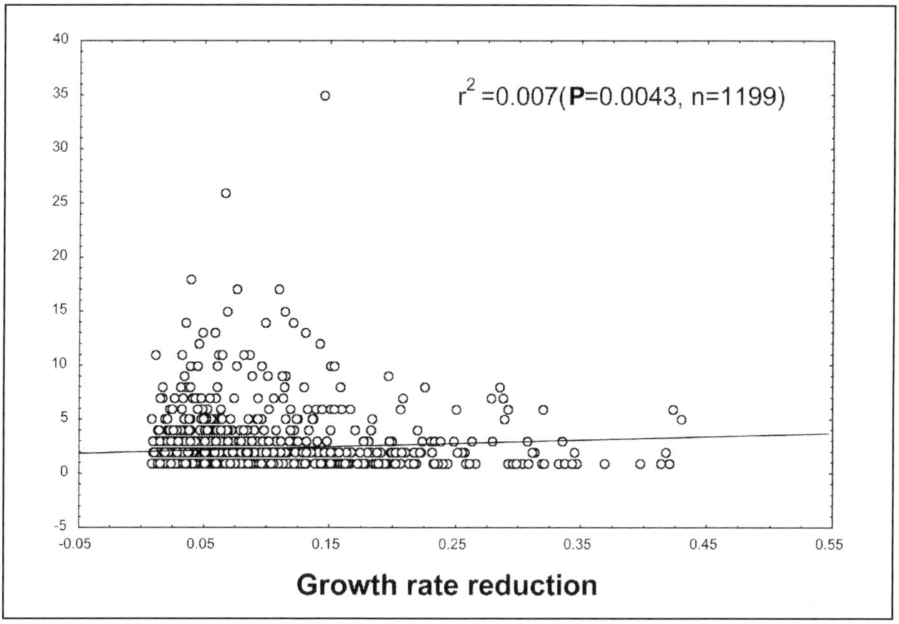

Figure 2. A weak but significant correlation between protein degree and gene knockout effect. Information on protein degrees shown here was obtained by pooling data from three independent sources, two large-scale protein interaction studies,[38,42] and a public data base of protein interactions[39] from which all interactions generated with the yeast two-hybrid assay had been eliminated. The horizontal axis shows the difference in the growth rate of a gene knock-out strain between the growth medium (among five different media) in which the strain grew at the highest rate, and the medium in which it grew at the lowest rate, as reported by Steinmetz and collaborators.[28] Growth rates are measured relative to a large pool of yeast gene deletion strains.[28] For most genes, the growth rate difference is an indicator of the largest gene knockout effect among the tested growth media. An analogous analysis using the growth rate change of a gene knockout mutation in only rich medium (YPD) yields the same results (not shown).

the network as a whole? The answer is yes.[34] Such an explanation may still involve natural selection, but on a *local* instead of a *global* scale. For example, whenever a mutation causes a new interaction between two proteins to occur, natural selection may determine whether this interaction becomes fixed in a population or eliminated from it, depending on whether the interaction is beneficial, neutral, or deleterious. However, this is selection acting on individual interactions rather than global properties of an entire network.

In a previous contribution, I have proposed an explanation of the protein interaction network's degree distribution from purely local processes such as gene duplications and mutations that generate new interactions and cause others to disappear.[34] The rate at which some of these processes occur can be roughly estimated from available protein interaction data, and based on these estimates, one can establish a quantitative mathematical model that explains the network's structure. This explanation falls within a class of models for network evolution that involve preferential attachment, that is, highly connected proteins are more likely to evolve new interactions than other proteins. Empirical data supports the notion that preferential attachment occurs in protein interaction networks, as shown in Figure 3. Others have also proposed models of protein network evolution,[35] models that differ in important details but that have one key commonality: They do not require natural selection on a global network feature,

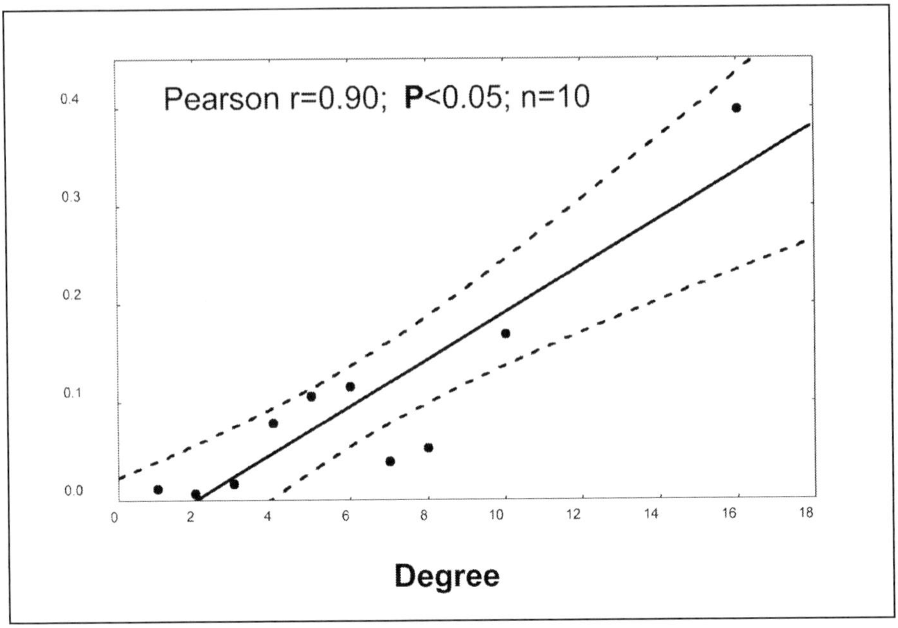

Figure 3. Preferential attachment in protein interaction networks. The horizontal axis shows protein degree *d*. The vertical axis shows the likelihood P_d that a protein of degree *d* evolves new interactions. This likelihood can be estimated from the number of newly evolved interactions between products of paralogous genes, as detailed in reference 34. For all member genes of a paralogous gene pair with a newly evolved interaction since their duplication, I determined the number I_d of those genes whose encoded proteins had *d* interactions to proteins different from its paralogue. To account for the fact that proteins of different degree occur at different frequencies in the network, I then divided this number by the relative frequency f_d of proteins of degree *d* in the network, and normalized the resulting quantity to obtain P_d, i.e., $P_d = (I_d/f_d))/\Sigma d\,(I_d/f_d)$. There is a strong, approximately linear association between protein degree and the likelihood to evolve new interactions. From Figure 5 in reference 34.

but they explain the network's structure from evolutionary events on the small, local scale of individual proteins.

Many models of network evolution based on preferential attachment predict that highly connected network nodes should be old nodes, nodes that were added very early in a network's history.[36] They should have arisen early in the evolution of the network. Because the protein interaction network shows preferential attachment (Fig. 3), the question arises whether such an association between protein age and connectivity exists. Specifically, one can ask whether highly connected proteins are phylogenetically old. Phylogenetically old proteins should have a wider taxonomic distribution than more recently arisen proteins. In two complementary analyses, I thus asked whether highly connected proteins have a wider phylogenetic distribution than less highly connected proteins.

Connectivity and Protein Age

For the first of these analyses, I used the fully sequenced genomes of six maximally diverse species. They represent fungi (*Schizosaccharomyces pombe*), protists (*Plasmodium falciparum*), plants (*Arabidopsis thaliana*), animals (*Drosophila melanogaster*), eubacteria (*Escherichia coli*), and archaea (*Methanococcus janaschii*). For each of the proteins in the protein interaction network of baker's yeast (*Saccharomyces cerevisiae*) I used gapped BLAST[37] to ask how many of

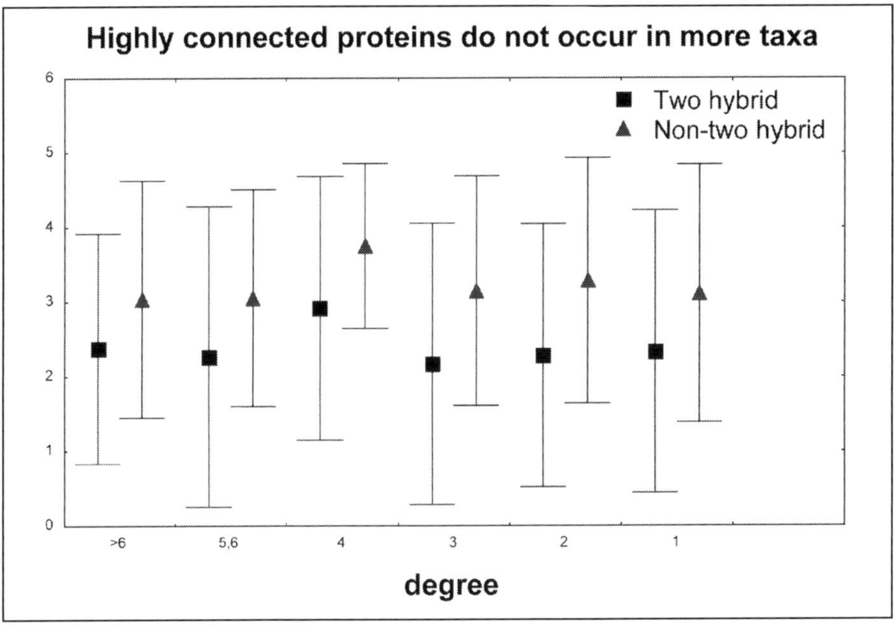

Figure 4. The vertical axis shows the average number of genomes (± one s.d.) among six fully sequenced genomes that contain at least one protein homologous to proteins whose degree is indicated on the horizontal axis. The analysis is based on two different data sets on yeast protein interactions, one ('two hybrid') from a high-throughput experiment using the yeast-two hybrid assay to identify such interactions,[38] the other ('non-two hybrid') from a publicly available database on protein interactions from which I eliminated all data generated with the two-hybrid assay.[39] Protein comparisons are based on the following six maximally diverse fully sequenced and publicly available genomes: *Schizosaccharomyces pombe* (www.sanger.ac.uk), *Plasmodium falciparum* (www.plasmodb.org), *Arabidopsis thaliana* (www.tigr.org), *Drosophila melanogaster* (www.fruitfly.org), *Escherichia coli* K12-MG1655 (www.tigr.org), *Methanococcus janaschii* DSM2661 (www.tigr.org). I used gapped BLAST[37] with a threshold protein alignment score of $E < 10^{-5}$ to identify homology. Results (not shown) are qualitatively identical for threshold scores of $E < 10^{-2}$ and $E < 10^{-10}$.

these six species contain a recognizable homologue of the yeast proteins. The data in Figure 4 show the results of this analysis for a BLAST protein alignment score threshold of $E < 10^{-5}$ to identify homology. Specifically, the figure shows the average number of taxa that contain at least one homologue to a yeast protein (vertical axis) plotted against the degree of this protein in the protein interaction network. The analysis shown is based on two different data sets of yeast protein interactions.[38,39] If highly connected proteins are phylogenetically old, then highly connected proteins should occur in significantly more of the six taxa than lowly connected proteins. The data of Figure 4, however, does not support this pattern. Figure 5 shows a complementary analysis, where I plotted average protein degree against the number of the six taxa in which a protein's homologue is found. If more widely distributed proteins are more highly connected, then they should have a higher degree. The data does not support this association either. Alignment score thresholds of $E < 10^{-2}$ and $E < 10^{-10}$ yield the same conclusion (data not shown).

In a second analysis, I cast my net wider than just the above six fully sequenced genome. I arbitrarily chose 15 highly connected proteins (degree > 4) and 15 proteins with low connectivity (degree one) from the yeast protein interaction network.[38] For each of these thirty proteins, I asked whether it has at least one homologue in any of six broad taxonomic groups:

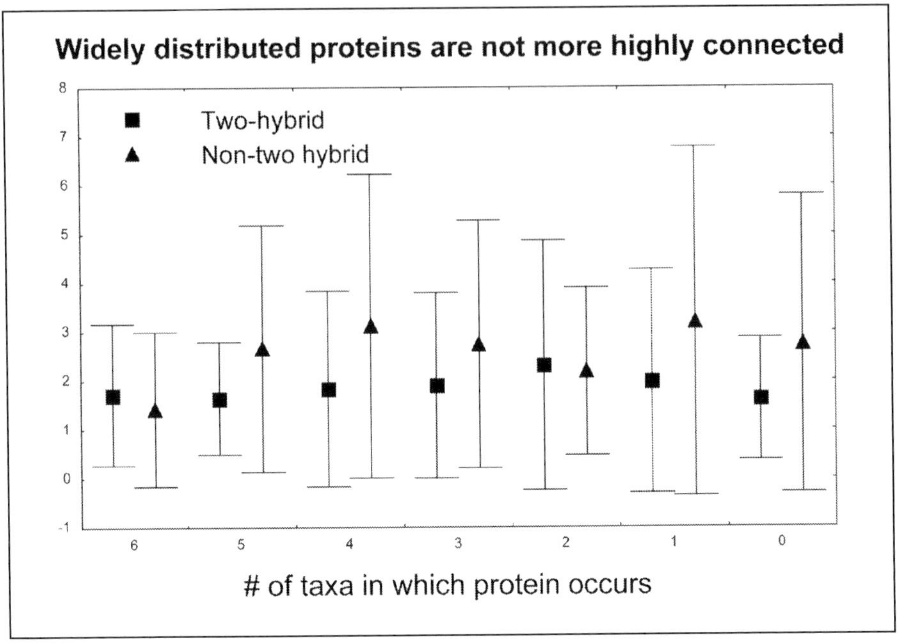

Figure 5. The vertical axis shows the average degree (± one s.d.) of proteins in the yeast protein interaction network as a function of the number of genomes—among six fully sequenced genomes—in which these proteins contain homologues, as shown on the horizontal axis. The analysis is based on two different data sets on yeast protein interactions, one ('two hybrid') using the yeast-two hybrid assay to identify such interactions,[38] the other ('non-two hybrid') a publicly available database on protein interactions from which I eliminated all data generated with the two-hybrid assay.[39] Protein comparisons are based on the following six maximally diverse fully sequenced and publicly available genomes: *Schizosaccharomyces pombe* (www.sanger.ac.uk), *Plasmodium falciparum* (www.plasmodb.org), *Arabidopsis thaliana* (www.tigr.org), *Drosophila melanogaster* (www.fruitfly.org), *Escherichia coli* K12-MG1655 (www.tigr.org), *Methanococcus janaschii* DSM2661 (www.tigr.org). For the data shown, I used gapped BLAST[37] with a threshold protein alignment score of $E < 10^{-5}$ to identify homology. Results (not shown) are qualitatively identical for threshold scores of $E < 10^{-2}$ and $E < 10^{-10}$.

metazoa, plants, protists, fungi (exclusive *Saccharomyces* spp.), eubacteria, and archaea. Table 1 summarizes the results. Seven out of 15 highly connected proteins and six out of 15 proteins with degree one have homologues in all eukaryotes. The same proportion (12 out of 15) of highly connected proteins and proteins with degree one have homologues in fungi outside the genus *Saccharomyces*. The same holds also for proteins that have no homologues outside this genus (3 out of 15 proteins). Based on this data, it appears that highly connected yeast proteins are not phylogenetically older than proteins of low degree.

While this finding is at first sight puzzling, the following analysis suggests a mundane explanation. This explanation emerges from a stochastic model of how the number of a protein's interaction partners changes over time. Consider one protein in a protein interaction network and denote as D_t the number of proteins this protein interacts with. If time t is measured in suitable discrete units, such as million years, then the change of this variable over time can be represented by a first order Markov process.[40] Specifically, designate as p_i the probability that the protein gains an interaction, that is, that its degree increases by one (through a mutation that has become fixed in a population). Formally $p_i = \text{Prob}(D_t = i+1 | D_{t-1} = i)$. Similarly, denote

Table 1. Taxonomic distribution of proteins with different connectivity in the yeast protein interaction network

High Degree Proteins

Name	Deg	M	Pr	P	F	B	Ar
GLC1	12	+++	+++	+++	+++	-	-
CDC7	11	+++	+++	+++	+++	-	-
PHO85	10	+++	+++	+++	+++	++	-
LSM4	9	+++	-	+	+	-	-
SAP4	8	-	-	-	+++	-	-
CSM1	8	-	-	-	-	-	-
YCK2	7	+++	+++	+++	+++	+	-
YIL105C	7	-	-	-	++	-	-
MET30	7	+++	++	++	+++	++	-
YDL012C	7	-	-	-	-	-	-
CLB2	6	+++	++	+++	+++	-	-
CVT19	6	-	-	-	-	-	-
ERF2	6	+++	+	+++	+++	-	-
CUP2	6	-	-	-	++	-	-
RPC19	5	++	-	-	++	-	-

Low Degree Proteins

Name	Degree	M	Pr	P	F	E	Ar
VPS4	1	++	+++	+++	+++	+++	+++
RHO1	1	++	+++	+++	+++	-	-
KRE6	1	-	-	-	+++	-	-
SMK1	1	++	+++	+++	+++	+	-
RLF2	1	-	-	-	-	-	-
YPR011C	1	+++	++	+++	++	-	-
YPR008W	1	-	-	-	++	-	-
APM1	1	+++	-	+++	+++	-	-
VIK1	1	-	-	-	-	-	-
HRR25	1	+++	+++	+++	+++	+	-
MKK2	1	+++	+++	+++	+++	-	-
YPL110C	1	+++	-	++	+++	-	-
MET31	1	-	-	-	-	-	-
YPL019C	1	-	++	-	+++	-	-
GDH1	1	-	+++	+++	+++	+++	-

The upper and lower parts of the table show the phylogenetic distribution of 15 arbitrarily chosen high and low degree proteins from publicly available yeast protein interaction data.[38] Gapped BLAST[37] was used to search for homologs to these yeast proteins in the GenBank database (www.ncbi.nlm.nih.gov). Columns in the table correspond to the following broad taxonomic groups. Metazoa (M), Protists (Pr), Plants (P), Fungi (F, exclusive of the genus Saccharomyces), Eubacteria (E) and Archaea (Ar). A '+' indicates that the respective protein has at least one putative homologue within the respective taxonomic group with a BLAST amino acid alignment score of $E < 10^{-10}$. '++' and '+++' indicate at least one homologue with $E < 10^{-20}$ and $E < 10^{-30}$, respectively.

the probability that the protein loses an interaction by q_i (q_i = Prob($D_t = i\text{-}1|D_{t\text{-}1} = i$)). Finally, let r_i denote the probability that D_t does not change between t-1 and t. This simple framework can capture a variety of observations. For instance, in an earlier contribution I suggested that the rate at which interactions get added and eliminated from the network must be approximately balanced, because of the high observed rate of interaction turnover.[34] This translates into $p_i \approx q_i$ for all i. In addition, the observation that proteins with more interaction partners show a greater turnover of interactions (Fig. 3) can be captured as a dependency of p_i on i, e.g., $p_i = i \times c$, where c is some constant.

A quantity of interest in this stochastic process is the expected waiting time until a protein first returns to the state $D_t = i$, i.e., $m_i = E(T_i|D_0 = i)$, where E indicates the expected value of the random variable T_i: = min{t>0: $D_t = i$}, which measures the time until the protein first visits state i. For $i = 0$, this expected time m_i is closely related to the residence time of a protein in the network, that is, the time during which a protein has a degree greater than zero. Quantities like m_i are difficult to calculate because we do not know how p_i, q_i and r_i depend on i, especially for large i. However, it is noteworthy that if the above assumptions held for arbitrarily large i, then this stochastic process would belong in the class of null-recurrent Markov processes,[41] whose expected waiting time to return to any state (not only $i = 0$) is infinite, and can thus not be calculated. We can, however, calculate related quantities that may explain why highly connected proteins are not necessarily phylogenetically old. Consider a protein with degree 1. What is the expected time until such a protein loses this interaction—and thus ceases to be part of the network—assuming that this protein never attains a degree higher than one? If we denote as τ the random variable measuring this time, then its distribution is given by Prob($\tau = k$) = $q_1 r_1^{k-1}$, which is essentially a geometric distribution. Its mean and variance are given by E(τ) = $q_1/(1-r_1)^2$, and Var(τ) = $r_1 q_1/(1-r_1)^3$. Order-of-magnitude estimates for upper bounds on the probabilities p_1 and q_1 suggest that they are of the order of 6×10^{-4} per protein and million year.[34] Using these values, E(τ) calculates as 416 million years, and its standard deviation as 588 million years. In other words, even a protein of low degree that does not acquire any further interactions through mutations takes more than an expected 400 million years to lose its only interaction, with an enormous standard deviation. For proteins that acquire more interactions in the course of evolution, this expected time would be much larger. Considering the standard deviation in and by itself, it is then hardly surprising that we can not distinguish proteins of different degrees by their phylogenetic distribution. The time for which even low degree proteins reside in the network can vary over an enormous range, a range greater than the time elapsed since the Cambrian radiation. A statistical test could not distinguish between the age of high and low-connectivity proteins if their residence time in a network can vary so widely.

Conclusions

In sum, I have reviewed evidence pertaining to the hypothesis that natural selection acts on the global structure of cellular networks and is responsible for their broad-tailed degree distribution. While associations between gene knock-out effects and protein degree weakly support this hypothesis for protein interaction networks, evolutionary studies and explanations of network structure based on purely local processes argue against it. I showed that the great dispersion of time for which proteins may reside in a network can obscure expected differences in the taxonomic distribution of highly and lowly connected proteins. Similar to metabolic reaction networks, where chemistry itself is an important factor shaping a network's structure, the minor role for natural selection in optimizing a network's degree distribution suggests an important role for protein chemistry in determining this distribution. Which of a protein's chemical features, such as domain composition or surface properties, renders some proteins highly connected? What aspect of protein chemistry is responsible for the observation that highly connected proteins show a greater evolutionary turnover of interactions? The answers to these and other questions are contained in accumulating structural data on thousands of proteins.

Acknowledgements

I would like to thank the Santa Fe Institute for its continued support, as well as the NIH for its support through grant GM063882.

References

1. Rzhetsky A, Gomez SM. Birth of scale-free molecular networks and the number of distinct DNA and protein domains per genome. Bioinformatics 2001; 17(10):988-996.
2. Wuchty S. Scale-free behavior in protein domain networks. Mol Biol Evol 2001; 18(9):1694-1702.
3. Wuchty S. Interaction and domain networks of yeast. Proteomics 2002; 2(12):1715-1723.
4. Koonin E, Wolf Y, Karev G. The structure of the protein universe and genome evolution. Nature 2002; 420(6912):218-223.
5. Branden C, Tooze J. Introduction to protein structure. New York: Garland, 1999.
6. Nagano N, Orengo C, Thornton J. One fold with many functions: The evolutionary relationships between TIM barrel families based on their sequences, structures and functions. J Mol Biol 2002; 321(5):741-765.
7. Li W-H. Molecular Evolution Massachusetts: Sinauer, 1997.
8. Bornberg-Bauer E. How are model protein structures distributed in sequence space? Biophys J 1997; 73(5):2393-2403.
9. Barabasi A-L, Albert R, Jeong H. Mean-field theory for scale-free random networks. Physica A 1999; 272(1-2):173-187.
10. Albert R, Barabasi A-L. Statistical mechanics of complex networks. Reviews of Modern Physics 2002; 47(1):47-94
11. Albert R, Jeong H, Barabasi AL. Error and attack tolerance of complex networks. Nature 2000; 406(6794):378-382.
12. Jeong H, Tombor B, Albert R et al. The large-scale organization of metabolic networks. Nature 2000; 407:651-654.
13. Jeong H, Mason SP, Barabasi A-L et al. Lethality and centrality in protein networks. Nature 2001; 411:41-42.
14. Wagner A, Fell D. The small world inside large metabolic networks. Proc Roy Soc London Ser B 2001; 280:1803-1810.
15. Fell D, Wagner A. The small world of metabolism. Nat Biotechnol 2000; 18:1121-1122.
16. Cascante M, Melendez—Hevia E, Kholodenko BN et al. Control analysis of transit—time for free and enzyme—bound metabolites - physiological and evolutionary significance of metabolic response—times. Biochem J 1995; 308:895-899.
17. Easterby JS. The effect of feedback on pathway transient response. Biochem J 1986; 233:871-875.
18. Schuster S, Heinrich R. Time hierarchy in enzymatic-reaction chains resulting from optimality principles. J Theor Biol 1987; 129(2):189-209.
19. Gleiss PM, Stadler PF, Wagner A et al. Small cycles in small worlds. Advances in Complex Systems 2001; 4:207-226.
20. Benner SA, Ellington AD, Tauer A. Modern metabolism as a palimpsest of the RNA world. Proc Natl Acad Sci USA 1989; 86:7054-7058.
21. Wachtershauser G. Before enzymes and templates: Theory of surface metabolism. Microbiol Rev 1988; 52:452-484.
22. Kuhn H, Waser J. On the origin of the genetic code. FEBS letters 1994; 352:259-264.
23. Morowitz HJ. Beginnings of Cellular Life. New Haven: Yale University Press, 1992.
24. Taylor BL, Coates D. The code within the codons. Biosystems 1989; 22:177-187.
25. Waddell TG, Bruce GK. A new theory on the origin and evolution of the citric acid cycle. Microbiologia Sem 1995; 11:243-250.

26. Lahav N. Biogenesis. New York: Oxford University Press, 1999.
27. Giaever G, Chu AM, Ni L et al. Functional profiling of the Saccharomyces cerevisiae genome. Nature 2002; 418(6896):387-391.
28. Steinmetz L, Scharfe C, Deutschbauer A et al. Systematic screen for human disease genes in yeast. Nat Genet 2002; 31(4):400-404.
29. Winzeler EA, Shoemaker DD, Astromoff A et al. Functional characterization of the S. cerevisiae genome by gene deletion and parallel analysis. Science 1999; 285(#5429):901-906.
30. Hahn M, Conant GC, Wagner A. Molecular evolution in large genetic networks: Does connectivity equal constraint? J Mol Evol. 2004; 58(2):203-11.
31. Fraser HB, Wall DP, Hirsh AE. A simple dependence between protein evolution rate and the number of protein-protein interactions. BMC Evol Biol 2003; 3:11.
32. Jordan IK, Wolf YI, Koonin EV. No simple dependence between protein evolution rate and the number of protein-protein interactions: Only the most prolific interactors tend to evolve slowly. BMC Evol Biol 2003; 3:1.
33. Jordan IK, Wolf YI, Koonin EV. Correction: No simple dependence between protein evolution rate and the number of protein-protein interactions: Only the most prolific interactors evolve slowly. BMC Evol Biol 2003; 3:5.
34. Wagner A. How large protein interaction networks evolve. Proc R Soc Lond B Biol Sci 2003; 270:457-466.
35. Sole RV, Pastor-Satorras R, Smith ED et al. A model of large-scale proteome evolution. Advances in Complex Systems 2002; 5:43-54
36. Albert R, Barabasi AL. Statistical mechanics of complex networks. Reviews of Modern Physics 2002; 74(1):47-97.
37. Altschul SF, Madden TL, Schaffer AA et al. Gapped blast and psi-blast : A new generation of protein database search programs. Nucleic Acids Res 1997; 25(17):3389-3402.
38. Uetz P, Giot L, Cagney G et al. A comprehensive analysis of protein-protein interactions in Saccharomyces cerevisiae. Nature 2000; 403(6770):623-627.
39. Mewes HW, Heumann K, Kaps A et al. MIPS: A database for genomes and protein sequences. Nucleic Acids Res 1999; 27:44-48.
40. Karlin S. A first course in stochastic processes. New York: Academic Press, 1975.
41. Kulkarni VG. Modeling and analysis of stochastic systems. New York: Chapman & Hall, 1995.
42. Ito T, Chiba T, Ozawa R et al. A comprehensive two-hybrid analysis to explore the yeast protein interactome. Proc Natl Acad Sci USA 2001; 98(8):4569-4574.

Chapter 5

The *Drosophila* Protein Interaction Network May Be neither Power-Law nor Scale-Free

J.S. Bader*

Abstract

Scale-free networks have become a topic of intense interest because of the potential to develop theories universally applicable to networks representing social interactions, internet connectivity, and biological processes. Scale-free topology is associated with power-law distributions of connectivity, in which most network components have only few connections while a very few components are extremely highly-connected. Here we investigate the power-law and scale-free properties of the network corresponding to protein-protein interactions in *Drosophila melanogaster*. We examine power-law behavior with a standard statistical technique designed to distinguish whether a power-law fit is adequate to describe the vertex degree distribution. We find that the degree distribution for the entire network, consisting of baits and preys, decays faster than power law. This fit may be confounded by artifacts of the screening procedure. The prey-only degree distribution is less likely to be confounded by the screening procedure, and is fit adequately by a power-law. When only the biologically relevant interactions are considered, however, the degree distribution again decays faster than power-law. Thus, power-law behavior may reflect interactions that are observed in vitro but not in vivo. We next describe an algorithm that may be able to extract the true distribution from the incomplete data. Finally, we investigate scale-free properties by characterizing organizational patterns over increasing spatial scales. We provide evidence for the existence of a length-scale that characterizes organization in the network. The existence of such a correlation length stands in contrast to scale-free networks, in which no length scale is special. These results suggest that the *Drosophila* protein interaction network may not be power-law and is not scale-free.

Introduction

Technological advances now permit the elucidation of biological networks on a genome scale. A recent report described using the two-hybrid method to identify the protein-protein interactions that underlie the protein complexes and multi-complex pathways in *Drosophila melanogaster*.[1] This was the first large-scale protein-protein interaction network determined for a metazoan and builds on earlier screens conducted for *Saccharomyces cerevisiae*.[2,3] Protein interaction networks have also been probed using mass spectrometry of protein complexes.[4,5] Chromatin immunoprecipitation experiments provide analogous data to support the identification of protein-DNA interactions and transcriptional regulatory networks.[6]

*J.S. Bader—Department of Biomedical Engineering, Johns Hopkins University, 201C Clark Hall, 3400 N. Charles St., Baltimore, Maryland 21218, U.S.A. Email: joel.bader@jhu.edu

Power Laws, Scale-Free Networks and Genome Biology, edited by Eugene V. Koonin, Yuri I. Wolf and Georgy P. Karev.

The advent of these large-scale data sets has stimulated interest in developing theories that explain how biological networks are organized, and how they continue to be shaped by evolution. Biological networks are examples of small-world networks, which occupy the middle ground between completely regular networks and random networks.[7] Like regular networks, small-world networks have clusters of interconnected vertices; like random networks, most pairs of vertices are connected by a short path of links.

A notable feature of biological networks, also represented in social networks and air travel networks, is the existence of hubs, a colloquial term for vertices with a high number of connections compared to typical vertices. In many examples of networks, the vertex degree distribution shows a decay much slower than the Gaussian-like distribution for random networks; the tail end of this distribution corresponds to hubs. A preferential or rich-get-richer model, in which new connections are biased towards vertices that are already highly connected, leads to a power-law vertex degree distribution and also leads to scale-free self-organization.[8]

Much is known about the statistical physics of self-organized networks and self-organized criticality. If biological networks are a realization of self-organized criticality, then universal results and scaling laws from physics should apply. Alternately, if biological networks have properties that differ from scale-free networks, then new theoretical developments may be required to describe their properties and behavior. Thus, there is great interest in determining whether evolution has shaped biological networks to resemble self-organized, scale-free networks.

Two testable hypotheses of a scale-free model are a power-law distribution of connections per vertex and the lack of a characteristic length scale for network organization. Here we examine whether the topology of the *Drosophila* protein interaction network follows these hypotheses.

One possible test for power-law behavior is to calculate the empirical vertex degree distribution, then check whether a power-law functional form provides a better fit than an alternate functional form, typically exponential or normal. This type of test does not confirm that the distribution follows a power law; instead, it indicates that a power law fits less poorly than other functional forms. We describe how a vertex degree distribution may be fit by a family of functions with terms corresponding to power-law decay, exponential decay, and even faster decay. We use standard statistical procedures to decide whether the optimal fit is a power law, or whether it is faster than power law.

These statistical tests are not straightforward due to experimental limitations in sampling the *Drosophila* network. First, at most 96 colonies were sequenced for each bait, which artificially limits the vertex degree observed for a bait protein. Next, some prey libraries were often obtained from mRNA libraries with power-law distributions of transcript abundances, which influences the bait-prey combinations that are sampled. Finally, analysis of the entire network is questionable as only 25% of the network was judged to be high-confidence for biological relevance. We address each of these factors in turn in an analysis of the *Drosophila* vertex degree distribution. We then describe an approach to predict the true vertex degree distribution from the incomplete distribution derived from partial sampling of the network.

We investigate structure in the network by characterizing motifs that represent order. A simple motif is the existence of a triangle, three vertices connected one to the next. The ratio of the number of observed to expected triangles is synonymous with the standard definition of the clustering coefficient for a small world network. This statistic is sensitive to organization over short length scales. To investigate organization over longer length scales, we investigate the distribution of longer cycles. This distribution may be measured for an empirical network. We introduce a simple mathematical model for a network organized to have one level of clustering and show that this model is sufficient to explain the observed cycle distribution. Thus, there is no need to invoke a continuous distribution of length scales. Moreover, the one-level model immediately yields a characteristic, testable scaling length for the network, which again stands in contrast to scale-free behavior.

We conclude with a discussion of models that may provide an improved theoretical framework for understanding the properties of biological networks and the evolutionary forces that have shaped them.

Observed Vertex Degree Distribution

The empirical network that serves as the basis of this study was obtained by a two-hybrid screen for protein-protein interactions in *Drosophila melanogaster*.[1] This work describes 20,405 pair-wise interactions involving 7046 proteins. One of the difficulties in analyzing topological properties of this network is that some properties are biased by the screening procedure. For example, the number of preys for each bait is limited by the number of colonies sequenced for each mating, typically 96 (see ref. 1 and references therein for background on the two-hybrid system). Furthermore, screens conducted with a prey library obtained directly from mRNA may be biased for highly-expressed proteins. Finally, interactions identified by the two-hybrid method often have questionable biological relevance.[9-11]

To address each of these points, we constructed a series of three vertex degree distributions. The first degree distribution considered all of the interactions observed for *Drosophila*, excluding a small number of self-interactions. This network included 20,278 interactions between 7000 proteins.

Next, we addressed the limited number of colonies sequenced for each bait by considering only the degree distribution for prey proteins. While each bait participates in at most 96 interactions due to the limited sampling, there should be no such limitation for the number of times a prey is observed as an interaction partner.

One possible limitation on observing a prey, however, is that it is not represented in the prey library. Or, if it is present, its abundance may be low. These preys may be systematically under-represented in two-hybrid screens that use prey libraries obtained directly from mRNA isolated from cells. Indeed, mRNA abundances may themselves follow a power-law distribution, with a few highly-represented species being responsible for the majority of the mRNA mass.

The screens in reference 1 attempted to avoid this limitation by conducting two-hybrid screens with two independent prey libraries. One library was obtained by isolating mRNA from *Drosophila* embryonic and adult developmental stages, then using these transcripts to generate a prey library. The second library was obtained by individually amplifying every predicted *Drosophila* gene from a cDNA library, with a 75% success rate in generating a prey with verified insert sequence and size. The resulting 10,787 preys were then pooled to yield a nearly perfectly normalized library. This pool was mated to each of 10,623 baits, yielding 31,270 bait-prey pairs whose sequences could be mapped to release 3.1 of the gene annotations from the Berkeley Drosophila Genome Project. After removing a small number of self-interactions, these 31,270 pairs corresponded to10,161 unique prey-bait pairs between 3001 preys and 2657 baits.

One of the challenges in interpreting two-hybrid data is that many of the interactions observed are spurious, with questionable biological relevance. The biological relevance of the interactions reported in reference 1 was modeled statistically. Each interaction was assigned a confidence score in the range from 0 to 1, with 0.5 as the approximate dividing point between low-confidence (< 0.5) and high-confidence (> 0.5) of biological relevance. Starting with the preys from the normalized screen described above, we obtained a third degree distribution by considering only the high-confidence interactions. This network corresponded to 3574 unique prey-bait pairs between 2093 preys and 2130 baits.

Vertex Degree Distributions and Power-Law Fits

We write $N(k)$ as the number of proteins in a network with exactly k neighbors. A typical procedure used to assess power-law behavior is to fit $N(k)$ by a power-law, exponential, or Gaussian decay, each corresponding to different random models,

$$\hat{N}(k) = \begin{Bmatrix} \exp(A + a_0 \log k) \\ \exp(A + a_1 k) \\ \exp(A + a_1 k + a_2 k^2) \end{Bmatrix}$$

where A in each case is an appropriate normalization constant. Typically, the fit is performed on a log-scale to minimize the quantity χ,[2]

$$\chi^2 = \sum_k \left[\log N(k) - \log \hat{N}(k)\right]^2,$$

and the functional form giving the smallest χ^2 is accepted as describing the decay. In certain cases, it may be preferable to normalize each term by its anticipated variance $1/\hat{N}(k)$ or construct bins on a logarithmic scale to avoid small counts.

Rather than comparing three separate fits, a standard statistical procedure is to assess the significance of a series of models of increasing complexity. The log-scale, exponential, and Gaussian functional forms for $N(k)$ can be considered as the first three terms in a Taylor series expansion,

$$\hat{N}(k) = \exp\left(A + a_0 \log k + a_1 k + a_2 k^2 + \ldots\right).$$

Using forward regression, we can build models of increasing complexity. The first model (power-law decay) fits uses only the terms A and a_0, with all higher order coefficients set equal to 0; the second model (power-law truncated by exponential decay) uses A, a_0, and a_1, with all higher order coefficients set to 0; and so on. We then assess the significance of each model relative to the preceding model using analysis of variance, an F test of the reduction of χ^2.

We used bins of width 1 for simplicity. To account for bins with a small number of counts, including empty bins where log $N(k)$ is undefined, we performed a series of fits. First, we excluded bins with 0 counts. Next, we excluded bins with 0 or 1 counts and refit the model. Next, we excluded bins with 0, 1, or 2 counts and refit the model. We reasoned that power-law behavior is typically defined only when at least 3 orders of magnitudes of power-law decay are observed. Thus, shaving off the tail of the distribution should not affect our ability to define a power-law exponent. Equivalently, a robust power-law fit should not require inclusion of the bins with the fewest number of counts.

The empirical vertex degree distributions are depicted in Figure 1A. The values estimated for the power-law decay parameter a_0 and the exponential decay parameter a_1 are depicted in Figures 1B,C. We note first that the degree distribution for the entire network has a highly significant exponential decay component when the bins with count 1 are excluded from the fit. The estimate for the exponent is approximately -0.03. The inverse of this exponent is of the same magnitude as the 48 to 96 clones sequenced for each bait, which supports the hypothesis that the experimental design has contributed to a decay that is faster than power law.

In contrast to baits, preys do not have an interaction count that is limited by the experimental design. Thus, we hypothesize that the degree distribution of preys (for each prey, the number of unique baits that identified it as an interaction partner) is less affected by sampling limitations. The vertex degree distribution for preys appears to be a power-law distribution (Fig. 1A). This appearance is borne out statistically with a power-law parameter ranging from -2.0 to -2.3 (Fig. 1B) and an exponential decay parameter that is indistinguishable from 0 at a p-value of 0.05 (Fig. 1C). These parameter estimates are stable over a range of minimum bin

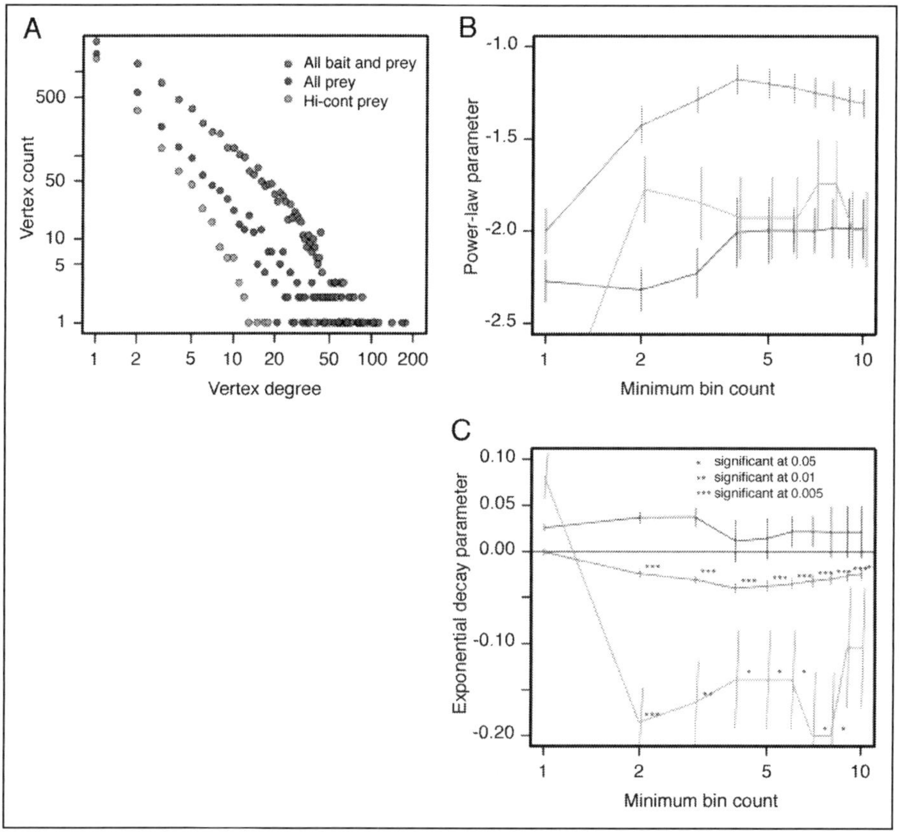

Figure 1. Vertex degree distributions and fits.

counts used for the fit. Thus, the network obtained from highly-normalized preys is described well by power-law decay.

The interactions that contribute to the prey degree distribution are not all biologically relevant. Some may reflect assay-specific artifacts having to do with the two-hybrid reporter system; other false-positive interactions may arise from weak or nonspecific interactions; other interactions may be highly reproducible in vitro but involve proteins that are restricted to different developmental stages or tissues and never interact in vivo. We hypothesize that the power-law distribution may reflect overall properties of the distribution of binding constants between proteins rather than the number of biologically-relevant interaction partners for a given protein. To test this hypothesis, we eliminated low-confidence interactions from the prey vertex degree distribution and refit the distribution. While the power-law decay parameter is close to the previous estimate of -2 (Fig. 1B), the exponential-decay parameter is now significantly different from 0. The value of the exponential-decay parameter ranges from -0.15 to -0.2 depending on the minimum bin count, at a p-value of 0.05 to 0.005.

Synthesizing these results, we propose that the power-law distribution reflects the distribution of binding constants observed between proteins in vitro, while an exponential distribution reflects the number of interaction partners that are relevant in vivo. We note that if the preferential attachment model is modified to restrict the potential interaction partners of a node, an exponential-decay degree distribution may result.[12] Restriction may occur naturally in a biological system due to temporal or spatial restrictions to protein expression.

We temper the strength of our conclusion by noting that the number of interaction partners of a protein was used as one of the explanatory variables in deriving the statistical confidence scores.[1]

Bait and Prey Distributions Reconciled

How far are we in identifying the complete set of protein-protein interactions that constitute the *Drosophila* protein interaction network? Answering this question has practical importance for estimating the cost of generating complete maps for other metazoan species, including human. Knowledge of the estimated complexity of metazoan protein interaction networks provides necessary input and tests for developing theories of biological network evolution.

Ideally, we would want to see consistency for the list of interaction partners for a protein when used as a bait and when used as a prey. A discrepancy in the interaction partners may indicate assay artifacts, for example DNA-binding activity in a protein that is used as part of the activation-domain fusion in the two-hybrid system. Unfortunately, when the number of interaction partners identified for a bait protein is limited by the number of clones selected for sequencing, it is not possible to compare the list of bait interaction partners and prey interaction partners directly. Even the simple summary statistic of the number of interaction partners may not be comparable.

Here we describe an approach that may be successful in reconciling the counts of interaction partners for a protein that is used as both a bait and a prey by inferring the true distribution of interaction partners for a protein when used as a bait based on the limited experimental evidence. Using the subscript i to label the bait, we sequence k_i clones that correspond to x_i unique prey proteins. Our goal is to estimate the total number of interaction partners m_i from which the xi observed prey species have been drawn. Thus, we wish to estimate

$$\Pr(\mathbf{m}|\mathbf{x},\mathbf{k}) = \sum_{\theta} \Pr(\mathbf{m},\theta|\mathbf{x},\mathbf{k}) = \sum_{\theta} \Pr(\mathbf{m}|\theta,\mathbf{x},\mathbf{k}) \Pr(\theta|\mathbf{x},\mathbf{k})$$

where $\mathbf{m}$ represents the underlying interaction counts $\{ m_i \}$ for each bait, x represents the observed counts $\{ x_i \}$, $\mathbf{k}$ represents the number of clones sequenced for each bait, and θ represents the set of parameters describing the vertex degree distribution.

To simplify the following discussion, will make the assumption that the probability distribution for θ is highly peaked near its maximum likelihood estimate θ^{ML},

$$\theta^{ML} = \arg\max \Pr(\theta|\mathbf{x},\mathbf{k}) = \arg\max \Pr(\mathbf{x}|\theta,\mathbf{k}),$$

which corresponds to a flat prior distribution for θ. In this case,

$$\Pr(m_i|x_i,k_i) \approx \Pr(m_i|x_i,k_i,\theta^{ML}) = \frac{\Pr(x_i|m_i,k_i)\Pr(m_i|\theta^{ML})}{\sum_{m'=1}^{\infty} \Pr(x_i|m',k_i)\Pr(m'|\theta^{ML})}.$$

The maximum likelihood estimate for q can itself be obtained as

$$\theta^{ML} = \arg\max_{\theta} \prod_{i} \sum_{m_i=1}^{\infty} \Pr(x_i|m_i,k_i) \Pr(m_i|\theta).$$

Once functional forms for $\Pr(m|\theta)$ and $\Pr(x|m,k)$ have been specified, the maximum likelihood estimate for θ may be found by direct maximization or expectation maximization; then, $\Pr(\mathbf{m}|\mathbf{x},\mathbf{k})$ is readily calculated.

Guided by the results for the prey vertex degree distribution, we suggest that an appropriate functional form for $\Pr(m|\theta)$ is

$$\Pr(m|\theta) = \exp(-\alpha_0 \ln m - \alpha_1 m) / \left[\alpha_1^{\alpha_0 - 1} \Gamma(1 - \alpha_0, \alpha_1)\right],$$

where θ is defined by the pair of parameters (α_0, α_1), we have moved to a continuous distribution for $m \geq 1$, and we have recognized that the normalization constant is simply related to the standard definition of the incomplete gamma function.

Finally, we require the probability distribution $\Pr(x_i|m_i,k_i)$. We again make a simplifying assumption, that each of the m_i interaction partners is equally likely to yield a clone. While this assumption is unlikely to be true even for a normalized prey library, it provides a necessary starting point for more advanced analysis. We make a second simplifying assumption that the presence or absence of each of the m_i prey species in the k_i clones is independent. In this case, $\Pr(x_i|m_i,k_i)$ is given by a binomial distribution,

$$\Pr(x|m,k) = \frac{m!}{x!(m-x)!}\left[1-\left(1-m^{-1}\right)^k\right]^x \left[1-m^{-1}\right]^{m-x}.$$

We speculate that the parameters α_0 and α_1 determined by the approach outlined above for the baits should agree with the power-law decay and exponential decay parameters obtained by fitting the vertex degree distribution for the preys. We further suggest that a discrepancy between the number of interaction partners estimated for a bait and the number observed when the same protein is used as a prey could signal an assay-dependent artifact. The value of the formulas provided above is that they provide a quantitative method for making such a determination. These formulas also provide a link between the amount of work done, measured by the parameter k, and the completeness of the map, measured by the factor x/m.

Note also that the prey vertex degree distribution is also affected by the finite sampling of each bait. The approach described here for the bait distribution could be modified to yield an improved estimate for the number of interaction partners for each prey.

Determining the Length Scale of the Network

We move now from examining the properties of the vertices to more global measures of network organization. A defining property of small-world networks is clustering: a pair of vertices connected to a third vertex has an enhanced likelihood of being connected to each other. This property has been used to infer unobserved connections in protein interaction networks.[13]

Clustering as typically defined measures the ratio of the number of triangles in a network (three vertices connected together) to the number of triangles observed in an equivalent randomized network. To examine clustering over longer length scales, we defined a more generalized measure by counting the number of higher-order cycles in a network, and comparing this count to the distribution observed in an equivalent randomized network.

Solutions to the cycle-count distribution may also be obtained from mathematical models of random networks. The mathematical models we describe below permit closed-form analytic solutions for the cycle-count distribution. The key simplifying assumption of the mathematical models are simplified vertex degree distributions. As described below, we check these assumptions by also performing simulation studies of an ensemble of randomized networks. Agreement between theory and simulation bolsters the credibility of the theory and suggests that the cycle-count distribution may be insensitive to certain details of the vertex degree distribution.

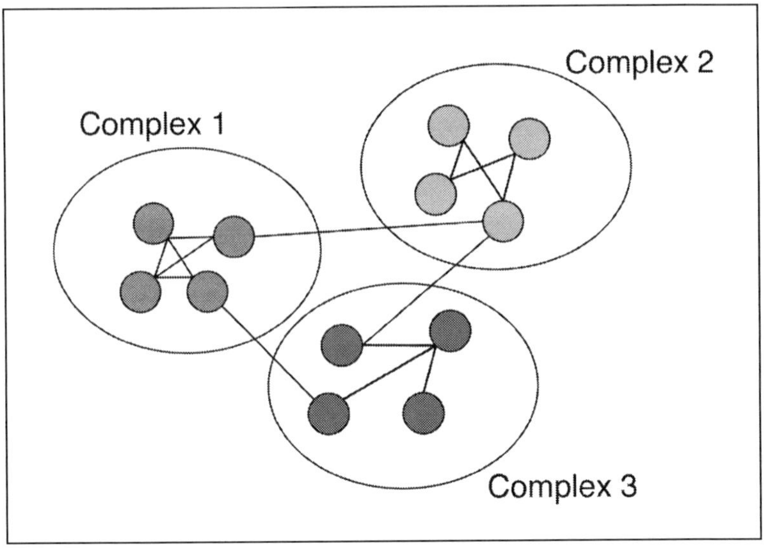

Figure 2. Illustration of enhanced connectivity within protein complexes.

We start with a mathematical model in which pairs of proteins in a network with N total proteins are connected with probability $J/(N-1)$. When J is much smaller than N, which is expected for biological networks, this yields a Poisson vertex degree distribution with mean J. The number of cycles of length L, $N(L)$, is equal to the number of ways to select L proteins times the probability that each is connected to the next, divided by the symmetry factor $2L$ for a closed loop of length L,

$$N(L)=\frac{N!}{(N-L)!}\left(\frac{J}{N-1}\right)^L\frac{1}{2L}.$$

The initial combinatorial factor is, $N^L\cdot[1+O(L^2/N)]$, where O is the symbol for asymptotic order, and the factor involving J is $(J/N)^L\cdot[1+O(L/N)]$. The simplified result for the cycle-count distribution for a random network is

$$N(L)=\left(J^L/2L\right)\cdot\left[1+O\left(L^2/N\right)\right].$$

We anticipate, however, that biological networks will be characterized by structure corresponding to protein complexes, with enhanced connectivity for proteins within a complex. This picture is illustrated in Figure 2. We incorporated this behavior in a random model in which each protein is assigned to one of several protein complexes, and the probability of an interaction is enhanced for proteins residing in the same complex.

To make the model explicit, we define K complexes with P proteins in each complex, giving $N = KP$ total proteins. Proteins within a complex are connected with probability, $J_W/(P-1)$ yielding J_W within-complex neighbors on average, and proteins in different complexes are connected with probability, $J_B/(N-K)$ yielding J_B between-complex neighbors on average.

Cycles in this model can exist entirely within a single complex, or can cross between complexes. We first calculate the cycle-count distribution for single-complex cycles,

$$N_w(L)=K\frac{P!}{(P-L)!}\left(\frac{J_W}{P-1}\right)^L\frac{1}{2L}.$$

Here we keep the first two terms for the combinatorial factor,

$$\frac{P!}{(P-L)!} = P^L \prod_{i=0}^{L-1}\left(1-\frac{i}{P}\right) = P^L\left[1-\frac{L(L-1)}{2P}+O\left(\frac{L^4}{P^2}\right)\right].$$

$$= P^L\left[\exp\left(-\frac{L^2-L}{2P}\right)+O\left(\frac{L^4}{P^2}\right)\right]$$

Multiplying this expression with the remaining terms yields the final expression

$$N_W(L) = K\frac{J_W^L}{2L}\left[\exp\left(-\frac{L^2-L}{2P}\right)+O\left(\frac{L^4}{P^2}\right)\right]$$

for the within-complex cycles.

The expected number of between-complex cycles may be estimated as

$$N_B(L) = \frac{N!}{(N-L)!}\left[\frac{P-1}{N-1}\cdot\frac{J_W}{P-1}+\frac{N-K}{N-1}\cdot\frac{J_B}{N-K}\right]^L$$

$$= \frac{N!}{(N-L)!}\left(\frac{J_W+J_B}{N-1}\right)^L = \frac{(J_W+J_B)^L}{2L}\left[1+O\left(\frac{L^2}{N}\right)\right]$$

We have assumed that a pair of proteins in the cycle is from the same complex with probability $(P\text{-}1)/(N\text{-}1)$ and from different complexes with probability $(N\text{-}K)/(N\text{-}1)$. To lowest order, the final combined expression for the cycle-count distribution is

$$N(L) = N_W(L)+N_B(L) = K\frac{J_W^L}{2L}\exp\left(-\frac{L^2-L}{2P}\right)+\frac{(J_W+J_B)^L}{2L}.$$

These mathematical models for unstructured and structured random networks were used to describe high-confidence protein-protein interaction networks obtained experimentally for *Drosophila* and also for *Saccharomyces* (Fig. 3). In both cases, the unstructured model failed to fit the experimental data, while the structured model provided an excellent fit.

Details of the fit are available in the original publications.[1,11] In particular, the approximations used in deriving the analytical formulas for the mathematical models were checked computationally by simulating a series of random structured and unstructured networks, and calculating the cycle-count distribution for an ensemble of random networks. The numerically-converged simulation results were indistinguishable from the analytic theory.[11]

The existence of structure in the network immediately suggests the existence of a length-scale that is characteristic of the structure. The elements of structure are K protein complexes, each with an average of P proteins and an average of J_W within-complex neighbors. From a central protein in a complex, there are approximately J_W proteins within 1 link, J_W^2 proteins within 2 links, and J_W^d proteins within d links. As d approaches the typical number of links separating pairs of proteins in a complex, J_W^d should approach the total number P of proteins in the complex. This relationship leads to the scaling law

$$d \sim \log_{J_W} P$$

for an appropriate topological correlation length in the network. The meaning of the correlation length is that proteins within d links of each other are likely to be in the same complex, and hence should have correlated properties.

We have checked this behavior for the *Saccharomyces* protein interaction network for correlations described by protein annotations and gene expression measurements.[11] The cycle-count

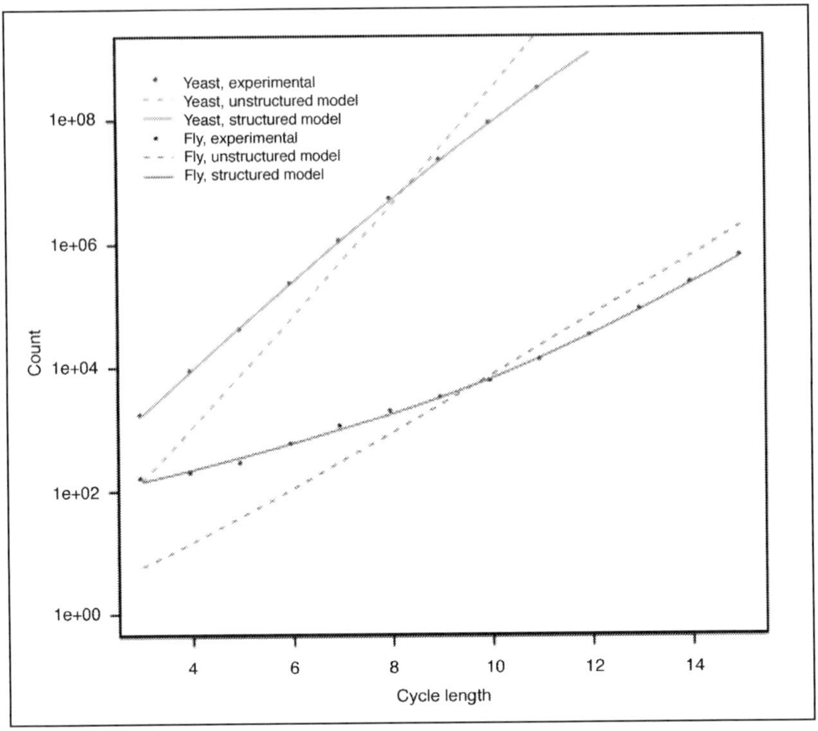

Figure 3. Structured network models provide an excellent fit to cycle-count distributions for experimental protein-protein interaction networks; unstructured network models do not fit the experimental data at all.

distribution suggested that typical protein complexes had 13-17 proteins, with 8-9 within-complex neighbors as determined by two-hybrid interactions or coimmunoprecipitation. This leads to a topological correlation length of 1.2 to 1.3 links. Next, correlations between all pairs of proteins in the network were calculated and averaged as a function of the number of links separating the pair. Correlations derived from database annotations and transcript profiling decayed exponentially with a decay constant of 1.2-2.0 links, which suggests that the topological properties of the network reflect biologically-relevant organization.

Although the cycle-count distribution may be fit adequately using a simple model for network structure, this does not rule out more complicated structure. A more elaborate model may be generated by assuming additional levels of structure with sub-complexes intermediate between proteins and complexes, or pathways represented by aggregates of complexes. Evidence for sub-complexes is provided by genetic interactions, which decay over a length scale that is about half the topological correlation length.[11] The levels of organization may also become quasi-continuous, as in fractal models for network organization.[14]

Conclusion

We have described approaches for a quantitative analysis of the topological properties of biological networks, with a focus on protein interaction networks revealed by recent large-scale experimental screens.

One of the much-discussed properties complex networks is an apparent power-law distribution of connectivity. We have employed a standard statistical approach to determine whether a vertex-degree distribution follows a power law, or whether it decays more rapidly. Employing this method, we show that the full network does not follow a power law, likely because the experimental design truncates the power-law decay prematurely. We then show that a subset of the data, in which the connectivity is less likely to be truncated by the experimental design, is described by power-law decay.

Before concluding that the biological network is characterized by a power-law degree distribution, we then focus on the subset of interactions predicted to be biologically-relevant based on a statistical model. The vertex-degree distribution again decays faster than a power law, and is described instead by power-law decay truncated by exponential decay.

We suggest that a reason for this behavior is that the experimental data contains a mixture of in vitro binding events, which follow a power-law distribution, and biologically-relevant binding events, which follow a faster-decaying distribution. This type of mixture model was, in fact, used in developing the statistical model that predicted biological relevance.

We have also provided a method for extracting the true vertex degree distribution (perhaps including interactions corresponding to false-positive artifacts from the in vitro method) from the limited experimental observations. We are currently attempting to use this method to predict how much of the actual network has been revealed by the experiments to date.

Finally, we have described a property, the existence of cycles in a network, that is sensitive to large-scale organization. We have used this property to extract a topological correlation length from protein interaction networks. Correlations based on biological properties appear to decay exponentially according to the same correlation length. The existence of a topological length scale would appear to exclude scale-free models of biological organization, which by definition lack a characteristic length scale. A better description may be given by models in which a constraint limits the evolution of a network,[12,15] or in scale-rich models in which the cost of satisfying a constraint depends on its scale.[16,17]

References

1. Giot L, Bader JS, Brouwer C et al. A protein interaction map of Drosophila melanogaster. Science 2003; 302:1727-1736.
2. Uetz P, Giot L, Cagney G et al. A comprehensive analysis of protein-protein interactions in Saccharomyces cerevisiae. Nature 2000; 403:623-627.
3. Ito T, Chiba T, Ozawa R et al. A comprehensive two-hybrid analysis to explore the yeast protein interactome. Proc Natl Acad Sci USA 2001; 98:4569-4574.
4. Gavin AC, Bosche M, Krause R et al. Functional organization of the yeast proteome by systematic analysis of protein complexes. Nature 2002; 415:141-147.
5. Ho Y, Gruhler A, Heilbut A et al. Systematic identification of protein complexes in Saccharomyces cerevisiae by mass spectrometry. Nature 2002; 415:180-183.
6. Lee TI, Rinaldi NJ, Robert F et al. Transcriptional regulatory networks in Saccharomyces cerevisiae. Science 2002; 298:799-804.
7. Watts DJ, Strogatz SH. Collective dynamics of 'small-world' networks. Nature 1998; 393:440-442.
8. Barabasi AL, Albert R. Emergence of scaling in random networks. Science 1999; 286:509-512.
9. von Mering C et al. Comparative assessment of large-scale data sets of protein-protein interactions. Nature 2002; 417:399-403.
10. Deane CM, Salwinski L, Xenarios I et al. Protein interactions: two methods for assessment of the reliability of high throughput observations. Mol Cell Proteomics 2002; 1:349-356.
11. Bader JS, Chaudhuri A, Rothberg JM et al. Gaining confidence in high-throughput protein interaction networks. Nat Biotechnol 2004; 22:00-00.
12. Mossa S, Barthelemy M, Eugene Stanley H et al. Truncation of power law behavior in "scale-free" network models due to information filtering. Phys Rev Lett 2002; 88:138701.
13. Goldberg DS, Roth FP. Assessing experimentally derived interactions in a small world. Proc Natl Acad Sci USA 2003; 100:4372-4376.

14. Ravasz E, Somera AL, Mongru DA et al. Hierarchical organization of modularity in metabolic networks. Science 2002; 297:1551-1555.
15. Jin EM, Girvan M, Newman ME. Structure of growing social networks. Phys Rev E Stat Nonlin Soft Matter Phys 2001; 64:046132.
16. Carlson JM, Doyle J. Complexity and robustness. Proc Natl Acad Sci USA 2002; 99(Suppl 1):2538-2545.
17. Newman ME, Girvan M, Farmer JD. Optimal design, robustness, and risk aversion. Phys Rev Lett 2002; 89:028301.

Chapter 6

Birth and Death Models of Genome Evolution

Georgy P. Karev, Yuri I. Wolf and Eugene V. Koonin*

Abstract

Gene duplication is the primary avenue of genome evolution. The gene repertoire of any species can be described as an ensemble of paralogous gene families, ranging in size from one to large numbers that amount to a substantial fraction of genes in the respective genome. Evolution of such an ensemble is naturally represented by a birth-and-death process, the birth of a gene being duplication, and death being gene inactivation and elimination. In addition to gene duplication and loss, evolution of gene families involves "true" innovation, i. e., appearance of genes new to the given lineage through horizontal gene transfer, emergence of genes from noncoding sequences, and change of preexisting genes beyond recognition. Assuming these three elementary processes, we developed a simple theoretical framework for analysis of genome evolution, the Birth, Death and Innovation Models (BDIMs). Comparison of the predictions made by different versions of BDIMs with empirical distributions of paralogous family size in genomes allows one to choose the adequate models. Stable family size distributions can evolve only under balanced BDIMs, in which duplication and deletion rates are asymptotically equal up to the second order. The linear BDIM, in which there is almost no dependence between the family size and birth-death rates, readily approximates the observed family size distribution at equilibrium. However, the stochastic version of this model yields unrealistic times for evolution of the large paralogous families that were detected in all genomes. In order to produce reasonable rates of family evolution, one needs to turn to nonlinear higher-degree BDIMs, which imply "interactions" between paralogs. These interactions may be interpreted as a proxy for natural selection, which should drive evolution of large paralogous families if their emergence is to be viewed as an adaptive reaction.

Power Laws, Scale-Free Networks, and Models of Genome Evolution

Power law distributions appear in an enormous variety of fundamentally different contexts. These distributions are described by the simple function $P(i) \cong ci^{-\gamma}$ where $P(i)$ is the frequency of nodes with exactly i connections or sets with exactly i members, γ is a parameter which typically assumes values between 1 and 3, and c is a normalization constant. Obviously, in double-logarithmic coordinates, the plot of P as a function of i is close to a straight line with a negative slope. The utility of these distributions was first noticed by Pareto who employed them to characterize the spread of wealth in society and later by Zipf for the description of word usage in texts.[1,2] However, it took another century after Pareto's groundbreaking work for

*Corresponding author: Eugene V. Koonin—National Center for Biotechnology Information, National Library of Medicine, National Institutes of Health, Bethesda, Maryland 20894, U.S.A. Email: koonin@ncbi.nlm.nih.gov.

Power Laws, Scale-Free Networks and Genome Biology, edited by Eugene V. Koonin, Yuri I. Wolf and Georgy P. Karev. ©2006 Eurekah.com and Springer Science+Business Media.

the power laws to take off in earnest. This happened largely in the context of network analysis, which was brought to prominence by the explosive growth of the Internet and rapidly moved into other areas, above all, biology and sociology.[3,4] The important biological examples include metabolic, expression, and protein-protein interaction networks.[5,6-9] The crucial property shared by the Internet, social, and biological networks is their scale-free character, i.e., independence of the node degree distribution on scaling.[10-12] Because the node degrees in a scale-free network are distributed according to a power law, such a network is resistant to error (i.e., random elimination of nodes) but sensitive to attack (i.e., directed removal of a highly connected node).[13] Biological networks, unlike social ones, are unlikely to be deliberately attacked, so this property seems to ensure their robustness.

Once multiple complete genomes of organisms from diverse walks of life have been sequenced, genome analyses showed that the distributions of a broad variety of genome-associated quantities followed the power laws, at least within some approximation.[5-7] Examples include the distribution of the number of transcripts per gene, the number of interactions per protein, the number of genes or pseudogenes in paralogous families, the number of connections per node in metabolic networks, and more .[6,14-16] Barabasi and coworkers realized that the mode of network evolution leading to the power law distributions and, accordingly, scale-free networks is preferential node attachment, whereby, at any moment in a network evolution, the probability of a node acquiring a new connection is proportional to the number of connections this node already has. Metaphorically (and in the spirit of the classical work of Pareto), this principle may be described as "the rich get richer" or, if one implies that nodes get highly connected through an evoluitonary process involving selection, "the fit get fitter".[3,4,9]

However, preferential attachment, being an extremely general (and important) principle associated with power law type distributions and scale-free phenomena, does not actually "explain" the emergence of these phenomena. In a sense, this explanation is almost tautological: preferential attachment seems to be the "nature" of the systems with power law distributions, which is independent of the physical identity of these systems or any specific evolutionary mechanisms. A genuine physical or biological explanation involves deciphering these mechanisms or at least identifying the simplest models of evolution that include realistic elementary events and are compatible with the observations.

Under this logic, families of paralogous genes seem to represent a perfect object for evolutionary modeling. Indeed, for these families, elementary evolutionary processes are defined naturally. By definition, paralogous families evolve by gene duplication. It has been long suspected and, with the advent of genomics, established beyond reasonable doubt that genome evolution proceeds largely by duplication of genes, gene segments, and even long genomic segments or entire genomes.[17-22] All sequenced genomes contain numerous paralogous genes, and in more complex genomes, the majority of genes have at least one paralog.[23,24] Duplication is followed by mutational diversification and gradually leads to functional differentiation of the paralogs. It is thought that such differentiation occurs via the routes of neofunctionalization (emergence, in one of the paralogs, of a new function nonexistent in the ancestral gene)[18] and, perhaps predominantly, subfunctionalization, i.e., distribution of the partitioning of subfunctions of the ancestral genes among the paralogs.[25,26] Hence, **duplication** obviously is the first elementary process of genome evolution. Genomes and gene families not only grow but often shrink or, probably most of the time, persist in equilibrium. Therefore, duplication is counter-balanced by the opposite elementary process, **gene loss**. Again, comparative genomics has shown that gene loss occurs in all species and may be extensive in certain lineages, particularly in parasites.[27-29] Finally, genes new to a given lineage may emerge either as a result of a dramatic change after duplication obliterating all "memories" of a gene's origin, or via horizontal gene transfer, or by evolution of a protein-coding gene from a noncoding sequence (rare as this latter process might be). Collectively, the contribution of these processes to genome evolution may be termed *innovation.* It seems plausible that gene duplication, gene loss, and innovation might comprise a reasonable minimal set of elementary events for modeling genome evolution. The only potential major addition could be gene rearrangement whereby genes accrete or lose domains. However, at least for first approximation modeling, these changes could fit either under duplication if they do not yield new genes without detectable relationships to preexisting families, or under innovation if they do. We should further note that evolutionary analysis of paralogous gene families can be naturally construed as a study of the evolution of genomes themselves if all genes are viewed as members of paralogous families, ranging in size (number of members) from 1 to N (the size of the largest family).

A natural framework for modeling evolution of gene families is a birth-and-death process, a concept well explored in many physical and chemical contexts.[30] Duplication constitutes a gene birth, and gene loss is a death event; innovation also can be readily incorporated in this context. The birth-and-death approach has been applied to modeling the evolution of paralogous genome family sizes,[16,31,32] the distribution of folds and families in the entire protein universe,[33] and protein-protein interaction networks.[34,35]

In a series of recent studies, we explored a very general class of models of genome evolution, which we dubbed BDIMs, after **b**irth-**d**eath-**i**nnovation **m**odels.[32,36,37] We found that BDIMs comprise a flexible and rich theoretical framework which allows one to explain both static (distribution of family) size and dynamic (the time required or the evolution of families of the observed size) aspects of genome evolution. Most importantly, it turned out to be possible to distinguish between different versions of BDIMs in terms of their agreement with the data.

Definitions, Assumptions and Empirical Data

We treat a genome as a "bag" of genes (gene fragments), coding for protein domains, which we will simply call **domains** for brevity (see ref. 32 for additional details and rationale). Domains are treated as independently evolving units disregarding the dependence between domains that tend to belong to the same multidomain protein. Each domain is considered to be a member of a family, which may have one or more members. Three classes of elementary events are considered:

i. domain **birth** which generates a new member in the same family as a result of gene duplication
ii. domain **death**, i.e., inactivation and/or deletion, and
iii. **innovation** which generates a new family with one member. Innovation may occur via domain evolution from a noncoding sequence or a sequence of a nonglobular protein, via horizontal gene transfer from another species, or via radical modification of a domain following a duplication. The rates of elementary events are considered to be independent of time (only homogeneous models are considered) and of the nature (structure, biological function, and other features) of individual families.

The data on the size of domain families in sequenced genomes were from the previous work.[32] Briefly, the domains were identified by comparing the CDD library of position-specific scoring matrices (PSSMs), which includes the domains from the Pfam and SMART databases, to the protein sequences from completely sequenced eukaryotic and prokaryotic genomes (http://www.ncbi.nlm.nih.gov/entrez/query.fcgi?db=Genome) using the RPS-BLAST program.[38]

We assume that: (i) time is continuous and more than one elementary event is unlikely to occur in a short time interval, (ii) all elementary events are independent of each other, and (iii) the rates of domain birth and death, respectively, λ_i and δ_i; depend on family size i only.

In a finite genome, the maximal number of domains in a family obviously cannot exceed the total number of domains and, in reality, is probably much smaller. Let N be the maximal possible number of domain family members (note that almost all of the results below are valid with $N=\infty$ under certain well defined conditions, which provide the existence of the ergodic distribution of the birth-and-death process). Let f_i be the number of domain families in i-th class, i.e., families that have exactly i domains in the given genome, $i = 1,2,\ldots N$. Originally,[32] we examined exclusively the deterministic version of BDIM:

$$d f_1(t)/dt = \nu-(\lambda_1 +\delta_1) f_1(t)+ \delta_2 f_2(t)$$

$$d f_i(t)/dt = \lambda_{i-1} f_{i-1}(t) -(\lambda_i +\delta_i) f_i(t)+ \delta_{i+1} f_{i+1}(t) \text{ for } 1<i<N, \quad (1)$$

$$d f_N(t)/dt = \lambda_{N-1} f_{N-1}(t) -\delta_N f_N(t).$$

The innovation rate, which we designate ν, is considered constant for a given genome in this model.

In the new version of the model, we also consider "virtual" families consisting of 0 domains. In this model, newborn domains are drawn from the 0 class and "dead" domains return to it. Introduction of the 0 class "closes" the model and, accordingly, transforms it into a Markov process. This provides for the possibility to explore the stochastic properties of the system. In these stochastic models, innovation was not introduced explicitly but is implied in the form of emergence of domains from the 0 class.

Let $p_i(t)$ be the frequency of a domain family of size i. Then $p_i(t)$ satisfy a well known system of forward Kolmogorov equations for birth-and-death process (see, e.g., ref. 39) which actually differs from model (1) only in the first equation:

$$d p_0(t)/dt = -\lambda_0 p_0(t)+\delta_1 p_1(t),$$

$$d p_1(t)/dt = \lambda_0 p_0(t)-(\lambda_1+\delta_1)p_1(t)+\delta_2 p_2(t),$$

$$d p_i(t)/dt = \lambda_{i-1} p_{i-1}(t)-(\lambda_i +\delta_i)p_i(t)+\delta_{i+1} p_{i+1}(t) \text{ for } 1<i<N, \quad (2)$$

$$d p_N(t)/dt = \lambda_{N-1} p_{N-1}(t)-\delta_N p_N(t).$$

The evolution of individual trajectories of the birth-and-death process $X(t)$, whose state probabilities satisfy the system (2), can be described as follows. At the starting time, the system is in some initial state x_0. The time axis $\{t \geq 0\}$ can be subdivided to intervals $[0,\tau_1)$, $[\tau_1, \tau_2)$, $[\tau_2, \tau_3)\ldots$ such that $X(t)$ is a constant on each of these intervals. If, at the moment τ_n, the system occupied the point $x_n = i$, then, at the moment τ_{n+1}, it moves either to the state i+1 with the probability $\beta_i = \lambda_i /(\lambda_i + \delta_i)$ or to the state i-1 with the probability $\mu_i = \delta_i /(\lambda_i + \delta_i)$. The *sojourn time* $t_i = \tau_{n+1} - \tau_n$ from arrival at point $x_n = i$ to exit from this point is a random variable independent of the previous history of the system and is distributed according to the exponential law $P\{t_i \geq x\} = \exp(-(\lambda_i + \delta_i)x)$. Note that the mean sojourn time in the state i is $E(t_i) = 1/(\lambda_i + \delta_i)$.

It is well known that process (2) has a unique stationary ergodic distribution $p_0,\ldots, p_N$ defined by the equalities $dp_i(t)/dt = 0$ for $0 \leq i \leq N$:

$$p_i = p_0 \prod_{j=1}^{i} \left(\lambda_{j-1} / \delta_j\right) \text{ for all } i = 1,\ldots N, \quad (3)$$

We consider also the variant of the model (2) without the 0-state; this model describes evolution of the size of a domain family that includes an indispensable (essential) gene and is not allowed to go extinct. Mathematically, the system (2) describes the state probabilities of well-known birth-and-death processes with a finite number of states and reflecting boundaries. Although this classical process has been studied in detail, it has not been previously noticed that it is a natural source of the power-law distributions.

Asymptotic Behaviors of the Ergodic Distribution of the Model

The ergodic distribution (2) is globally stable and is approached exponentially with respect to time from any initial state. We proved[32] that the asymptotic behavior of this distribution is completely defined by the "asymptotic balance" between the birth and death rates, λ_i and δ_i, as functions of i. Let us suppose that, for large i, the following expansion is valid:

$$\lambda_{i-1}/\delta_i = i^s\theta\,(1 - \gamma/i + O(1/i^2)) \tag{4}$$

where s and γ are real numbers and θ is positive. Then

i. *if* $s \neq 0$ *(nonbalanced BDIM), then* $p_i \sim \Gamma(i)^s\theta^i\, i^{-\gamma}$ where $\Gamma(i)$ is the Γ-function;
ii. *if* $s = 0$ *and* $\theta \neq 1$ *(first-order balanced BDIM), then* $p_i \sim \theta^i i^{-\gamma}$;
iii. *if* $s = 0$; $\theta = 1$ *and* $\gamma \neq 0$ *(second-order balanced BDIM), then* $p_i \sim i^{-\gamma}$;
iv. *if* $s = 0$; $\theta = 1$ *and* $\gamma = 0$ *(high-order balanced BDIM), then* $p_i \sim 1$.

Then, equilibrium frequencies of a BDIM have a power asymptotic behavior if and only if the BDIM is second-order balanced. Precise formulas for p_i can be obtained for specific forms of λ_i and δ_i (see refs. 32, 37 for details) and several of them will be considered below.

If the per-family birth and death rates linearly depend on the number of domains in a family

$$\lambda_i = \lambda(i+a),\ \delta_i = \delta(i+b) \text{ for } i>0,\ a \text{ and } b \text{ are constants} \tag{5}$$

(linear BDIM), the equilibrium distribution of domain family sizes is defined by:

$$p_i = p_0\frac{\Gamma(1+b)\,\lambda_0}{\Gamma(1+a)\,\lambda}\theta^i\frac{\Gamma(i+a)}{\Gamma(i+1+b)} \sim \theta^i i^{-\gamma},\ \text{where } \theta = \lambda/\delta,\ \gamma = 1+b-a. \tag{6}$$

A linear BDIM is, by definition, at least first-order balanced; if $\lambda = \delta$ (so that $\theta = 1$), the resulting second-order balanced linear BDIM has a power asymptotic with $\gamma = 1 + b - a$. Thus, the linear BDIM is the simplest model which yields provides the power asymptotic of the stationary state.

Previously, we applied deterministic BDIMs to approximate the observed distribution of protein domains in a variety of prokaryotic and eukaryotic genomes by minimizing the χ^2 value for the observed and predicted distributions. The simplest model that resulted in a good fit to the observed domain family size distributions for all analyzed genomes was the second-order balanced linear BDIM (Fig. 1). For all analyzed genomes, $P(\chi^2)$ for this model was >0.05, i.e., no significant difference between the model predictions and the observed data was detected.

Since the deterministic, linear, second-order balanced BDIM gave an excellent fit to the stationary distribution of gene family sizes for all analyzed genomes, we expanded the study by exploring the dynamic behavior of ensembles of gene families using stochastic versions of BDIM. Specifically, the following problems were analyzed:

1. probability of formation of the largest family from a singleton before getting to extinction;
2. mean and variance of the time required for formation of a family of a given size from a singleton, particularly, the largest identified family;
3. mean and variance of extinction time for a family of a given size;
4. mean and variance of the number of elementary events (gene duplication and elimination) prior to extinction or formation of a family of a given size.

These problems are readily solved within the framework of birth-and-death processes (see, e.g., refs. 39,40; all relevant formulas are compiled in ref. 41 (Mathematical Appendix)). Naturally, we first addressed these problems using the linear, second order balanced, stochastic BDIM, the direct counterpart of the model that successfully explained the stationary distribution of family sizes. However, as discussed below, we found that this model yielded evolutionary parameters incompatible with empirical data, which prompted us to examine more complex versions

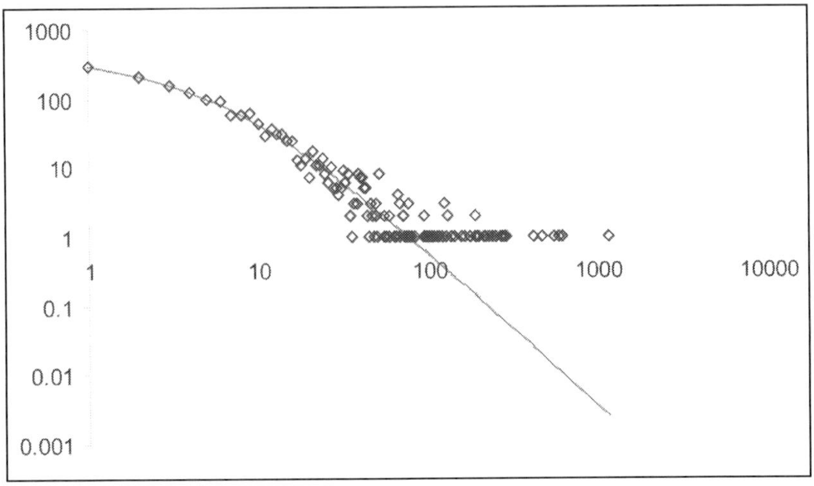

Figure 1. Fit of empirical domain family size distributions to the second-order balanced linear BDIM: *Homo sapiens.*

of BDIMs. We explored polynomial, rational, and logistic BDIMs with the aim of selecting the model that is best compatible with the data under a critical constraint: the stationary ergodic distribution of all models should be the same as it is for the original linear BDIM.

It follows from (3) that the following modification of any form of BDIM:

$$\lambda^*_i = \lambda_i g(i),\ \delta^*_i = \delta_i g(i-1) \qquad (7)$$

where $g(i)$, $i = 1,\ldots N$, is a positive function, $g(0) = 1$, results in a BDIM with the same ergodic distribution of the family sizes as the original one. We studied the class of modification (7) for the linear second order balanced BDIM with $\lambda_i = \lambda(i + a)$, $\delta_i = \lambda(i + b)$ for $i > 0$, which produce the stationary distribution $p_i \sim i^{-\gamma}$, where $\gamma = 1 + b - a$. In particular, modifications of a linear BDIM with $g(i) = (i + 1)^{d-1}$ or $g(i) = (i + 1)^{d-1}(1 - i/(N + c))$ define, respectively, broad classes of rational or logistic BDIMs with the same stationary distribution as the original linear BDIM, but with very different dynamic properties.

All stochastic models of genome evolution face an important "time unit" problem. If models (1), (2) are second order balanced, such that $\lambda = \delta$, then λ is a time-scaling constant and the models have a natural "innate" time scale measured in $1/\lambda$ units (hereinafter *internal time units*). However, if we wish to measure the time in real time units, such as years, we must estimate the parameter λ using available estimations of the duplication rate. For this purpose, we choose the average duplication rate, r_{du}. An estimate of the average duplication rate was produced by Lynch and Conery[26] by counting the number of recent duplicates in three eukaryotic genomes and dividing this number by the estimated rate of silent nucleotide substitutions. They obtained the value $r_{du} \sim 2\mathrm{x}10^{-8}$ duplications/gene/year, which we used for our calculations. The estimations of λ based on the empirical average duplication rate vary for different nonlinear BDIMs. Indeed, in terms of the model (2), the average duplication rate is, by definition,

$$r_{du} = \sum_{i=1}^{N-1} p_i \lambda_i / i.$$

Let us introduce coefficient $c_{du} = r_{du}/\lambda$, which connects the internal model parameter λ with the empirical value of r_{du} such that

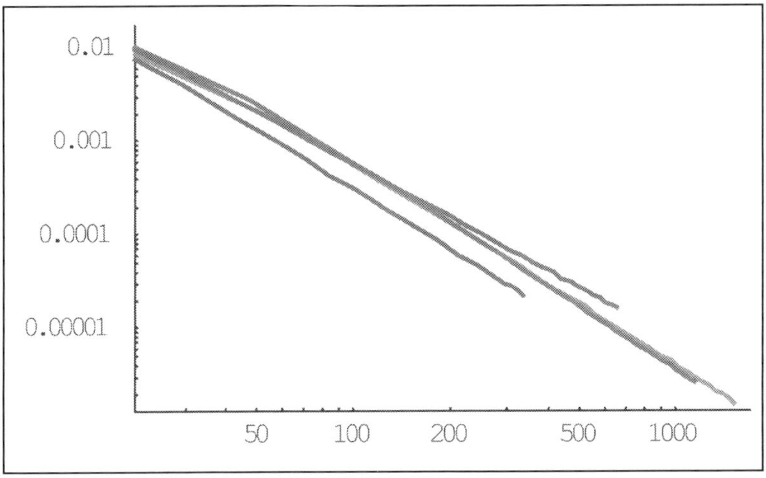

Figure 2. Probabilities of formation of families starting from a singleton, $P^1(1,n)$, versus family size (n) for the linear BDIM .The plot is in double logarithmic scale. The model parameters are for *D. melanogaster* (blue), *C. elegans* (purple), *H. sapiens* (red), *A. thaliana* (green). A color version of this figure is available online at http://www.Eurekah.com.

$$\lambda = 2 \times 10^{-8} / c_{du} . \qquad (8)$$

For all transformations (7) of the linear BDIM, the stationary probabilities p_i are the same as for the original linear model, but the birth rates λ_i and, accordingly, c_{du} vary. We show that the *internal time unit* becomes smaller with the increase of the "model degree" which results in some interesting effects discussed below.

Linear Stochastic BDIM and Its Applications

The probability of formation of a family of size n starting from a family of size i before getting to extinction can be computed with the help of known formulas for the birth-and-death process to reach state n before reaching state 0.

For the linear 2nd order balanced BDIM, the probability that a singleton expands to a family of size n before dying, $P^1(1,n)$, is

$$P^1(1,n) \cong \frac{\gamma\Gamma(1+b)}{\Gamma(1+a)} n^{-\gamma}$$

where $\gamma = 1 + b - a$, with the same power γ as the equilibrium frequencies of the families.[37] The values of probabilities $P^1(1,n)$ for different species are shown in Figure 2. These probabilities are rather small, $P^1(1,N) \sim 10^{-5}$.

The random birth-and-death process (2) certainly visits the state 0 in the course of time; this means that any domain family will eventually go extinct (and then formally can be "reborn", returning from the 0 class). The mean time of extinction of the largest family is an important characteristic of the evolutionary process described by these models. The plot of $E^1{}_n$, the mean time of extinction of the family of initial size n for the linear 2nd order balanced BDIM (measured in internal time units), versus n for different species is shown in Figure 3.

The formation time of a family of a given size was computed for the version of BDIM (2) that describes evolution of an essential gene (no 0-state). For the linear BDIM, plots of the

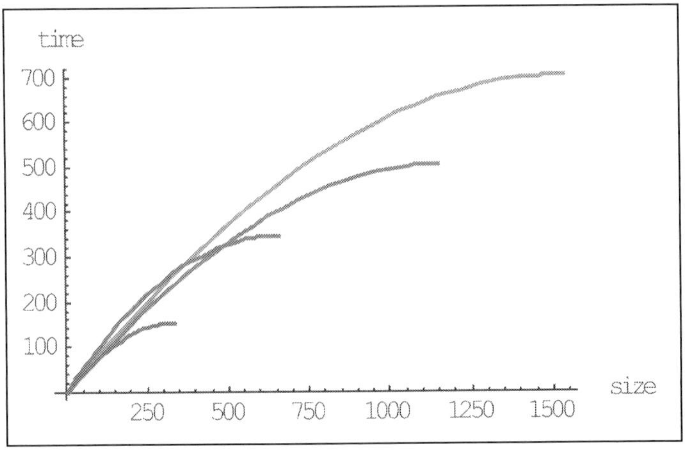

Figure 3. Mean time of extinction ($E^1{}_n$) depending on family size (n) for the linear BDIM. Time is in 1/ λ units. The model parameters are for *D. melanogaster* (blue), *H. sapiens* (red), *A. thaliana* (green), *C. elegans* (purple). A color version of this figure is available online at http://www.Eurekah.com.

mean time of formation $M^1{}_n$ (in internal time units) are shown in Figure 4. The times of formation and extinction of a family of a given size under stochastic BDIM are random variables. Thus, the questions remains how well do the **mean** values represent these variables. To address this issue, we calculated the variances and coefficients of variation of the extinction and formation times, $s^{(1)}{}_N$, and $\sigma^{(1)}{}_N$, for the linear BDIM. It should be noticed that coefficient of variation does not depend on the model parameter λ and therefore is an important and informative characteristic of the process. We found that both coefficients are very large, e.g., $s^{(1)}{}_{335}$ = 194.11 and $\sigma^{(1)}{}_{335}$ = 81.79 for *D. melanogaster*, and $s^{(1)}{}_{1151}$ = 649.8 and $\sigma^{(1)}{}_{1151}$ = 308.4 for *H. sapiens*.

Summarizing the results obtained for the stochastic characteristics of the linear BDIM, we found that, firstly, the probability of formation of a large family from a singleton is quite small ($\sim 10^{-6}$ for large genomes), and, secondly, the ratio of the mean times of formation and extinction of the largest families is very large ($\sim 0.5 \div 1 \times 10^3$). Thirdly, the coefficient c_{du} is in the range of $1 \div 3$ for the linear model and all considered species (e.g., c_{du} = 1.8 for *Dme* and c_{du} = 2.7 for *Hsa*). Using the values of this coefficient and the available estimates of gene duplication rates[26] to estimate the internal time unit, 1/λ, with formula (8), gives the mean time of formation of the largest families $M^1(1;N) \sim 10^{13} - 10^{14}$ yrs, which is three to four orders of magnitude greater than the current estimate for the age of the Universe.[42] Thus, the mean family formation times given by the linear BDIM would become realistic only if the recent analyses underestimated the gene duplication rate by a factor of $\sim 10^4$, which does not seem plausible. Accordingly, the linear BDIM cannot provide an adequate description of genome evolution, at least when only the mean time of family formation is considered. As mentioned above, the coefficient of variation of the family formation time is extremely large ($\sim$100), so large deviations from the mean time, up to 2 orders of magnitude, are not improbable. At the end of this chapter (see *Conclustions and Perspective*), this issue is addressed with an alternative approach, namely computer simulations, which exploit the large number of families in evolving genomes and the substantial variance of the times of their formation. First, however, we consider nonlinear, higher order models that have the potential to yield faster evolution, allowing for the formation of large families observed in complex genomes.

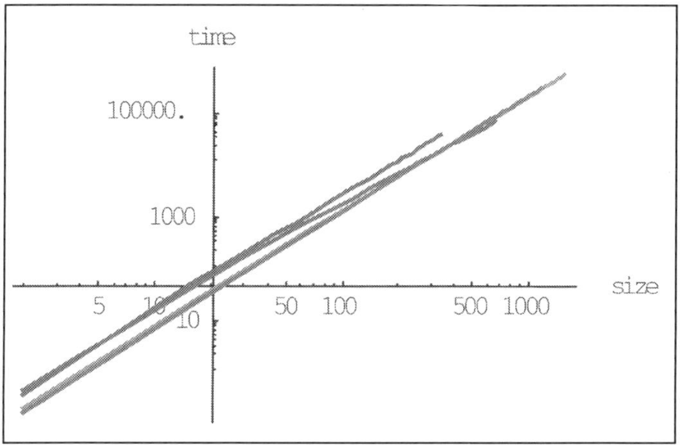

Figure 4. Mean time of formation M^1_n (in $1/\lambda$ units) depending on family size (n) for the linear BDIM (in double logarithmic scale). The model parameters are for *D. melanogaster* (blue), *H. sapiens* (red), *A. thaliana* (green), *C. elegans* (purple). A color version of this figure is available online at http://www.Eurekah.com.

Nonlinear Modifications of the Model: Polynomial BDIM

Now our goal is to modify the linear BDIM in such a way that:

i. the stationary distribution of the family sizes stays the same
ii. the modified models account for much more rapid evolution of family sizes for realistic values of duplication rates
iii. the ratio of the mean times of family formation and extinction is substantially greater than it is under the linear model.

To provide for fast evolution of gene families, the mean sojourn times t_i in each state i, $t_i = 1/(\lambda_i + \delta_i)$, should be substantially shorter than those in the linear model. The appropriate models can be constructed through the transformation of birth and death rates (7); it should be emphasized that the values of parameters a and b that have been previously determined for the linear BDIM to fit empirical data for different species can be employed for the modified models.

We show that nonlinear BDIM modifications with the function $g(i) = (i + 1)^{d-1}$ satisfy the requirements (i) and (ii) and partially solve the problem (iii). Informally, polynomial BDIMs with $d = 2,3$, etc. can be introduced as follows. Under the linear BDIM, the dependence of the birth and death rates on family size is very weak such that the growth rate is almost proportional to the family size (and asymptotically tends to exact proportionality for large i) and there is no significant feedback between the family size and growth rate. In contrast, the quadratic model ($d = 2$) includes dependence of birth and death rates of individual domains on pairwise interactions, whereas higher order models imply more complex interactions. In general, if interactions of the order d are postulated, then the second order balanced BDIM has λ_i and δ_i, which are polynomials on i with the same degree d and the same higher coefficients. Nonlinear polynomial BDIMs predict evolutionary rates that are dramatically greater then those for the linear BDIM; in particular, the **quadratic** BDIM with birth and death rates defined as

$$\lambda_i = \lambda(i + a)(i + 1),\ \delta_i = \lambda(i + b)i, \tag{9}$$

is close to the best (in an exact sense explained below) modification of the initial linear BDIM for achieving the fastest possible evolution rate. The probability of formation of a family of size

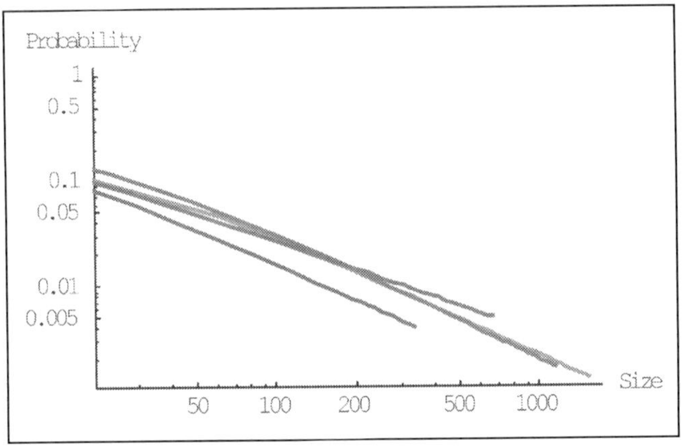

Figure 5. Probability of formation of families starting from a singleton, $P^{(2)}(1,n)$, versus family size (n) for the quadratic BDIM. The plot is in double logarithmic scale. The model parameters are for *D. melanogaster* (blue), *C. elegans* (purple), *H. sapiens* (red), *A. thaliana* (green). A color version of this figure is available online at http://www.Eurekah.com.

n from a singleton before extinction is much greater for the quadratic model than for the linear model, $P^{(2)}(1,N)\sim10^{-3}$ (Fig. 5). The mean time of extinction for the quadratic BDIM, $E^{(2)}{}_{n}$, measured in *internal time units*, is much shorter than that for the linear model (Fig. 6). The mean time of formation of families from an essential singleton, $M^{(2)}{}_{n}$, is also much shorter than that for the linear model (Fig. 7). For the largest family, the mean time of formation is approximately two orders of magnitude greater than the mean extinction time; thus, the transition from the linear to the quadratic BDIM lowers the ratio of extinction to formation time by about an order of magnitude, thus partially solving the problem (iii).

The coefficient c_{du} for the quadratic model is in the range 5 ÷ 25 for all considered species (e.g., $c_{du} = 11.67$ for *Dme* and $c_{du} = 24.48$ for *Hsa*). Using these values to estimate the value of the internal time unit, $1/\lambda$, with formula (8), gives the mean time of formation of the largest

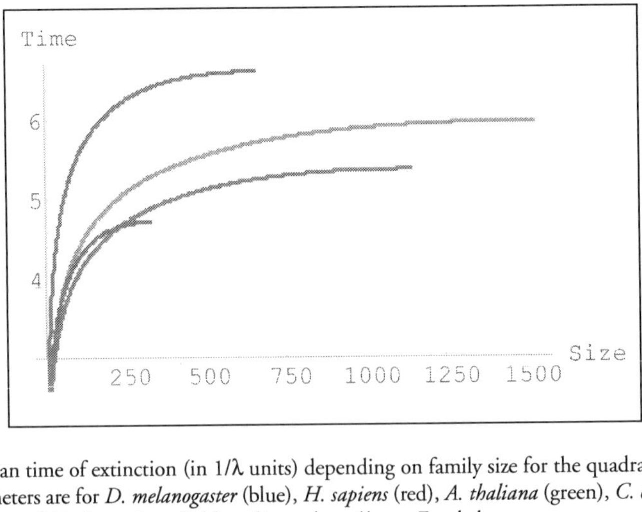

Figure 6. Mean time of extinction (in $1/\lambda$ units) depending on family size for the quadratic BDIM. The model parameters are for *D. melanogaster* (blue), *H. sapiens* (red), *A. thaliana* (green), *C. elegans* (purple). A color version of this figure is available online at http://www.Eurekah.com.

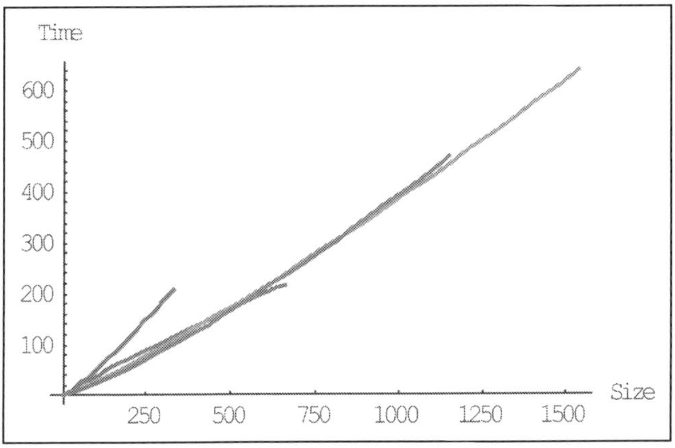

Figure 7. Mean time of formation (in $1/\lambda$ units) depending on family size for the quadratic BDIM. The model parameters are for *D. melanogaster* (blue), *H. sapiens* (red), *A. thaliana* (green), *C. elegans* (purple). A color version of this figure is available online at http://www.Eurekah.com.

families $M^{(2)}(1;N) \sim 10^{11}$ yrs [120.4 Ga (billion years) for *Dma* and 573.9 Ga for *Hsa*), which is still two orders of magnitude greater than the actual evolution time (~1 billion years). Although the variance of family formation time for the quadratic BDIM was significantly lower than for the linear BDIM, the coefficients of variation of the formation time and extinction time for the largest family under the quadratic BDIM were ~1.3 - 1.5 times greater than the respective coefficients for the linear BDIM. Due to the large variation coefficients, the actual values of the formation/extinction times of the largest family could differ from the mean value by up to two orders of magnitude with a relatively high probability (see discussion below).

We further examined the cubic BDIM and showed that this model is characterized by extremely high, apparently unrealistic evolutionary rates compared to atively with the linear and even the quadratic models.[37] Thus, the optimal degree of the model probably lies between 2 and 3. We investigated the stochastic behavior of the system and its characteristics within the broader framework of rational BDIMs where birth and death rates, $\lambda(i)$ and $\delta(i)$, are rational functions of the family size i.

Nonlinear Rational BDIM

We will examine models represented as transformed linear BDIM with

$$\lambda_i = \lambda(i + a)(i + 1)^{d-1}, \ \delta_i = \lambda(i + b)i^{d-1}, \tag{10}$$

where $d \geq 1$ is the model degree (the "degree of interactions"). Let us recall that the highest degrees and the corresponding coefficients of the birth and death rates must be equal to provide for the power asymptotics of the stationary distribution, $P(i) \sim i^{-\gamma}$. The power γ of this distribution is completely determined by the degree d and the coefficients at i^{d-1} (a and b in (10)). Thus, the model (2), (10) is representative of all rational BDIMs of the degree d with a given power asymptotic of the stationary distribution; this distribution for model (10) is exactly the same as for the corresponding linear model with $\lambda_i = \lambda(i + a)$, $\delta_i = \lambda(i + b)$ as described in reference 32. In this section we analyze the dependence of the main stochastic characteristics of model (10) on the model degree d. The probability of formation of the largest family from a singleton before extinction increases along with the model degree (Fig. 8). Conversely, the mean times of extinction and formation of the largest family (in *internal time units*) decrease with the increase of model degree (Figs. 9, 10).

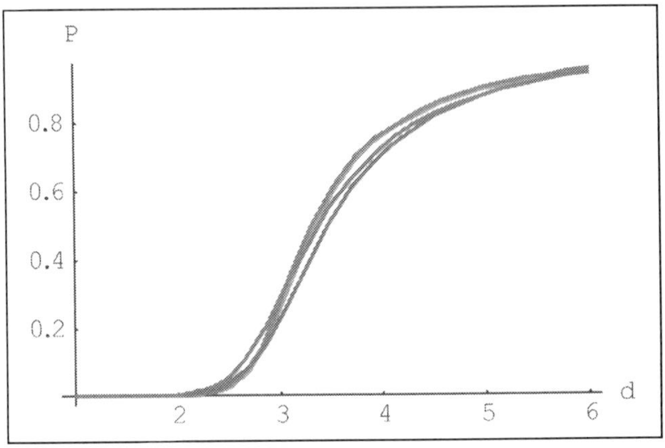

Figure 8. Probability of formation of the largest family starting from a singleton, $P^d(1,N)$, for rational BDIMs depending on model degree *d*. The model parameters are for *D. melanogaster* (blue), *H. sapiens* (red), *A. thaliana* (green), *C. elegans* (purple). A color version of this figure is available online at http://www.Eurekah.com.

A comparison of the mean times of formation and extinction for rational BDIMs reveals an interesting property of nonlinear BDIMs: for any given family of size *n*, there exists such a model degree that the times of family formation and extinction are equal. Accordingly, at higher model degrees, the mean time of formation becomes shorter than the mean time of extinction. For the rational model with model parameters taken for *H. sapiens*, this ratio is <1 for all $d > 3.1$ (Fig. 11).

One would expect that increasing the degree (the "order of interaction") of BDIM should result in much faster family evolution; this is, indeed, the case *under a fixed value of the parameter* λ.[37] However, we have also shown that this effect is offset by the rapid growth of the

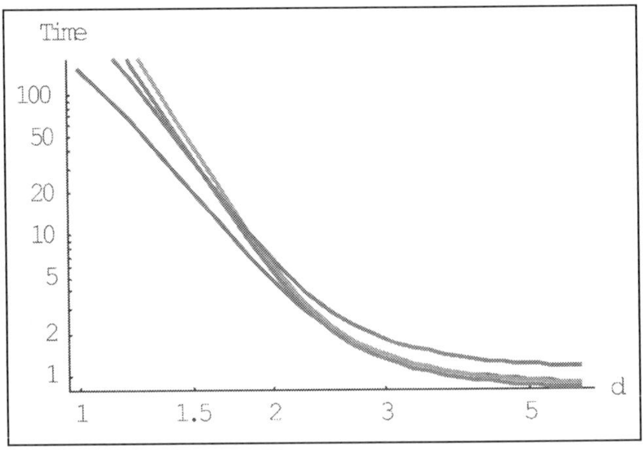

Figure 9. Mean time of extinction (in $1/\lambda$ units) of the largest family for the rational BDIM depending on the model degree *d*. The plot is in double logarithmic coordinates. The model parameters are for *D. melanogaster* (blue), *C. elegans* (violet), *H. Sapiens* (red) and *A. thaliana* (green). A color version of this figure is available online at http://www.Eurekah.com.

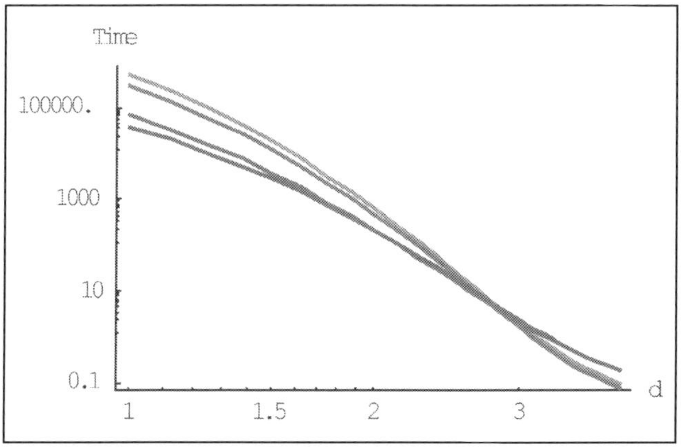

Figure 10. Mean time of formation (in $1/\lambda$ units) of the largest family for the rational BDIM depending on the model degree *d*. The plot is in double logarithmic coordinates. The model parameters are for *D. melanogaster* (blue), *C. elegans* (violet), *H. apiens* (red) and *A. thaliana* (green). A color version of this figure is available online at http://www.Eurekah.com.

internal time unit. Accordingly, $1/\lambda = c_{du}/r_{du} = c_{du}/20$ Ga (assuming $r_{du} = 2 \times 10^{-8}$, after ref. 26) increases with the model degree (Fig. 12). For the model parameters taken for *H. sapience*, the internal time unit is 0.136 Ga for $d = 1$, 1.22 Ga for $d = 2$, and 146.9 Ga for $d = 3$. Thus, although the mean times of formation (and extinction), when measured in internal time units dramatically decrease with the increase of the model degree, the evolution time *in years* does not decrease indefinitely, but rather passes through a minimum at *d* between 2 and 3 (Fig. 13). Specifically, the *d* values resulting in the fastest gene family evolution are 2.67 for *D. melanogaster* and 2.71 for *H. sapiens.* Even the minimum mean time of the largest family formation achievable with the rational BDIMs of the optimal degree is on the order of 10^{11} years (56.55 Ga for *D. melanogaster* and 204 Ga for *H. sapiens,*[37]), which is incompatible with the age of life on Earth. Thus, a rational BDIM of any degree cannot provide an adequate description of genome

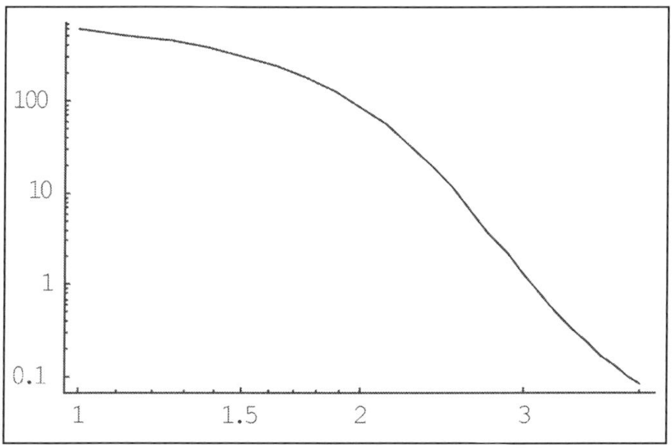

Figure 11. Ratio of the mean times of extinction and formation of the largest family depending on the model degree; the model parameters are for *H. sapiens.* The plot is in double logarithmic coordinates.

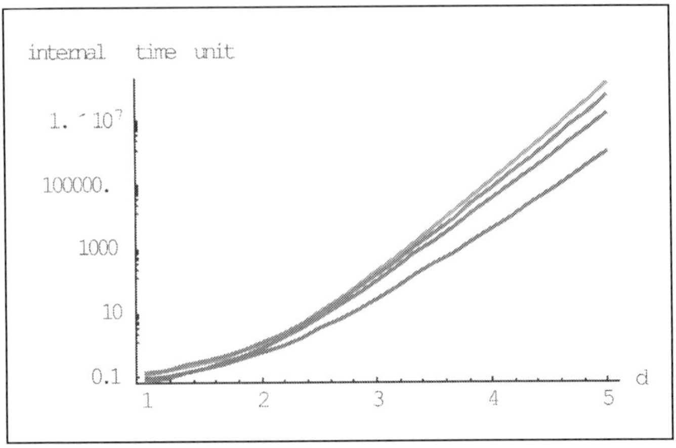

Figure 12. Internal time unit depending on model degree (logarithmic scale). The model parameters are for *D. melanogaster* (blue), *C. elegans* (violet), *H. Sapiens* (red) and *A. thaliana* (green). A color version of this figure is available online at http://www.Eurekah.com.

evolution when only the mean time of family formation is considered. Accordingly, for assessing the feasibility of the formation of the largest families under a given model, the coefficient of variation of the formation time should be taken into account. As shown above, this coefficient is quite large (≥100 for all considered genomes) such that the actual formation time of the largest family could differ from its mean value by two orders of magnitude or more, which would bring the time required for the formation of families of the observed size close to realistic spans of ~10^9 yrs. In order to investigate this problem further, we turned to computer simulation analysis.

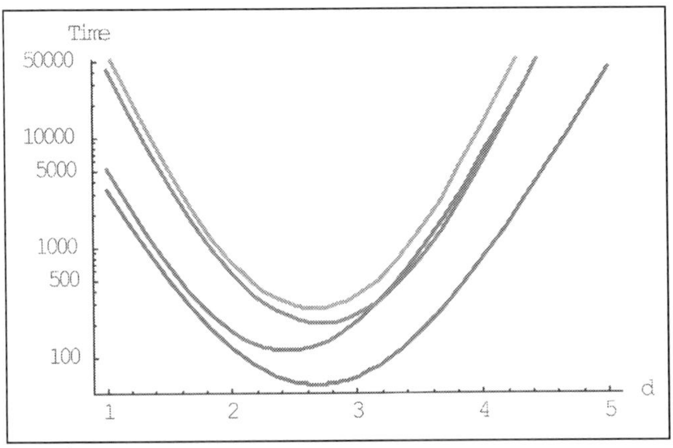

Figure 13. Dependence of the time (in years,Ga) required for the formation of the largest family on the model degree *d* for the rational BDIM. The plot is in semi-logarithmic coordinates. The model parameters are for *D. melanogaster* (blue), *C. elegans* (violet), *H. Sapiens* (red) and *A. thaliana* (green). A color version of this figure is available online at http://www.Eurekah.com.

Simulation of Gene Family Evolution under BDIMs of Different Degrees

In the previous section, we determined the mean time of family formation for BDIM of different degrees and found that even the shortest mean time obtained with the optimal model degree was substantially greater than the time available for genome evolution. However, for assessing the feasibility of the formation of the largest families during the evolution of real genomes, the more relevant value is not the mean but the minimum time of family formation over the entire ensemble of genes. Given the large variance of the family formation time estimates, this minimum value is likely to be much less than the mean. Analytic determination of this value is hard so we resorted to Monte Carlo simulation analysis. Model parameters estimated for human genome evolution were employed for this analysis.

The simulated evolution started from 3000 families of size one and continued until the largest family reached 1024 members (a convenient number to approximate the size of the largest family in eukaryotic genomes). The time scale was adjusted such that $r_{du} = 2 \times 10^{-8}$ duplications/gene/year.[26] A series of simulations was performed for nonlinear rational BDIMs of different degrees.

At each discrete time step, for each family, a birth or death of a domain belonging to the family was simulated by (respectively) increasing or decreasing the family size counter; additionally, a new family of size 1 was created with the probability proportional to the innovation rate ν (the resulting process is analogous to the classical model of Karlin and McGregor[43]). The probabilities of a birth or death for a given family of size k were, respectively, proportional to λ_k and δ_k.

A series of simulations with rational BDIM of different degrees was run until the largest family reached 1024 members. For the linear BDIM, the median time required to produce the first family of this size was 49.5 Ga and the mean (± standard deviation) was 52.6 ± 21.1 Ga. The quadratic BDIM reached this level much faster, with the median time of 2.52 Ga and the mean of 2.64 ± 0.78 Ga. Not unexpectedly, these values are orders of magnitude smaller than the mean values estimated above.

As shown in Figure 14, the time at which the largest family in a genome reaches 1024 members depends on d in a similar fashion as the mean time for a single family, i.e., there is a clear minimum at a specific value of d. At the optimal value of $d \approx 2.2$, the model reaches this family size in 2.2 ± 0.5 Ga, which is compatible with the timescale of evolution of eukaryotes.[44,45] Compared to the minimal evolution time predicted for a single family, the genome-size ensemble of gene families reached the threshold size much faster (by 1.5-2.5 orders of magnitude), and the optimum values of d was lower by ~0.5 (Fig. 14).

The Mean Number of Elementary Events before Family Extinction and Formation

Comparing the mean family formation and extinction times predicted by BDIMs with the actual evolutionary timescale allowed us to choose the model with the best fit to the empirical data. The *number* of elementary evolutionary events, namely, gene duplications and deletions, predicted by BDIMs is of interest in itself as an approximation of an important characteristic of genome evolution.

To calculate the mean number of elementary events during evolution of gene families, we employed the *embedding* chains $\{Y(n)\}$ instead of the original BDIM. The embedding chain $\{Y_n\}$ for a particular BDIM is a random walk with discrete time on the same set of states and transition probabilities $p_{i,i+1} = \beta_i = \lambda_i/(\lambda_i + \delta_i)$, $p_{i,i-1} = \mu_i = \delta_i/(\lambda_i + \delta_i)$ and $p_{ij} = 0$ for all other cases. The transition from state i to state $i + 1$ ($i - 1$) corresponds to duplication (deletion) of a domain in a family of size i. The only difference between the original

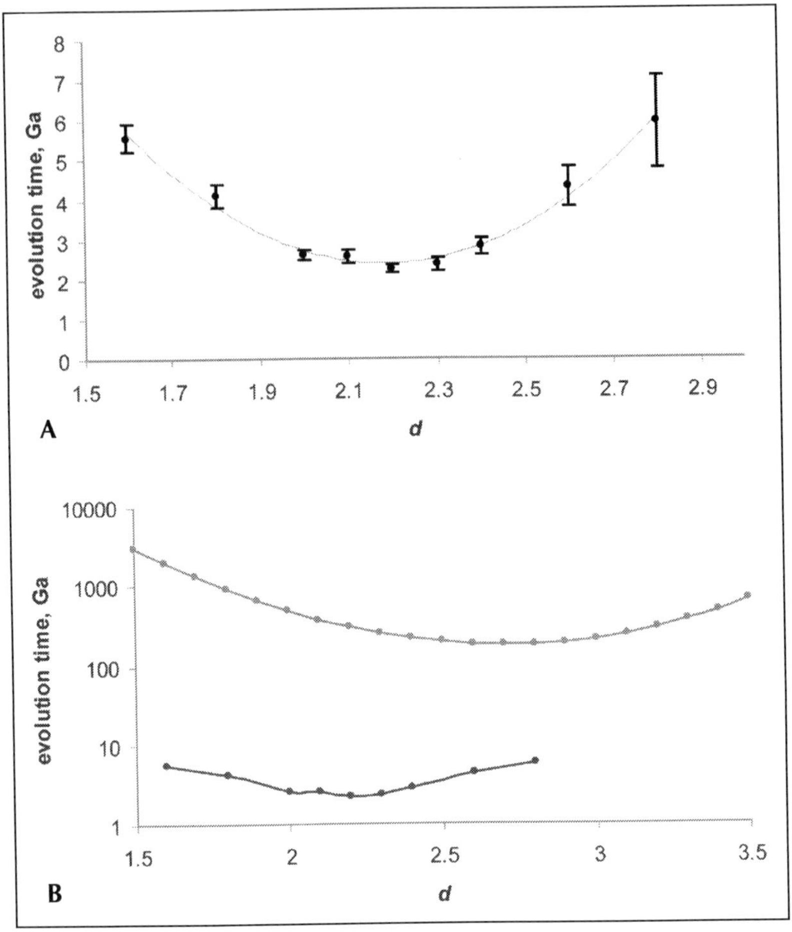

Figure 14. Simulation of gene family evolution depending on model degree. A) The minimal time required for the formation of a family with 1024 members determined by Monte Carlo simulation starting from an ensemble of 3000 singletons. The standard deviation is shown for each point. B) The minimal time required for the formation of a family with 1024 members. The upper curve shows the prediction of the respective BDIM and the lower curve shows the simulation results (the same as in (A) but in the logarithmic scale).

birth-and-death process and the embedding chain is that the sojourn time for the embedding chain is equal to 1 for any state i instead of a random variable which is exponentially distributed with the mean equal to $1/(\lambda_i + \delta_i)$. The ratio β_i/μ_i $(= \lambda_i/\delta_i)$ characterizes the trend of family evolution from the state i, i.e., is the family more likely to grow or to shrink; for a **symmetric** random walk, $\beta_i/\mu_i = 1$ for all i. For the rational models, $\beta_i/\mu_i \cong 1$ for large i. Using embedded chains, we calculated the mean number of elementary events occurring before the formation of a family of the given size, $f_n^{(d)}$ depending on the rational BDIM degree, d (Fig. 15). As expected, these numbers drop with the increase of the model degree but remain quite large even for the optimal degree. For example, for *D. melanogaster* $f_{335}^{(1)} = 734725$, $f_{335}^{(2)} = 127567$, $f_{335}^{(3)} = 60755$, and for *H. Sapiens*, $f_{1151}^{(1)} = 1.29*10^7$, $f_{1151}^{(2)} = 1.68*10^6$, $f_{1151}^{(3)} = 756238$. The coefficient of variation $\Sigma_N^{(d)}$ of the number of events before

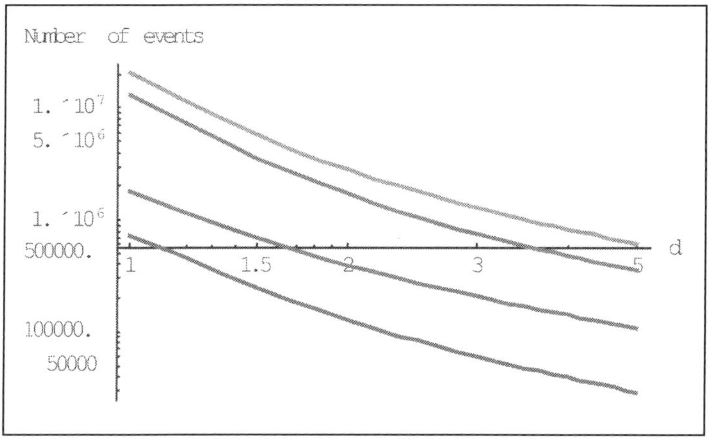

Figure 15. The mean number of elementary events (duplications and deletions) required, under the rational BDIM, for the formation of the largest family, dependent on the model degree. The model parameters are for *D. melanogaster* (blue), *C. elegans* (violet), *H. sapiens* (red) and *A. thaliana* (green). A color version of this figure is available online at http://www.Eurekah.com.

the formation of a family of the largest size is large and only slightly decreases when the model degree increase; for example, $\Sigma_{335}{}^{(1)} = 87$, $\Sigma_{335}{}^{(2)} = 86.6$ and $\Sigma_{335}{}^{(3)} = 80$ for for *D. melanogaster* and $\Sigma_{1151}{}^{(1)} = 299.31$, $\Sigma_{1151}{}^{(2)} = 296.4$ and $\Sigma_{1151}{}^{(3)} = 276.23$ for *H. sapiens*. The mean number of elementary events occurring before the extinction of a family of the given size, $e^{(d)}{}_n$, is always greater than $f_n{}^{(d)}$ but the ratio $e^{(d)}{}_n/f_n{}^{(d)}$ is close to 1 (<1.1, in a sharp contrast to the ratio of the mean times of formation and extinction discussed above) for all genomes and all model degrees rapidly and monotonically tends to 1 as *d* increases.

Given that all the analyzed BDIMs are balanced, i.e., the birth and death rates are asymptotically equal, it was not unexpected that the mean number of events required for the formation of a large family (or the number of events preceding the extinction of such a family) was orders of magnitude greater than the size of the family. This suggests a highly dynamic picture of genome evolution where numerous duplications counterbalanced by gene losses are typically involved in the evolution of large families. However, the number of events required for the formation of a family of the given size quickly drops with the increase of a model degree (Fig. 15), which may be construed as reflection of positive selection leading to proliferation of families that are of adaptive value to the organism.

Conclusions and Perspective

Here and in the previous publications,[32,36,37] we describe a rather general class of models, which are based on the classical concept of a birth-and-death process and seem to be applicable to the genome evolution process. Similar, although not identical and apparently less general, modeling approaches have been considered by others.[16,31,46] Even earlier, evolution of gene families has been modeled within the distinct mathematical framework of multiplicative processes.[47]

The utility of birth-and-death type models in evolutionary genomics in itself is not a trivial matter and stems from fundamental features of genome evolution which, in part, have been presciently envisaged by classic geneticists and, in part, became apparent after the advent of genomics. As captured in the title of Ohno's famous book,[18] although foreseen even in the early days of genetics,[17,48] gene duplication probably is **the principal** mechanism of genome evolution. Of course, genomes cannot grow ad infinitum and, through most of the evolutionary history, the number of genes within a given phylogenetic lineage probably remains roughly constant. Hence duplication is intrinsically coupled to gene loss. The results of comparative genomics further show that many genes in each lineage cannot be obviously linked to other genes through duplication. Without necessarily specifying the biological mechanisms (these could involve rapid change after duplication, gene acquisition via horizontal transfer, and possibly, birth of genes from noncoding sequences), it is reasonable to view these unique genes as resulting from innovation. For genomes to maintain equilibrium, the combined rates of duplication and innovation over the entire ensemble of gene families should equal the rate of gene loss, at least when averaged over long time spans. Furthermore, the observed distribution of family sizes, which asymptotically tends to a power law, dictates a much more specific connection between the gene birth and death rates, namely, the second order balance (4).

The incentive to examine these models in detail stems from at least three rather fundamental questions: (i) are the above elementary evolutionary mechanisms sufficient to account for the empirically observed characteristics of genomes, (ii) what is the contribution of natural selection to the general quantifiable features of genomes, such as the size distribution of gene families, and (iii) how similar or how different are the models describing evolution of phylogenetically distant genomes, such as those of prokaryotes and eukaryotes. The analysis of BDIMs starts to provide some answers, although it is premature to consider these final in any sense. The critical observation made in the course of BDIM analysis was that different versions of these models could be readily distinguished on the basis of goodness of fit to the empirical data. This being the case, we found that the simplest possible model in which all paralogs are considered independent does not explain the data well. Thus, turning to the first of the above questions, we have to conclude the "something else" is required to model genome evolution, on top of the three elementary processes. This "something" is dependence or "interaction" between gene family members which results in self-accelerating family growth. In order to account for the observed stationary distribution of family sizes, it is sufficient to introduce a very weak dependence as embodied in the linear BDIM. However, when we switched from the deterministic to the stochastic version of BDIMs which provide for the possibility of analysis of the dynamics of the systems evolution, we found that evolution under the linear BDIM was much too slow to account for the emergence of the large families of paralogs found in all genomes during the time of life's evolution. Only higher order BDIMs, with degrees between 2 and 3, i.e., with "strong interactions" between family members were found to provide for sufficiently fast evolution to be compatible with the real biological timescale.

Obviously, these findings beg the question: what is the nature of the mysterious "interactions" between paralogs? This brings us to the second of the above major problems. BDIMs do not explicitly include the notion of selection. However, the simplest interpretation of the interactions implied by the higher order BDIMs seems to be that these reflect adaptive evolution of gene families driven by positive selection. Should that be the case, we are justified to conclude that very weak selection would suffice to explain the stationary distribution of family sizes, but much stronger selective pressure is needed to account for the dynamics of genome evolution. However, the interpretation of BDIM degree as a manifestation of selection is, at this point, no more than a guess. One of the further developments of genome evolution modeling involves introducing selection explicitly and determining whether the resulting more sophisticated models will be equivalent to the higher order BDIMs explored here.

BDIMs worked well in describing evolution of all analyzed genomes, from the smallest prokaryotic ones to the most complex genomes of plants and animals. However, the parameters of the resulting models, i.e., the duplication, deletion, and innovation rates differed significantly, suggesting some tantalizing answers to the third of the questions posed above. In particular, we found that the innovation rates in prokaryotes were an order of magnitude greater than those in eukaryotes.[32] An optimistic interpretation of this difference is that the relatively high innovation rates detected for prokaryotes reflect rampant horizontal gene transfer, an increasingly recognized defining feature in the evolution of bacteria and archaea.[49-51] Should that be the case, we might be justified to conclude that BDIMs are telling us something new regarding the extent of this phenomenon. However, it would be premature to rule out the pessimistic explanation, i.e., that the observed differences are due to some cryptic modeling artifacts. The issue definitely deserves further investigation, through refined modeling approaches and analysis of additional comparative-genomic data.

In conclusion, it makes sense to ask the $64K question: do the models discussed in this chapter (and similar ones) reveal something new about biology? So far we seem to have only rather equivocal answers. Earlier in this section, we discuss some interesting hints on new aspects of the role of selection in genome evolution and on distinct regimes of evolution in different domains of life. Realistically, however, the principal conclusions seem to be quite general and mostly methodological. Indeed, it was observed in these and related analyses that important aspects of genome evolution can be realistically modeled with simple, straightforward approaches. Perhaps more importantly, the work summarized here makes the next step by showing (to paraphrase Einstein's famous aphorism) that models of genome evolution should be as simple as possible but not simpler and that we seem to be able to identify the minimal required level of complexity. Future developments will show whether or not a path exists from these general findings to new biology.

References

1. Pareto V. Cours d'Economie Politique. Paris: Rouge et Cie 1897.
2. Zipf GK. Human behaviour and the principle of least effor. Boston: Addison-Wesley, 1949.
3. Barabasi AL. Linked: The New Science of Networks. New York: Perseus Pr, 2002.
4. Mendes JF, Dorogovtsev SN. Evolution of Networks: From Biological Nets to the Internet and Www. Oxford: Oxford University Press, 2003.
5. Gisiger T. Scale invariance in biology: Coincidence or footprint of a universal mechanism? Biol Rev Camb Philos Soc 2001; 76:161-209.
6. Luscombe N, Qian J, Zhang Z et al. The dominance of the population by a selected few: Power-law behaviour applies to a wide variety of genomic properties. Genome Biol 2002; 3:(research 0040.0041-0040.0047).
7. Koonin EV, Wolf YI, Karev GP. The structure of the protein universe and genome evolution. Nature 2002; 420:218-223.
8. Kuznetsov VA. Distribution associated with stochastic processes of gene expression in a single eukariotic cell. EUROSIP Journal on Applied Signal Processing 2001; 4:285-296.
9. Barabasi AL, Oltvai ZN. Network biology: Understanding the cell's functional organization. Nat Rev Genet 2004; 5:101-113.
10. Barabasi AL, Albert R. Emergence of scaling in random networks. Science 1999; 286:509-512.
11. Bilke S, Peterson C. Topological properties of citation and metabolic networks. Phys Rev E Stat Phys Plasmas Fluids Relat Interdiscip Topics 2001; 64:036106.
12. Dorogovtsev SN, Mendes JF. Scaling properties of scale-free evolving networks: Continuous approach. Phys Rev E Stat Phys Plasmas Fluids Relat Interdiscip Topics 2001; 63:056125.
13. Albert R, Jeong H, Barabasi AL. Error and attack tolerance of complex networks. Nature 2000; 406:378-382.
14. Jeong H, Tombor B, Albert R et al. The large-scale organization of metabolic networks. Nature 2000; 407:651-654.
15. Jeong H, Mason SP, Barabasi AL et al. Lethality and centrality in protein networks. Nature 2001; 411:41-42.

16. Qian J, Luscombe NM, Gerstein M. Protein family and fold occurrence in genomes: Power-law behaviour and evolutionary model. J Mol Biol 2001; 313:673-681.
17. Fisher RA. The possible modification of the response of the wild type to recurrent mutations. Am Nat 1928; 62:115-126.
18. Ohno S. Evolution by gene duplication. Berlin, Heidelberg, New York: Springer-Verlag, 1970.
19. Henikoff S, Greene EA, Pietrokovski S et al. Gene families: The taxonomy of protein paralogs and chimeras. Science 1997; 278:609-614.
20. Jordan IK, Makarova KS, Spouge JL et al. Lineage-specific gene expansions in bacterial and archaeal genomes. Genome Res 2001; 11:555-565.
21. Lespinet O, Wolf YI, Koonin EV et al. The role of lineage-specific gene family expansion in the evolution of eukaryotes. Genome Res 2002; 12:1048-1059.
22. Kellis M, Birren BW, Lander ES. Proof and evolutionary analysis of ancient genome duplication in the yeast Saccharomyces cerevisiae. Nature 2004; 428:617-624.
23. Chervitz SA, Aravind L, Sherlock G et al. Comparison of the complete protein sets of worm and yeast: Orthology and divergence. Science 1998; 282:2022-2028.
24. Lander ES, Linton LM, Birren B et al. Initial sequencing and analysis of the human genome. Nature 2001; 409:860-921.
25. Lynch M, Force A. The probability of duplicate gene preservation by subfunctionalization. Genetics 2000; 154:459-473.
26. Lynch M, Conery JS. The evolutionary fate and consequences of duplicate genes. Science 2000; 290:1151-1155.
27. Aravind L, Watanabe H, Lipman DJ et al. Lineage-specific loss and divergence of functionally linked genes in eukaryotes. Proc Natl Acad Sci USA 2000; 97:11319-11324.
28. Katinka MD, Duprat S, Cornillot E et al. Genome sequence and gene compaction of the eukaryote parasite Encephalitozoon cuniculi. Nature 2001; 414:450-453.
29. Koonin EV, Fedorova ND, Jackson JD et al. A comprehensive evolutionary classification of proteins encoded in complete eukaryotic genomes. Genome Biol 2004; 5:R7.
30. Gardiner CW. Handbook fo Stochastic Models for Physics, Chemistry and the Natural Sciences. Berlin: Springer-Verlag, 1985.
31. Rzhetsky A, Gomez SM. Birth of scale-free molecular networks and the number of distinct DNA and protein domains per genome. Bioinformatics 2001; 17:988-996.
32. Karev GP, Wolf YI, Rzhetsky AY et al. Birth and death of protein domains: A simple model of evolution explains power law behavior. BMC Evol Biol 2002; 2:18.
33. Dokholyan NV, Shakhnovich B, Shakhnovich EI. Expanding protein universe and its origin from the biological Big Bang. Proc Natl Acad Sci USA 2002; 99:14132-14136.
34. Pastor-Satorras R, Smith E, Sole RV. Evolving protein interaction networks through gene duplication. J Theor Biol 2003; 222:199-210.
35. Wagner A. How the global structure of protein interaction networks evolves. Proc R Soc Lond B Biol Sci 2003; 270:457-466.
36. Karev GP, Wolf YI, Koonin EV. Mathematical modeling of the evolution of domain composition of proteomes: A birth-and-death process with innovation. In: Galperin MY, Koonin EV, eds. Computational Genomics: From Sequence to Function. Amsterdam: Horizon Press, 2002:3:261-314.
37. Karev GP, Wolf YI, Koonin EV. Simple stochastic birth and death models of genome evolution: Was there enough time for us to evolve? Bioinformatics 2003; 19:1889-1900.
38. Marchler-Bauer A, Panchenko AR, Shoemaker BA et al. CDD: A database of conserved domain alignments with links to domain three-dimensional structure. Nucleic Acids Res 2002; 30:281-283.
39. Bhattacharya R, Waymire E. Stochastic processes with applications. New York: Wiley, 1990.
40. Ross SM. Introduction to probability models. Boston: Academic Press, 1989.
41. Karev GP, Wolf YI, Berezovskaya FS et al. Gene family evolution: An in-depth theoretical and simulation analysis of nonlinear birth-death-innovation models. BMC Evol Biol 2004; 4:32.
42. Krauss LM, Chaboyer B. Age estimates of globular clusters in the Milky Way: Constraints on cosmology. Science 2003; 299:65-69.

43. Karlin S, McGregor J. The number of mutant forms maintained in a population. In: LeCam L, Neyman J, eds. Proc Fifth Berkeley Symposium on Mathematical Statistics and Probability. Berkeley: University of California Press, 1967.
44. Hedges SB, Chen H, Kumar S et al. A genomic timescale for the origin of eukaryotes. BMC Evol Biol 2001; 1:4.
45. Hedges SB. The origin and evolution of model organisms. Nat Rev Genet 2002; 3:838-849.
46. Reed WJ, Hughes BD. A model explaining the size distribution of gene and protein families. Math Biosci 2004; 189:97-102.
47. Huynen MA, van Nimwegen E. The frequency distribution of gene family sizes in complete genomes. Mol Biol Evol 1998; 15:583-589.
48. Bridges CA. Salivary chromosome maps. J Hered 1935; 26:60-64.
49. Doolittle WF. Lateral genomics. Trends Cell Biol 1999; 9:M5-8.
50. Koonin EV, Makarova KS, Aravind L. Horizontal gene transfer in prokaryotes: Quantification and classification. Annu Rev Microbiol 2001; 55:709-742.
51. Gogarten JP, Doolittle WF, Lawrence JG. Prokaryotic evolution in light of gene transfer. Mol Biol Evol 2002; 19:2226-2238.

CHAPTER 7

Scale-Free Evolution:

From Proteins to Organisms

Nikolay V. Dokholyan* and Eugene I. Shakhnovich

Introduction

One of the most intriguing problems in molecular biology is the origin of the vast population diversity of protein families.[1-4] Following the assumption that the protein families are populated at random, one would expect a multinomial distribution of the family populations.[5] However, it has been discovered[6-9] that distribution of the family populations is by far nonexponential, but has a long tail, which signifies that some specific mechanisms govern populations of protein families. To explain such diversity, there emerged two views of **convergent** and **divergent** evolution (Fig. 1).

In the **convergent** evolution scenario,[10] it is postulated that the present population distribution of protein fold families is the result of convergent processes in the course of evolution which were selectively populating folds. In the simplest scenario, it is presumed that evolution has reached equilibrium in the protein structural space.** Due to the underlying physical nature of evolutionary processes, i.e., the physical nature of amino acid interactions that underlie the properties of specific folds, the expected equilibrium distribution of family population follows the Boltzman distribution. Thus, more "designable" folds, that can be encoded by many sequences, have a higher representation in genomes.[11,13-18] This assumption, called the "designability principle", is based on phenomenological considerations[13] and on observations drawn from exhaustive enumeration of all sequences in simplified two- and three-dimensional lattice protein models. In the course of evolution more designable folds become more populated than less designable folds, which results in the uneven distribution of observed populations of protein families.[17,19]

There have been several arguments[10,13-15,20] based on various observations favoring convergent evolution. Teichmann et al proposed that structural similarities arise solely due to physical interactions that favor particular packing and chain topologies.[20] Functional pressure was proposed to be the paladin of protein structural convergence. One of the most striking example is that of the Ser/His/Asp catalytic triad,[10,21] which is found in a number of folds that have no significant sequence similarity. Antifreeze proteins (AFP) provide a crucial defense for

** Strictly speaking the equilibrium may have not been reached, nevertheless protein families can still be populated according to some "attractive" features such as designability.[11,12]

*Corresponding author: Nikolay V. Dokholyan—Department of Biochemistry and Biophysics, The University of North Carolina at Chapel Hill, School of Medicine, Chapel Hill, North Carolina 27599, U.S.A. Email: dokh@med.unc.edu

Power Laws, Scale-Free Networks and Genome Biology, edited by Eugene V. Koonin, Yuri I. Wolf and Georgy P. Karev.

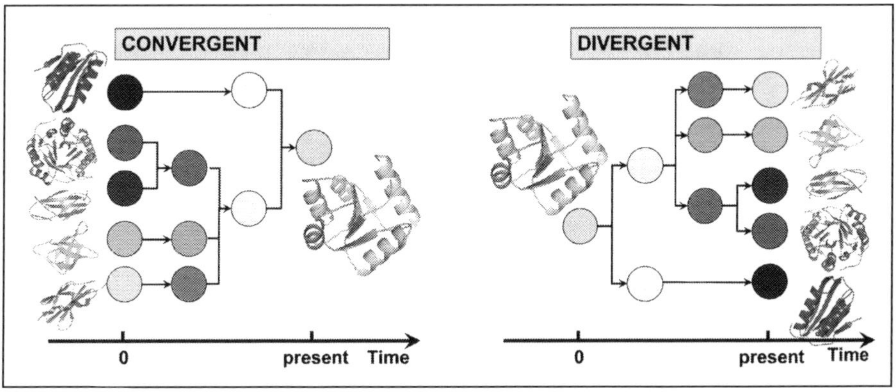

Figure 1. Schemes of the convergent and divergent evolution scenarios.

organisms against sub-zero temperatures. The beetles *Dendroides Canadensis* and *Tenebrio Molitor* AFP have dissimilar sequences from known plant and fish AFP,[22,23] indicating functional convergence of AFP proteins. Another striking example is of Zn-dependent carboxypeptidases that cleave off the C-terminal amino acid residues from proteins and peptides.[24,25] Two families of nonhomologous carboxypeptidases—thermolysin and mitochondrial processing peptidase[24]—show functional and structural similarity, although the topologies—the arrangement of the secondary structure elements in these folds—are different. In general, most of the observations of convergent evolution have relied on finding functional similarity (and structure of the active site) between proteins with low sequence similarity.[20,26-28]

The arguments of protein hereditary relation based solely on their sequence similarity are questionable. It was shown that a sequence is not a robust feature of proteins.[29] In fact, a protein structure often remains stable after a single amino acid substitution. Amino acid substitutions may accumulate in the course of evolution: some of them will be destabilizing, some will be stabilizing. However, as long as a protein itself is stable in its environment, there may be no reasons for it to be eliminated from the genome of a corresponding organism. One can argue that amino acid substitutions, that affect protein folding kinetics and function, may be more damaging for the viability of a protein in a cell. However, many single-domain proteins have just a few amino acids—**protein folding nucleus**[30-32]—that govern fast folding kinetics. In a lowest approximation, multi-domain protein folding kinetics is also governed by a few amino acids—he nuclei of each individual domain. Thus, even though the evolutionary may exert pressure to preserve important amino acids for protein folding kinetics, the small number of them may not prevent proteins diverge strongly in the course of evolution.

The evolutionary pressure to preserve functionally important amino acids (e.g., the active site) may depend on the number of alternative proteins in cells that are capable of performing the same function. Even if a protein were to lose completely its function, it may still survive in a cell and acquire a new function later on in the course of evolution. For example, both enoyl-CoA hydratase and 4-chlorobenzoyl-CoA dehalogenase show significant sequence and structure similarities, but they catalyze different reactions.[33] In addition, in proteins, that play a structural role in the cell, such as fibronectin,[34] functionally important amino acids are the same as those that stabilize these proteins. In proteins, with a binding site, the number of amino acids that constitute a binding site is small. Therefore, the evolutionary pressure to preserve functionally important amino acids may not affect strongly the ability of proteins to diverge during evolution.

It is possible that the evolutionary pressure to preserve a protein's sequence in a more "designable" fold family is not strong enough for protein sequences to diverge from each other up to the point when their sequence similarity becomes of the order of randomly chosen proteins. Thus, the sequence may not be a robust measure of hereditary relation between proteins.

In the **divergent** evolution scenario, the present day proteins are the pra-children of a small set of prebiotic proteins. They diverge from a few "original" proteins by duplication, deletion, and accumulating amino acid substitutions.[10,35-37] Crucial support for divergent evolution came when the gene duplication was documented.[38-40] The principal advantage of the divergent over convergent evolution scenario is that the former does not rely on the "designability" principle.* Of course, there still are examples of proteins[10] that have similar structure function but vastly different sequence, which are disputed to be an indication of the convergent evolution. However, since the sequence may not be indicative of hereditary relation, such argument is unsupported. If structure is more conserved in the course of evolution, it is important to retrace the evolutionary relation based on structure for those proteins that have sequence similarity below the accepted level for the homologous proteins (approximately 25%).

Gerstein and Levitt pioneered the structural census of protein sequences[41] and discovered that most popular folds—that are most often used in proteins of various organisms—constitute the largest fold families.[41,42] This fact, however, can also be interpreted from **both** convergent and divergent evolution scenarios. Based on convergent evolution scenario, those folds that are most adaptive to a new function are more populated than more "rigid" folds are populated. From the divergent evolution perspective, the more often a given fold is used in the cell, the more often it is expressed and, therefore, the more often it varies in the course of evolution.

The absence of the experimental crux makes it challenging tests strongly support one scenario versus another. Despite the large number of examples that favor divergent or convergent scenarios, there is no unified biophysical theory that would combine into a single theoretical framework for understanding apparently disconnected observations within a single evolutionary concept.

Protein Evolutionary Relationships from Structure Similarities

One step towards a unifying theory of protein evolution is the reconstruction of protein relationships based on their structural similarity. There have been several efforts made in quantifying structural similarities between proteins.[4,43,44] The ambiguity in all of these efforts arises from complications in rigorous quantitative definition of structural similarity. Semi-intuitive definitions of folds have been employed to construct two popular databases, SCOP[4] and CATH.[43] The main drawback of these databases is that they are somewhat subjective.

The FSSP database based on the DALI structure comparison algorithm[44] defines a quantitative measure of structural similarity, the Z-score. However, selection of the threshold value Z_{min} of the Z-score, beyond which proteins are considered structurally similar, also introduces an element of ambiguity into FSSP-based family classification. In a recent paper,[45] Getz and coauthors provided a quantitative relationship between FSSP, CATH and SCOP classifications. These authors noted that the matrix of pairwise Z-scores can be viewed as a weighted graph, where each two proteins that have similarity $Z > 2$ ($Z = 2$ is the minimal Z-score reported in FSSP) are connected by an edge that carries weight corresponding to the Z-score similarity between these two proteins. Getz et al[45] employed clustering algorithms, developed for weighted graphs, to identify fold families. However, clustering of weighted graphs is not exact as it may depend on the chosen algorithm and other factors. Another well-known problem with structural classification of whole proteins presented in FSSP is so-called "floats" where two structurally unrelated proteins having a common "promiscuous" domain are identified as structurally similar. It is, therefore, crucial to reconstruct protein structural relationships taking into account the problem of "floats".

* The divergent evolution and prevalence of more designable structures do not contradict each other.[12]

Protein Structure-Function Relations from an Evolutionary Perspective

Functional annotation of proteins is crucial for our understanding of how the cooperative organization of proteins in cells relates to the specific cell anatomy and function.[20] Understanding cell anatomy and function is, in turn, important for understanding the evolution of organisms. On a practical side, the ability to alter a cell's function and/or development may aid rational drug development. However, one of the challenging tasks of structural genomics is the determination of protein function based on its structure.

The determination of the function of a hypothetical protein is currently based on three strategies.[20] The first strategy is based on identifying any sequence similarity to known proteins. Even at low sequence similarities, there may be a set of conserved amino acids constituting an active site. These amino acids may indicate the function of a hypothetical protein.[46,47] The principal limitation of this strategy is the extent to which functionally important amino acids are conserved. It has been demonstrated on five various fold families,[29] that evolutionary pressure to preserve functionally important amino acids may not be as strong as the pressure to preserve amino acids responsible for protein stability in cells. Therefore, the determination of "true" conservation of amino acids due to their functional role may be arduous.

The second strategy for functional assignments of hypothetical proteins is the search for protein surface cavities using sequence and structural similarities to proteins with known function. As in the first strategy the extent of the success of this methodology depends strongly on the conservation of local sequence and structural motifs. The driving assumption for such strategy is the possible similarity of the active sites between proteins sharing the same or similar function.[48-50] There have been several mechanisms proposed to search for local functional motifs by comparison to libraries of three-dimensional structural templates[27,28,49,51,52] and the analysis of the physical properties of protein surfaces.[53] Teichmann et al[20] described two examples of structural genomics leading the functional annotation of hypothetical proteins: the HdeA protein from *Escherichia coli*[54] and the protein corresponding to gene 226 from *Methaococcus janaschii.*[55,56]

The third strategy is based on the crystallographic studies of bound cofactors in the native protein structure. The main limitation of this strategy is that it requires experimental reconstruction of the three-dimensional structure of protein-ligand complexes, which may be unsuccessful. Even in successful cases, the time scale for the experimental structure determination is much larger than that by using bioinformatics approaches described above.

Due to the severe limitations of all three strategies, it is, thus, crucial to develop a novel technique to rigorously relate protein structure to protein function. Shakhnovich et al proposed a strategy that is based on the assumption of evolutionary relation of proteins that may be so distant that neither structural nor sequence similarities directly are able to identify the function of a given protein. This strategy is to identify a divergent evolutionary pathway—a set of structurally similar proteins that link two dissimilar proteins.[57]

Protein Evolutionary Relations within and between Individual Proteomes

An overwhelming amount of various experimental observations, DNA sequencing data, and resolved protein structures in the past few decades open inviting opportunity to understand the cell machinery at a molecular level. This opportunity, however, is hampered by the fact that there is no unifying view that would serve as a framework for a theoretical basis to explain all available data from molecular to cellular levels of descriptions. Present knowledge offers us understanding of biological processes at various scales: from small molecules living at the Angstrom scale (10^{-10} m) to organisms living at the meter scale. It is an enticing challenge then to bridge these scales by developing a unifying theory.

One step to create a bridge between the nano- and hundred-nano-scales is to reconstruct cell organization at the molecular level. To construct such a bridge it is necessary to reconstruct the cell protein-protein interaction network. A large number of techniques have been developed for the systematic analysis of protein interactions,[58-60] such as yeast two-hybrid-based methods,[61,62] surface plasmon resonance biosensors,[63] isothermal titration calorimetry,[64] optical spectroscopy,[65] mass spectrometry of protein complexes,[66-68] protein chips,[69] and other methods that combine computational and experimental approaches.[70] These methods aim to reconstruct full-scale protein interaction networks in primitive organisms, such as yeast[66,67] and *Helicobacter pylori*.[62] These methods indeed offer novel insight on protein interactions, although their application is currently limited to the simpler unicellular organisms.

Computational methods are alternative approaches to experimental ones. Large amounts of available biological data and cost-effectiveness made computational approaches recently bloom. There have been undertaken several principal computational efforts. The phylogenetic profile method[71-73] is based on comparison of complete genomes of various organisms. Such comparison can be correlated with the set of specific functions present in one organism and absent in another. The principal drawbacks of this method are that (1) it can be used only for complete genomes, (2) some functions may be redundant and not represented by the same set of proteins, and (3) it can not be used for most common and essential proteins to most organisms. The conservation of gene neighborhoods has been utilized to predict functional genes in bacterial genomes.[74-76] The applicability of this approach though is limited to bacterial genomes. A search for domain fusion events[46,77-80] has been used to find the functional role of promiscuous domains incorporated in various larger proteins across the phyla. Such a search, though, is limited to multi-domain proteins. Other methods, such as mirrortree[81] and in silico two-hybrid methods[82] to search for protein interaction networks are sensitive to coverage of species under study, since they are dependent on multiple sequence alignments. It is crucial to develop a theoretical basis for techniques that would reconstruct the protein relations within individual proteomes and reconstruct the evolutionary relations between them based on available data.

Sequence Divergence

There are several principal facts about protein sequence-structure relation observed:[1,3,4,13,15,18,33,83-90] (i) proteins taken from various species and having sequence identity, *ID*, at least *ID* = 25-30% have similar three-dimensional structures (native state)[90-97] and are said to belong to the same fold family; (ii) some pairs of proteins sharing the same fold have sequence similarity as low as expected for random sequences *ID* ~8-9%;[44,87,98] (iii) within the same fold family, protein sequences have only 3-4% "anchored" amino acids.[87]

In 1987, eleven leading evolutionary biologists[99] made a statement asking the scientific community for the appropriate usage of the term "homology". Two proteins are said to be homologous if they possess a common evolutionary origin (e.g., ref. 100). Because many proteins that have high sequence similarity are homologous, this term has been used loosely in the discussion of any proteins with high sequence identity. Proteins that have no common ancestor, but possess structural similarity, are called analogs.

If the sequence identity of two structurally similar proteins is high (*ID* > 25-30%), there is a high probability that these proteins share a common ancestor, and thus, statistically, one would rarely be mistaken when calling these two proteins homologs. If the sequence similarity of two structurally similar proteins is low (*ID* < 25%), it is difficult to establish whether these proteins are homologs or analogs.[29,35] In fact, despite clever efforts,[100] it is still questionable whether there is a unique solution to the problem of determining whether two proteins with low sequence identity are homologs or analogs, i.e., whether they evolved in divergent or convergent evolution.

Two proteins are likely to be homologs that diverged from the same root if they still carry the same function (i.e., if the evolutionary time elapsed from their common divergence point is smaller than functional relaxation time τ_F). However, if two structurally similar proteins with low sequence identity have significantly different functions, then there is little information with which to identify them as homologs or analogs. These two proteins might be homologs, although one of them has evolved to possess a new function.[33] However, these two proteins can also be analogs and their similarity in structure is purely accidental or, for example, is due to a potential similarity of the structure of the binding site. The question then becomes – how can we retrace the history of these two proteins?

Our results[29] suggested that it may be impossible to retrace the history of two structurally similar proteins with low sequence identity based purely on sequence analysis. In this case, the ancestral relation classification terms—homologs and analogs—become meaningless. There are two reasons we believe this to be so. To explain these reasons, in reference 29 we proposed a model of evolution (Energy Gap Model) that attempts to reproduce the principal protein observations (i-iii) described above. The Energy Gap Model is based on the design of a set of structurally identical sequences by the *Z*-score minimization.[101-104] The idea is to find the similarities in the sequences of such a set and to recover those residues that are conserved across this set. The protein folding theory[105,106] suggests that *Z*-score minimization is equivalent to maximizing the energy gap between misfolded or unfolded conformations and the native state of a protein. It has been pointed out that such maximization results in stable and fast-folding proteins.[102,107] Thus, by designing sequences that have the same fold, we attempt to mimic evolution in diversifying protein sequences for the same fold family. In addition, the Energy Gap Model is a dynamical model, i.e., there is an implicit time scale that allows one to follow the evolution of sequences during the design procedure. The model is discussed in detail in reference.[29]

Why It May Be Impossible to Reconstruct Hereditary Relations between Proteins Based Solely on Their Sequence Similarity?

Firstly, the correlation function $C(\tau)$, which measures the probability of an amino acid not to be affected by mutations in time τ, decays exponentially, so that beyond the correlation function relaxation time one can not relate the sequences—original, and the one observed at time τ later. Secondly, it did not make a difference if we started our design procedure from one sequence or from two unrelated sequences. These sequences diverged so much from each other in a short design simulation time, that one could not identify which initial sequence we used in the design procedure. Furthermore, our results[29] suggested that some degree of homology may occur even between sequences that converged from unrelated root to the same structure, i.e., in clear analogs. The reason for that is that as we showed in reference 29 some positions may feature conserved residues due to physical requirement of stability of a common fold. Physical conservation of certain classes of amino acids at some positions in protein folds may be reflected on the genetic level due to the specifics of genetic code. Such conservation in some cases may be confused with homology due to the origin of sequences in divergent evolution. A rigorous definition of analogs and homologs can therefore come only either from the understanding of the correlation times τ between consecutive mutations or by reconstructing the actual structural and/or functional evolutionary pathways. If the time scale is smaller than the typical time scale for the formation of a family of homologs, τ_0, then the homology is well-defined: the homologous sequences in this case have high sequence similarity, while the analogous sequences have low sequence similarity. At a longer time scale $\tau >> \tau_0$, unless there is a high sequence similarity between sequences, the notion of homology and analogy becomes meaningless.

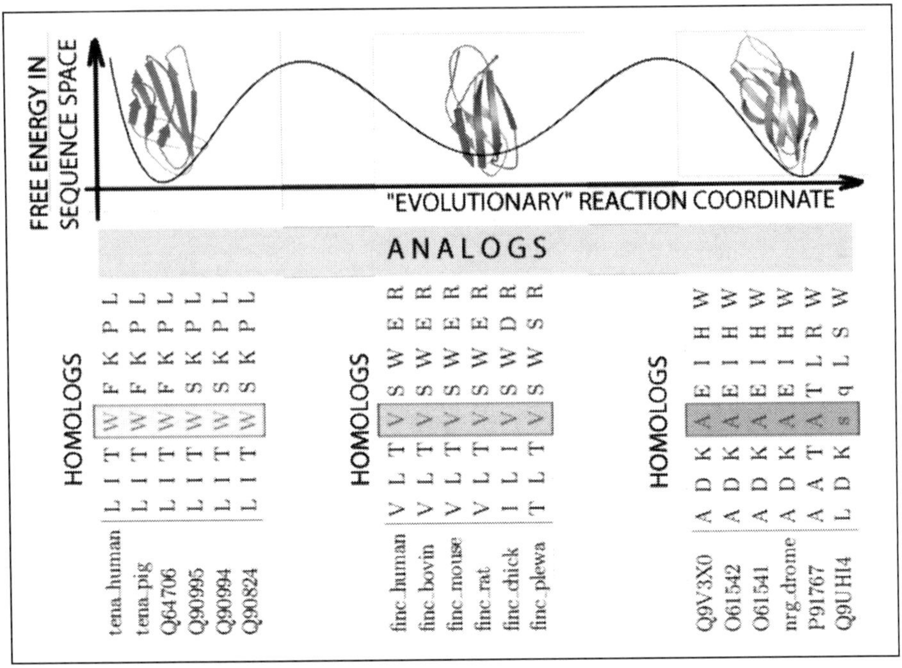

Figure 2. A schematic representation of the evolutionary processes that result in conservation patterns of amino acids. For a given family of folds, e.g., immunoglobulin (Ig) fold in this diagram, there are several alternative minima (3) in the hypothetical free energy landscape in the sequence space as a function of the "evolutionary" reaction coordinate (e.g., time). Each of these minima are formed by mutations in protein sequences at some typical time scales, τ_0, that do not alter the protein's thermodynamically and/or kinetically important sites, forming families of homologous proteins. Transitions from one minimum to another occur at time scales, $\tau=\tau_0 \exp(\Delta G/T)$ where ΔG is the free energy barrier separating one family of homologous proteins from another. At time scale τ mutations occur that would alter several amino acids at the important sites of the proteins in such a way that the protein properties are not compromised. At time scale τ the family of analogs is formed. In three minima we present three families of homologs (1TEN, 1FNF, and 1CFB) each comprised of six homologous proteins. We show 8 positions in the aligned proteins: from 18 to 28. It can be observed that at position 4 (marked by blocks) in each of the families presented in the diagram amino acids are conserved within each family of homologs, but vary between these families. This position corresponds to position 21 in Ig fold alignment (to 1TEN) and is conserved.

The Underlying Scenario of Protein Evolution

We conjecture that the hierarchical organization of structurally similar proteins may be the result of the separation of the evolutionary time scales, shown schematically in Figure 2. On a time scale τ_0, a set of mutations occur that do not affect those amino acids that play crucial thermodynamic, kinetic and/or functional roles. As a result, there is little variation in sequences at the important sites of proteins. If a mutation occurs at the thermodynamically, kinetically and/or functionally important sites, it usually substitutes amino acids with close physical properties so that core, nucleus and/or functional site are not disrupted and the protein folds into its family fold, is stable in this fold, and its function is preserved. At this time scale, a family of homologs is born.

Rarely, at time scale τ, correlated mutations or larger-scale sequence rearrangements occur[108-110] that modify **several** amino acids at the core, nucleus and/or functional site, so that the stability and kinetics of proteins are not altered. Such a set of mutations can drastically modify the sequence of the protein. However, within the time scale τ_0, a family of homologs is born within which there is conservation of (already new) amino acids in the specific (important) sites of homologous proteins. Although there are alternations in the specific sites of the proteins at the time scale τ, these sites are more preserved than the rest of the sequence. The proposed view of protein evolution is consistent with the observations of the hierarchical organization of structurally similar proteins in families of homologs. Sets of families of homologs are organized, in turn, in super-families of (possible) analogs. The evolutionary time in our analysis is associated with the number of mutations that accumulate in the course of evolution. Because the rates may vary between families and even proteins, the relation of evolutionary time to physical time is not straightforward. Evolutionary time can be rigorously defined statistically as the number of mutations that occur in a fold family, averaged over all family members. The real time for one family may be different from that of another. These considerations complicate interpretation of sequence-based approaches to organismic phylogeny and calls for more robust, structure based approaches to phylogeny (Deeds, Hennesey, Shakhnovich, in preparation).

Support for such a scenario comes from several studies reporting observations of correlated mutations in proteins in the course of evolution.[108,109,111] In addition, Axe et al[112] have demonstrated that random substitution of core residues in ribonuclease barnase by hydrophobic residues preserves the activity of barnase in a significant number of cases. They produced barnase mutants in which 12 of 13 hydrophobic core residues have together been randomly replaced by hydrophobic alternatives. A strikingly high proportion (23%) of mutants maintained structural integrity enough to support enzymatic activity of barnase.

Murzin[33] proposed an elegant scenario of the evolution of protein architecture while maintaining its function. He argued that protein folding pathways may be altered by mutations. As a result, a local free energy minimum of the wild type protein may become a global free energy minimum of a mutant protein. The conformations at these states—global free energy minima of mutant and wild type proteins—may have no structural resemblance. However, these states may maintain the same function. As an example, Murzin argued that catalytic domain of the carboxypeptidase G_2[113] is structurally similar to aminopeptidase from *Aeromonas Proteolytica.*[114] However, these enzymes fold into two topologically different topoisomers.

Reconstructing Evolutionary Relations between Proteins

To overcome difficulties of reconstructing the evolutionary relation of proteins based on their sequences, we resort to the analysis of structural relationships between proteins. We employed a graph representation of the protein domain universe, in which we considered only protein **domains** that do **not** exhibit pairwise sequence similarity in excess of 25% and each such protein domain represented a node of the graph.[8] We used protein domains as identified by Dietmann and Holm in the FSSP database of protein domains.[115] Structural similarity between each pair of protein domains was characterized by their DALI *Z*-score.[115] We defined a structural similarity threshold Z_{min} and connected any two domains on our graph that had DALI *Z*-score $Z \geq Z_{min}$ by an edge. This way we created the protein domain universe graph (PDUG). It is crucial to note that, in contrast to weighted graphs considered in reference 45, the PDUG, is an unweighted graph where each edge that made it above threshold is considered equally. Clustering of such an unweighted graph represents its partitioning into disjoint clusters which can be carried out exactly using the classical depth-first search algorithm.[116] Each disjoint cluster represents a family of structurally related proteins in which each protein is presented only once (Fig. 3). Disjoint PDUG clusters are, in principle, equivalent to the *fold* classification level of the SCOP database.[4]

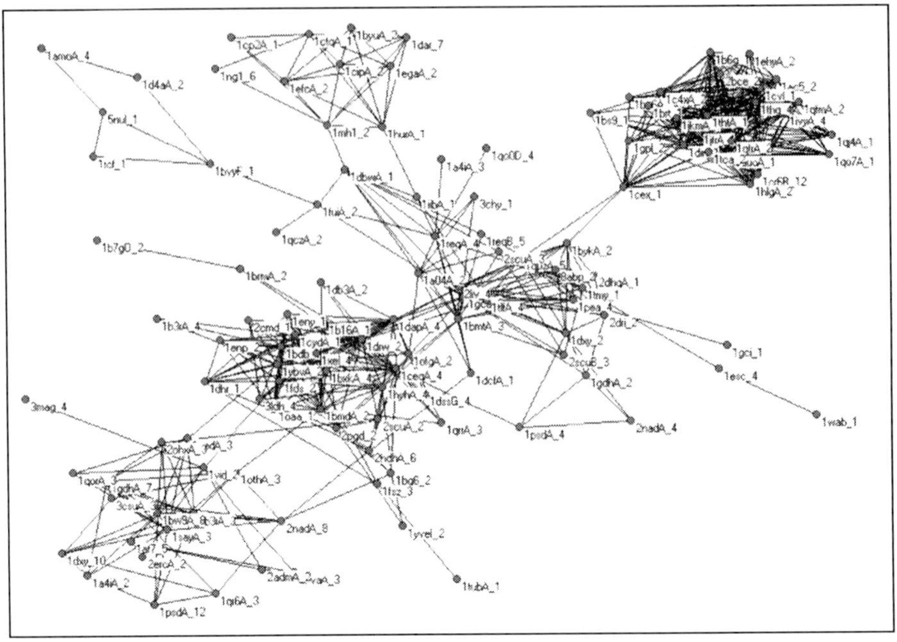

Figure 3. An example of a large cluster of TIM-barrel fold protein domains. Protein domains whose DALI similarity Z-score is greater than $Z_{min} = 9$ are connected by lines.

Properties of the Protein Domain Universe Graphs

We computed the size of the largest cluster in PDUG and random control graph as a function of Z_{min}.[8] We found a pronounced transition of the size of the largest cluster in PDUG at $Z_{min} = Z_c \approx 9$. The random graphs feature a similar transition, but at a higher value of $Z_{min} = Z_c \approx 11$. The distribution of cluster sizes depends significantly on whether $Z_{min} > Z_c$ or $Z_{min} < Z_c$ for both the PDUG and random graphs. We also found that the probability density $P(M)$ of cluster sizes M for both the PDUG and random graphs follows a power-law at their respective Z_c: $P(M) \propto M^{-2.5}$. The observed power-law behavior of $P(M)$ is simply a consequence of criticality at Z_c as it is featured prominently both for the PDUG and random graphs. The power-law probability density of cluster sizes is a **generic** percolation phenomenon that has been observed and explained in both percolation[117,118] and random graph theories.[119] Gerstein and coworkers also reported a power-law distribution for fold family sizes derived from the SCOP database[7] and attributed the observed power-law distribution to a certain evolutionary mechanism. However, we showed in reference 8 that random graphs featured the same power-law distribution for fold family sizes and were simply explained by percolation theory.[117-119]

In order to characterize the structural properties of the PDUG we computed the probability $\wp(k)$ of the number of edges per node k taken at $Z_{min} = Z_c$ for individual clusters. It is known that $\wp(k)$ distinguishes random graphs from various graphs observed in science and technology.[120] In drastic contrast with the equivalent random graph, the PDUG is scale-free with $\wp(k) \propto k^{-1.6}$ with a high degree of statistical significance (p-value less than 10^{-8}). The power law fit of $\wp(k)$ is most accurate at $Z_{min} \approx Z_c$, and noticeably deteriorates above and below Z_c. The fit at $Z_{min} > Z_c$ quickly becomes meaningless as the range of values of connectivity k rapidly diminishes as greater Z_{min} lead to mostly disconnected domains. At $Z_{min} < Z_c$ the power law fit also becomes problematic in the whole range of k because at large values of k (50-100) $\wp(k)$ shows some nonmonotonic behavior which can be interpreted as a maximum at large k (the data are insufficient to conclude that with certainty). However the remarkable property of a maximum $\wp(k)$ at $k = 0$ i.e., dominance of orphans remains manifest at all Z_{min} values. This is in striking contrast with random graph which is not scale-free at any value of Z_{min} and where $\wp(k)$ allows almost perfect Gaussian fit with a maximum at higher values of k.

The discovery of the scale-free character of the protein domain universe is striking. It has immediate evolutionary implication by pointing out the possible origin of all proteins from a single or a few precursor folds—a scenario parallel to the origin of the Universe from Big Bang. An alternative scenario, whereby protein folds evolved de novo and independently, would have resulted in random PDUG rather than the observed scale-free one.

The rigorous method of clustering protein **structures**[8] provides a number of insights. First of all, using graph theory for protein structure classification removes the ambiguities that are inherent in the highly useful, albeit manual, approaches to structural classification of proteins.[4,43] Perhaps not surprisingly we observed that the structure of the graph representing the protein domain universe depends on the Z_{min} threshold value of Z-score above which protein domains are considered structurally similar and are connected by an edge of the graph. However, at a certain critical value $Z_{min} = Z_c$, the structure of the PDUG becomes remarkably universal, simple and amenable to theoretical understanding from an evolutionary standpoint.

An important component of the analysis presented in reference 29 is random control where PDUG was compared with random graph. Our results showed that random weighted graph having the same weight (Z-score) distribution as PDUG featured same cluster size distribution. Since clusters in PDUG can be associated with fold level classification of protein structure, this observation suggested that nonuniform distribution of nonhomologous proteins over folds may not be due to special features of "most popular" protein folds as suggested previously by some researchers.[15,85] However that does not necessarily imply that observed protein folds are not selected based on their physical properties.[121] It is possible that the divergent evolution scenario described here occurs only on these selected folds while unfeasible ones are not observed in nature. However the analysis presented in reference 29 points out that explanation of the nonuniform distribution of nonhomologous proteins over observed folds does not require invoking the "designability principle"[15] or related conjectures about the nonuniform density of sequences in space of protein folds.[85]

We discovered that the structure of the PDUG is by far nonrandom, but rather represents a scale-free network featuring power-law distribution of the number of edges per node. The most striking qualitative aspect of the observed distribution is the much greater number of "orphans" (i.e., domains that are not structurally similar to any other domains) compared with random graph control. Importantly this qualitative feature remains prominent at any value of threshold Z_{min} despite the fact that power-law fits of $\wp(k)$ gets worse when Z_{min} deviates from Z_c. A natural explanation of this finding is from a divergent evolution perspective. The model of divergent evolution presented in reference 8 is in qualitative agreement with PDUG as it produces a large (compared with random graph) number of orphans (Fig. 4).

Besides reproducing the scale-free behavior of the PDUG, the divergent evolution model also quantitatively captures more specific graph properties of PDUG. In particular it was shown that the distribution of clustering coefficients[122] of nodes of PDUG is almost exactly matched by the divergent evolution model. This is in contrast to the random control where the scale-free PDUG has been randomly rewired while connectivity of each node is kept intact (Deeds and Shakhnovich, unpublished results).

Orphans are created in the model mostly via gene duplication and their subsequent divergence from a precursor. This may be meaningful biologically because duplicated genes may be under less pressure and hence prone to structural and functional divergence. The divergent

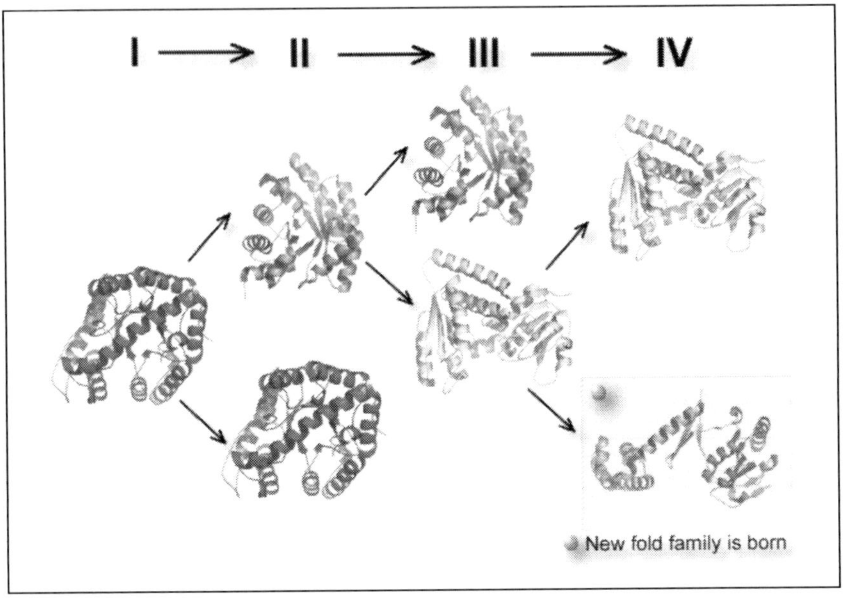

Figure 4. A cartoon representation of the divergent evolution model presented in reference 8. In this model, proteins diverge through a number of gene duplication events and point mutations (I→II→III→IV). While a single amino acid substitution may not significantly alter the protein structure, a number of them may result in drastic changes in protein structure. If these changes result in a functional and, most importantly, stable protein, a new fold family is born. In the model, each protein is represented by a node. Nodes representing proteins with significant structural similarity are connected by edges with a weight. If in the course of evolution an edge's weight lowers below the threshold value, the nodes become disconnected. At each evolutionary step, a randomly chosen node is duplicated and an edge with a weight (chosen from uniform distribution) is created that connects progeny to its parent. If the weight is below a threshold value, the nodes become disconnected and a new protein family is born. In addition, at each evolutionary step, after gene duplication a newly created node may become connected to its pra-parent.

evolution model presented in reference 8 is a schematic one as it does not consider many structural and functional details and its assumptions about the "geometry" of the protein domain space in which structural diffusion of proteins occurs may be simplistic. However, its success in explaining the qualitative and quantitative features of PDUG supports the view that all proteins might have evolved from a few precursors.

An important aspect of the model proposed in the reference 8 is that it provides only a conceptual framework for reconstructing protein structural space. The fine details of evolution contain crucial ingredients that underlie selective pressure in the model proposed in reference 8. Recently Deeds et al[123] uncovered how the features of an underlying protein structural space might impact protein structural evolution using lattice polymers as a completely characterized model of this space. In reference 123 we developed a measure of the structural comparison of lattice structures in analogy to the one used to understand structural similarities between real proteins. We used this measure of structural relatedness to create a graph of lattice structures and compared this graph (in which nodes were lattice structures and edges were defined using structural similarity) to the graph obtained for real protein structures. In reference 123 we found that the graph obtained from all compact lattice structures exhibited a distribution of structural neighbors per node consistent with a random graph. We also found that subgraphs of 3500 nodes chosen either at random or according to physical constraints, such as selective protein designability,[11] also represented random graphs. We developed a divergent evolution model based on the lattice space which produces graphs that were capable of recapitulating the scale-free behavior observed in similar graphs of real protein structures. Indeed, in contrast to this universal behavior, we observed subgraphs with power-law degree distributions only as the result of a very specific evolutionary sampling procedure. This not only demonstrated that scale-free graphs may be derived from such spaces but also that the rules underlying divergent graph evolution models are sufficient to produce this behavior.

Evolution of Proteins and Organisms

An important observation has been made by Deeds et al[124] who determined the structural content of 59 fully sequenced bacterial proteomes. Deeds et al identified structural proteomes—subgraphs of PDUG that belong to a specific organism through mapping of the PDUG representative domains on to a homologous domain of a given organism. Each such proteome contains a subset of domains from the PDUG. Strikingly, Deeds et al[124] found that these subgraphs are themselves scale-free networks (Fig. 5).

Deeds et al[124] explored two convergent evolutionary models to explain the scale-free organization of proteoms and concluded that such models were unlikely to explain the PDUG structural patterns. Addition of speciation events to a divergent model, however, resulted in model organisms that exhibit nonrandom subgraphs similar to those observed for real organisms. Deed et al[124] demonstrated that any divergent model must include some ingredient of speciation in order to account for the nonrandom overlap between structural proteomes. Such analysis of structural proteomes allowed authors to discount convergent models of structural evolution in favor of a specific divergent view that includes both organismal and structural evolution.

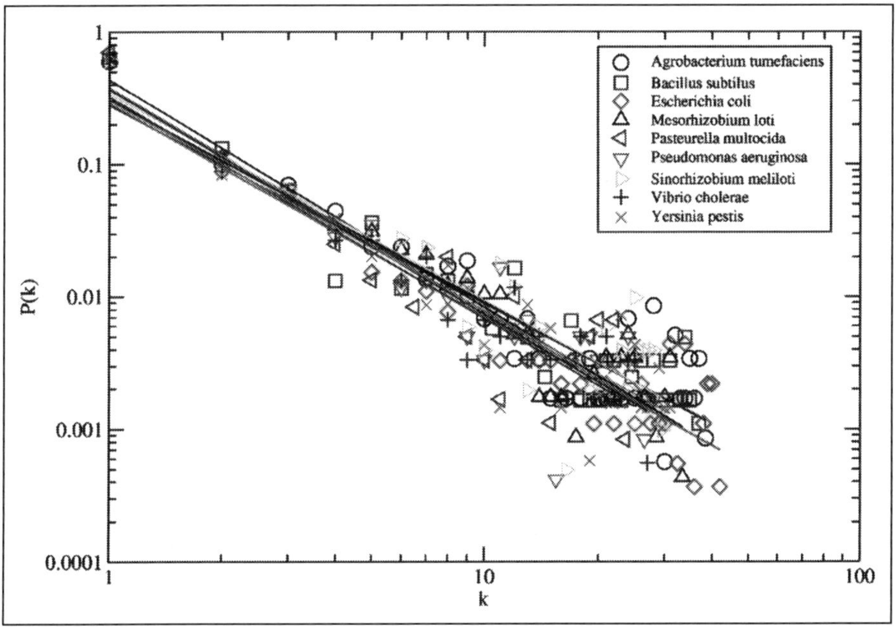

Figure 5. Degree distributions for nine bacterial subgraphs.[124] The degree distributions were shifted by a degree of 1 to allow display of orphan (degree 0) nodes on a log–log plot. Notably all proteomes exhibit similar to PDUG *P(k)* behavior: the power-law fits of these organisms yielded exponents that were approximately -1.6.

An important consequence of this study was the observed correlation between the Darwinian divergent evolution of organisms and the evolution of proteins. Such correlation couples two basic biological scales—microscopic (proteins) and macroscopic (organisms). The fact that these scales are coupled suggests a truly scale-free evolution of molecules and organisms. It also signifies of the single unifying law that governs evolution of proteins and organisms.

Reconstruction of Protein Structure-Function Relations

Most evolutionary models concede some form of relationship between protein structure and function.[125,126] Understanding this relationship is central not only to evolutionary biology, but also to structural genomics.[127] However, despite many efforts, the establishment of a clear relationship between structure and function has been elusive. This is in part due to the fact that neither structural nor functional relationships are well defined. The structure-function relationship between proteins can be understood in light of an evolutionary prospective. Two views on protein evolution have previously been suggested to account for the uneven distribution of sequences in fold space: that of convergent evolution[17,128,129] and divergent evolution.

Convergent evolution posits that different folds evolved independently and the same ("most popular") protein structures are recycled many times by proteins having different functions.[17] According to this model, new proteins may not be related by evolution to their orthologues as new proteins with similar function are rediscovered anew in many organisms. New proteins spawn by chance, but some structures are more populated than others because they are suggested to be more advantageous (thermodynamically, kinetically, evolutionarily). Such a scenario would suggest little relationship between structure and function.[130]

An alternative scenario is that of divergent evolution that suggests that a single or a few progenitor proteins give rise to many different, perhaps even unrelated offspring via processes of gene duplication and mutation.[131] These offspring can differ significantly from each other, either in sequence or structure, and can perform a varied array of functions many generations later.[132] It was shown recently that divergent evolution scenario implies important, observable structural relationships between domains: namely a scale-free organization of the protein universe that relays the history of how proteins are related to each other and would suggest a strong relationship between structure and function.[8]

It is important to note that divergent evolution implies a structure-function relationship that mirrors the structural hierarchy of PDUG. As protein structures diverged from progenitor proteins, so did functions. This relationship is necessitated by the requirement that the protein domain and all its descendants remain both functional[133] and structurally stable during the progression of evolution. Since evolution of function is similar in spirit and timeframe to that of structure, the structure-function relationship can also be observed in the context of a hierarchical functional annotation that allows comparison of protein functions at various levels of specificity of description. The level of hierarchical description is important, as it is the focal lens of functional evolution. Such hierarchical functional description is provided to the bioinformatics community by the Gene Ontology (GO) consortium.[134]

The main result of reference 135 is a striking finding that the corollary relationship between structural evolution and acquisition of new function by protein domains necessitated by a divergent evolution scenario can be **quantitatively** observed on the PDUG. Looking at PDUG through a hierarchical description of structural comparisons we find that we can characterize different clusters by the "functional fingerprint" that they display. A functional fingerprint is the distribution of functions within a particular cluster. We find that this distribution is quite unique to a given fold family at certain levels of functional annotation provided by GO. If we relax the Z_{min} threshold, we can also see an influx of protein domains into structurally similar clusters. These newly joined domains do not destroy the functional fingerprint of these clusters. This preservation of unique functional fingerprints through evolutionary dynamics further highlights the close relationship between structure and function necessitated by divergent evolution.

The Importance of Independent Functional Hierarchical Description

The simplistic divergent evolution model[8] that explains the nonrandom behavior of the PDUG is based solely on the premise that a protein has an ancestor that is its closest structural homologue. This model fits the data observed on the PDUG. The model characterizes the "oldest" proteins as those having the largest number of descendants and consequently the number of descendants for each protein depends on the protein's evolutionary age. We can therefore argue from our divergent evolution model that the older clusters and proteins are more populated and have more connections in PDUG. Of course, there is a significant stochastic component evolution of proteins that may drastically affect both family populations and their connectivity.

To detect mutual evolution between structure and function, in reference 135 we independently annotated proteins based on their function. By considering the function of all the proteins that are annotated and disregarding sequence homologues, we found that proteins have, in general, diverse functional descriptors. These descriptors are unique such as Methionine synthase, b12-binding domains or methylmalonyl-coa-mutase. On the other hand, all proteins can be broken up into just six or seven major functional categories such as enzyme, ligand binding, transporter. It seems apparent that the elucidation of a functional relationship between proteins depends on the system of description. Some medium specificity of functional description must be used if we are to quantitatively measure functional relationships between proteins. Since we do not know the coarseness of the needed annotation, we clearly need a hierarchical system.

A hierarchical system of functional annotation was recently developed by the GO consortium.[134] The GO system of annotation is well suited for measuring functional relationships between proteins because it defines a machine language where we can compare protein functions with little ambiguity based on their unique GO identifiers at different levels of specificity of annotation. The GO hierarchical language is organized as a directed acyclic graph. Each node in this graph is an annotation, a functional descriptor that we can assign to a gene or gene product. As the graph is traversed down, more precise functional descriptions populate the nodes. In this graph, the parent-leaf relationship of the nodes has an "all children are a subset of the parent" conjecture. For example, all adolases are enzymes as are CoA ligases because there is an edge from enzymes to both categories. In reference 135 we independently mapped protein function onto the whole of PDUG.

In order to carry out a completely machine based annotation, we used a direct mapping of the genes found in SwissProt Database that coded for the PDB entry of the protein domain in PDUG. We mapped the SwissProt entries to the curated annotation of SwissProt by the Gene Ontology Consortium. Each such annotation was mined independently by the GO consortium primarily from literature searches (http://www.geneontology.org). This yielded a nontrivial mapping from PDB to GO, thus giving each protein its functional assignment. The assignment is nontrivial because some SwissProt entries had many functional annotations corresponding to large, multifunctional, multi-domain proteins, from which our domain was only one. In this case, we kept all functional annotations. Working with domains alleviates the problems of "flow of structure" inside the clusters.[136] Flow of structure can happen when proteins A and B share a common domain C. Proteins A and B could then have highly nonrandom structural similarity, but different functions due to the noncommon domain being active. This way, domains may be erroneously classified as functionally equivalent while this may not actually be the case.

Divergent Evolution Observed

In reference 135 we presented strong evidence for divergent evolution of structure and function in protein domains by relating proteins' structures to their functions. We observed a homogeneity of function within structural clusters: the functional fingerprint. Functional fingerprints differ between the structural clusters. We observed the phenomenon of older, more populated clusters diffusing more in the functional space than newer, less populated clusters. For example, the largest structural cluster mainly populated by proteins with the distinctive Rossman fold, is mainly localized in the guanile nucleotide binding GO annotation.

When we considered less populated and therefore presumably younger clusters, we observed that the function is more localized. This is probably because the domain family had less time to diverge in structure and consequently function. The TIM barrel fold mainly has the function of hydrolases. Immunoglobulin (Ig) folds are very specialized folds performing mostly B-cell receptor functions. Interestingly, globins localize 95% into the "oxygen transporter" functional category. The probability that a set of randomly chosen proteins falls into one particular category at the fifth level of the GO ontology is diminishingly small. For example, for all globins this probability was found to be of the order of 10^{-80}.

It has been known for a long time that there are specialized folds. For example, the Ig folds are known to perform immunity/defense functions, and it is not surprising to find that its functional annotation differed significantly from all other structural clusters. We still found significant homogeneity even in more ubiquitous and less specialized folds such as TIM barrel and Rossman. Using the analysis of reference 135 it may be possible to identify a fold family by deciphering its functional fingerprint.

In reference 135 as we attached newly diverged protein domains to their ancestral clusters, the proteins attached with matching functional descriptions and complemented the functional fingerprint of their ancestors. We also noticed that there are functional categories that are more populated in the PDUG. These were the functions of many structurally similar, but sequentially different proteins. We therefore asked why some functions are more redundant than other ones. We speculated that these are the older functions (those that older proteins started with) that evolved much earlier and consequently have close descendant proteins performing similar function.

The PDUG functional annotations[135] revealed an interesting phenomenon related to the origin of orphans: as we increased the amount of structural evolutionary time, that we controlled by decreasing the threshold Z_{min} (higher Z_{min} represents a more recent snapshot of evolution), the orphans join ancestral clusters. Approximately half of the proteins are not orphans even at $Z_{min} = 9$. As we decreased Z_{min} from 9 to 2, we found that a half of the orphans join ancestral clusters, however the nucleus of the functional annotation within each cluster also grew almost proportionally. The functional nucleus is the collection of nonhomologous proteins that dominate functional annotations inside clusters. They are visibly seen as propagating together through the GO directed acyclic graph. This is in stark difference to random sampling of the protein domain universe where no such "nucleus" can be found and where we observe a more random distribution of functional annotation across all levels of GO. Notably, as we decreased Z_{min}, many functions peripheral to the nucleus diffuse into the fingerprint.

Conclusion

Refinement of the methodologies of protein structure determination yielded a massive amount of important information about protein structure. Due to the fundamental developments in the field of molecular evolution, this information unveiled a peculiar picture of the protein structural, sequence, and functional spaces. In particular, graph-theoretical approaches enable us to decipher specific characteristics of these spaces.

It has been suggested[29] that protein thermodynamics is one of the important evolutionary driving forces that shape the protein sequence space and govern the architecture of the protein structural space. This force relates protein sequence and structural spaces.

One striking observation is the scale-free organization of the PDUG[8]—protein structural space—which is signified by hierarchical relations between structurally similar proteins. The emergence of power-law scaling of the PDUG connectivity $\wp(k)$ is the result of evolutionary dynamics that is as robust at the scale of specific proteomes or at the scale of all organisms. The correlation between structural organization of proteomes and appearance of new organisms (speciation)[124] also suggest a truly universal "scale-free" evolutionary dynamics, whereby the appearance of new protein fold families is parallel to appearance of new species.

Distributions of function and structure over the PDUG act as two evolutionary lenses. It is evident that the evolution of structure and function is mutual and governed by the same underlying principles.[135] Since according to divergent evolution, aside from the biochemical consideration of function structure correlation, there is also biological pressure for proteins to retain close functional as well as structural similarity to their ancestors upon mutation and duplication. This implies a possibility to trace protein lineages via structural comparisons and further identify a possible function of putative proteins.

References

1. Levitt M, Chothia C. Structural patterns in globular proteins. Nature 1976; 261:552-558.
2. Koonin EV, Wolf YI, Karev GP. The structure of the protein universe and genome evolution. Nature 2002; 420:218-223.
3. Orengo CA, Jones DT, Thornton JM. Protein superfamilies and domain superfolds. Nature 1994; 372:631-634.
4. Murzin AG, Brenner SE, Hubbard T et al. Scop - A structural classification of proteins database for the investigation of sequences and structures. J Mol Biol 1995; 247:536-540.
5. Feller W. An introduction to probability theory and its applications. 1968.
6. Yanai I, Camacho CJ, Delisi C. Prediction of gene family distributions in microbial genomes: Evolution by gene duplication and modification. Phys Rev Lett 2000; 85:2641-2644.
7. Qian J, Luscombe NM, Gerstein M. Protein family and fold occurrence in genomes: Power-law behaviour and evolutionary model. J Mol Biol 2001; 313:673-681.
8. Dokholyan NV, Shakhnovich B, Shakhnovich EI. Expanding protein universe and its origin from the biological Big Bang. Proc Natl Acad Sci USA 2002; 99:14132-14136.
9. Karev G, Wolf Y, Rzhetsky A et al. Birth and death of protein domains: A simple model of evolution explains power law behavior. BMC Evol Biol 2002; 2:18.
10. Ponting CP, Russell RR. The natural history of protein domains. Annu Rev Biophys Biomol Struct 2002; 31:45-71.
11. England JL, Shakhnovich EI. Structural determinant of protein designability. Phys Rev Lett 2003; 90:art-218101.
12. England JL, Shakhnovich BE, Shakhnovich EI. Natural selection of more designable folds: A mechanism for thermophilic adaptation. Proc Natl Acad Sci USA 2003; 100:8727-8731.
13. Finkelstein AV, Gutun AM, Badretdinov AY. Why are the same protein folds used to perform different functions. FEBS Lett 1993; 325:23-28.
14. Govindarajan S, Goldstein RA. Why are some protein structures so common? Proc Natl Acad Sci USA 1996; 93:3341-3345.
15. Li H, Helling R, Tang C et al. Emergence of preferred structures in a simple model of protein folding. Science 1996; 273:666.
16. Rykunov DS, Lobanov MY, Finkelstein AV. Search for the most stable folds of protein chains: III. Improvement in fold recognition by averaging over homologous sequences and 3D structures. Proteins 2000; 40:494-501.
17. Taverna DM, Goldstein RA. The distribution of structures in evolving protein populations. Biopolymers 2000; 53:1-8.
18. Buchler NEG, Goldstein RA. Surveying determinants of protein structure designability across different energy models and amino-acid alphabets: A consensus. J Chem Phys 2000; 112:2533-2547.

19. Tiana G, Shakhnovich B, Dokholyan NV et al. Imprint of evolution on protein structures. Proc Natl Acad Sci USA, 2004; 101:2846-2851.
20. Teichmann SA, Murzin AG, Chothia C. Determination of protein function, evolution and interactions by structural genomics. Curr Opin Struct Biol 2001; 11:354-363.
21. Dodson G, Wlodawer A. Catalytic triads and their relatives. Trends Biochem Sci 1998; 23:347-352.
22. Duman JG, Li N, Verleye D et al. Molecular characterization and sequencing of antifreeze proteins from larvae of the beetle Dendroides canadensis. J Comp Physiol [B] 1998; 168:225-232.
23. Duman JG. Antifreeze and ice nucleator proteins in terrestrial arthropods. Annu Rev Physiol 2001; 63:327-357.
24. Makarova KS, Grishin NV. Thermolysin and mitochondrial processing peptidase: How far structure-functional convergence goes. Protein Sci 1999; 8:2537-2540.
25. Makarova KS, Grishin NV. The Zn-peptidase superfamily: Functional convergence after evolutionary divergence. J Mol Biol 1999; 292:11-17.
26. Chothia C, Hubbard T, Brenner S et al. Protein folds in the all-beta and all-alpha classes. Annu Rev Biophys Biomol Struct 1997; 26:597-627.
27. Wallace AC, Borkakoti N, Thornton JM. TESS: A geometric hashing algorithm for deriving 3D coordinate templates for searching structural databases. Application to enzyme active sites. Protein Sci 1997; 6:2308-2323.
28. Russell RB. Detection of protein three-dimensional side-chain patterns: New examples of convergent evolution. J Mol Biol 1998; 279:1211-1227.
29. Dokholyan NV, Shakhnovich EI. Understanding hierarchical protein evolution from first principles. J Mol Biol 2001; 312:289-307.
30. Abkevich VI, Gutin AM, Shakhnovich EI. Specific nucleus as the transition-state for protein-folding - evidence from the lattice model. Biochemistry 1994; 33:10026-10036.
31. Fersht AR. Nucleation mechanisms in protein folding. Curr Opin Struct Biol 1997; 7:3-9.
32. Dokholyan NV, Buldyrev SV, Stanley HE et al. Identifying the protein folding nucleus using molecular dynamics. J Mol Biol 2000; 296:1183-1188.
33. Murzin AG. How far divergent evolution goes in proteins. Curr Opin Struct Biol 1998; 8:380-387.
34. Pankov R, Yamada KM. Fibronectin at a glance. J Cell Sci 2002; 115:3861-3863.
35. Lupas AN, Ponting CP, Russell RB. On the evolution of protein folds: Are similar motifs in different protein folds the result of convergence, insertion, or relics of an ancient peptide world? J Struct Biol 2001; 134:191-203.
36. Russell RB, Ponting CP. Protein fold irregularities that hinder sequence analysis. Curr Opin Struct Biol 1998; 8:364-371.
37. Russell RB. Domain Insertion. Protein Eng 1994; 7:1407-1410.
38. Muller HJ. Bar Duplication. Science 1936; 83:528-530.
39. Ohno S. Evolution by Gene Duplication. Springer-Verlag: Berlin, 1970.
40. Ohno S, Wolf U, Atkin NB. Evolution from fish to mammals by gene duplication. Hereditas 1968; 59:169-187.
41. Gerstein M, Levitt M. A structural census of the current population of protein sequences. Proc Natl Acad Sci USA 1997; 94:11911-11916.
42. Qian J, Stenger B, Wilson CA et al. PartsList: A web-based system for dynamically ranking protein folds based on disparate attributes, including whole-genome expression and interaction information. Nucl Acids Res 2001; 29:1750-1764.
43. Orengo CA, Bray JE, Buchan DWA et al. The CATH protein family database: A resource for structural and functional annotation of genomes. Proteomics 2002; 2:11-21.
44. Holm L, Sander C. Protein-structure comparison by alignment of distance matrices. J Mol Biol 1993; 233:123-138.
45. Getz G, Vendruscolo M, Sachs D et al. Automated assignment of SCOP and CATH protein structure classifications from FSSP scores. Proteins 2002; 46:405-415.
46. Sprinzak E, Margalit H. Correlated sequence-signatures as markers of protein-protein interaction. J Mol Biol 2001; 311:681-692.
47. Dietmann S, Fernandez-Fuentes N, Holm L. Automated detection of remote homology. Curr Opin Struct Biol 2002; 12:362-367.
48. Russell RB, Sasieni PD, Sternberg MJE. Supersites within superfolds. Binding site similarity in the absence of homology. J Mol Biol 1998; 282:903-918.

49. Irving JA, Whisstock JC, Lesk AM. Protein structural alignments and functional genomics. Proteins 2001; 42:378-382.
50. Brocchieri L, Karlin S. Conservation among HSP60 sequences in relation to structure, function, and evolution. Protein Sci 2000; 9:476-486.
51. Bradley P, Kim PS, Berger B. TRILOGY: Discovery of sequence-structure patterns across diverse proteins. Proc Natl Acad Sci USA 2002; 99:8500-8505.
52. Andrade MA, Perez-Iratxeta C, Ponting CP. Protein repeats: Structures, functions, and evolution. J Struct Biol 2001; 134:117-131.
53. Jones S, Thornton JM. Prediction of protein-protein interaction sites using patch analysis. J Mol Biol 1997; 272:133-143.
54. Gajiwala KS, Burley SK. HDEA, a periplasmic protein that supports acid resistance in pathogenic enteric bacteria. J Mol Biol 2000; 295:605-612.
55. Hwang KY, Chung JH, Kim SH et al. Structurebased identification of a novel NTPase from Methanococcus jannaschii. Nat Struct Biol 1999; 6:691-696.
56. Stec B, Yang HY, Johnson KA et al. MJ0109 is an enzyme that is both an inositol monophosphatase and the 'missing' archaeal fructose1,6-bisphosphatase. Nat Struct Biol 2000; 7:1046-1050.
57. Shakhnovich BE, Harvey JM, Comeau S et al. ELISA: Structure-function inferences based on statistically significant and evolutionarily inspired observations. BMC Bioinformatics 2003; 4:34.
58. Lakey JH, Raggett EM. Measuring protein-protein interactions. Curr Opin Struct Biol 1998; 8:119-123.
59. Legrain P, Wojcik J, Gauthier JM. Protein-protein interaction maps: A lead towards cellular functions. Trends Genet 2001; 17:346-352.
60. Valencia A, Pazos F. Computational methods for the prediction of protein interactions. Curr Opin Struct Biol 2002; 12:368-373.
61. Fields S, Song OK. A novel genetic system to detect protein protein interactions. Nature 1989; 340:245-246.
62. Rain JC, Selig L, De Reuse H et al. The protein-protein interaction map of Helicobacter pylori. Nature 2001; 409:211-215.
63. Schuck P. Reliable determination of binding affinity and kinetics using surface plasmon resonance biosensors. Curr Opin Biotechnol 1997; 8;498-502.
64. Doyle ML. Characterization of binding interactions by isothermal titration calorimetry. Curr Opin Biotechnol 1997; 8:31-35.
65. Ahmadian MR, Hoffmann U, Goody RS et al. Individual rate constants for the interaction of Ras proteins with GTPase-activating proteins determined by fluorescence spectroscopy. Biochemistry 1997; 36:4535-4541.
66. Gavin AC, Bosche M, Krause R et al. Functional organization of the yeast proteome by systematic analysis of protein complexes. Nature 2002; 415:141-147.
67. Ho Y, Gruhler A, Heilbut A et al. Systematic identification of protein complexes in Saccharomyces cerevisiae by mass spectrometry. Nature 2002; 415:180-183.
68. Back JW, de Jong L, Muijsers AO et al. Chemical cross-linking and mass spectrometry for protein structural modeling. J Mol Biol 2003; 331:303-313.
69. Zhu H, Bilgin M, Bangham R et al. Global analysis of protein activities using proteome chips. Science 2001; 293:2101-2105.
70. Tong AHY, Drees B, Nardelli G et al. A combined experimental and computational strategy to define protein interaction networks for peptide recognition modules. Science 2002; 295:321-324.
71. Gaasterland T, Ragan MA. Microbial genescapes: Phyletic and functional patterns of ORF distribution among prokaryotes. Microb Comp Genomics 1998; 3:199-217.
72. Pellegrini M, Marcotte EM, Thompson MJ et al. Assigning protein functions by comparative genome analysis: Protein phylogenetic profiles. Proc Natl Acad Sci USA 1999; 96:4285-4288.
73. Bono H, Okazaki Y. Functional transcriptomes: Comparative analysis of biological pathways and processes in eukaryotes to infer genetic networks among transcripts. Curr Opin Struct Biol 2002; 12:355-361.
74. Tamames J, Casari G, Ouzounis C et al. Conserved clusters of functionally related genes in two bacterial genomes. J Mol Evol 1997; 44:66-73.
75. Dandekar T, Snel B, Huynen M et al. Conservation of gene order: A fingerprint of proteins that physically interact. Trends Biochem Sci 1998; 23:324-328.

76. Overbeek R, Fonstein M, D'Souza M et al. The use of gene clusters to infer functional coupling. Proc Natl Acad Sci USA 1999; 96:2896-2901.
77. Marcotte EM, Pellegrini M, Thompson MJ et al. A combined algorithm for genome-wide prediction of protein function. Nature 1999; 402:83-86.
78. Marcotte EM, Pellegrini M, Ng HL et al. Detecting protein function and protein-protein interactions from genome sequences. Science 1999; 285:751-753.
79. Enright AJ, Iliopoulos I, Kyrpides NC et al. Protein interaction maps for complete genomes based on gene fusion events. Nature 1999; 402:86-90.
80. Tsoka S, Ouzounis CA. Prediction of protein interactions: Metabolic enzymes are frequently involved in gene fusion. Nat Genet 2000; 26:141-142.
81. Goh CS, Bogan AA, Joachimiak M et al. Coevolution of proteins with their interaction partners. J Mol Biol 2000; 299:283-293.
82. Pazos F, Valencia A. In silico two-hybrid system for the selection of physically interacting protein pairs. Proteins 2002; 47:219-227.
83. Chothia C. Proteins - 1000 Families for the Molecular Biologist. Nature 1992; 357:543-544.
84. Finkelstein AV, Badretdinov AY, Gutin AM. Why do protein architectures have boltzmann-like statistics. Proteins 1995; 23:142-150.
85. Finkelstein AV, Gutin A, Badretdinov A. Why are some protein structures so common? FEBS Lett 1993; 325:23-28.
86. Davidson AR, Sauer RT. Folded proteins occur frequently in libraries of random amino-acid-sequences. Proc Natl Acad Sci USA 1994; 91:2146-2150.
87. Rost B. Protein structures sustain evolutionary drift. Fold Des 1997; 2:S19-S24.
88. Chothia C, Gerstein M. Protein evolution - How far can sequences diverge? Nature 1997; 385:579.
89. Grishin NV. Estimation of evolutionary distances from protein spatial structures. J Mol Evol 1997; 45:359-369.
90. Holm L. Unification of protein families. Curr Opin Struct Biol 1998; 8:372-379.
91. Sander C, Schneider R. Database of homology-derived protein structures and the structural meaning of sequence alignment. Proteins 1991; 9:56-68.
92. Flaherty KM, Mckay DB, Kabsch W et al. Similarity of the 3-dimensional structures of actin and the atpase fragment of A 70-Kda heat-shock cognate protein. Proc Natl Acad Sci USA 1991; 88:5041-5045.
93. Holmes KC, Sander C, Valencia A. A new ATP-binding fold in actin, hexokinase and Hsc70. Trends Cell Biol 1993; 3:53-59.
94. Orengo CA, Michie AD, Jones S et al. CATH - a hierarchic classification of protein domain structures. Structure 1997; 5:1093-1108.
95. Dodge C, Schneider R, Sander C. The HSSP database of protein structure sequence alignments and family profiles. Nucl Acids Res 1998; 26:313-315.
96. Sanchez R, Pieper U, Melo F et al. Protein structure modeling for structural genomics. Nat Struct Biol 2000; 7:986-990.
97. Pearl FMG, Lee D, Bray JE et al. Assigning genomic sequences to CATH. Nucl Acids Res 2000; 28:277-282.
98. Holm L, Sander C. An evolutionary treasure: Unification of a broad set of amidohydrolases related to urease. Proteins 1997; 28:72-82.
99. Reeck GR, de Haen C, Teller DC et al. "Homology" in proteins and nucleic acids: A terminology muddle and a way out of it. Cell 1987; 50:667.
100. Fitch WM. Distinguishing homologous from analogous proteins. Syst Zool 1970; 19:99-113.
101. Goldstein RA, Lutheyschulten ZA, Wolynes PG. Optimal protein-folding codes from spin-glass theory. Proc Natl Acad Sci USA 1992; 89:4918-4922.
102. Shakhnovich EI, Gutin AM. Engineering of stable and fast-folding sequences of model proteins. Proc Natl Acad Sci USA 1993; 90:7195-7199.
103. Abkevich VI, Gutin AM, Shakhnovich EI. Improved design of stable and fast-folding model proteins. Fold Des 1996; 1:221-230.
104. Shakhnovich EI. Theoretical studies of protein-folding thermodynamics and kinetics. Curr Opin Struct Biol 1997; 7:29-40.
105. Bryngelson JD, Wolynes PG. Spin-glasses and the statistical-mechanics of protein folding. Proc Natl Acad Sci USA 1987; 84:7524-7528.
106. Abkevich VI, Gutin AM, Shakhnovich EI. Theory of kinetic partitioning in protein folding with possible applications to prions. Proteins 1998; 31:335-344.

107. Shakhnovich EI. Protein design: A perspective from simple tractable models. Fold Des 1998; 3:R45-R58.
108. Altschuh D, Vernet T, Berti P et al. Coordinated amino-acid changes in homologous protein families. Protein Eng 1988; 2:193-199.
109. Thomas DJ, Casari G, Sander C. The prediction of protein contacts from multiple sequence alignments. Protein Eng 1996; 9:941-948.
110. Mirny LA, Shakhnovich EI. Universally conserved positions in protein folds: Reading evolutionary signals about stability, folding kinetics and function. J Mol Biol 1999; 291:177-196.
111. Pazos F, Helmer-Citterich M, Ausiello G et al. Correlated mutations contain information about protein-protein interaction. J Mol Biol 1997; 271:511-523.
112. Axe DD, Foster NW, Fersht AR. Active barnase variants with completely random hydrophobic cores. Proc Natl Acad Sci USA 1996; 93:5590-5594.
113. Rowsell S, Pauptit RA, Tucker AD et al. Crystal structure of carboxypeptidase G(2), a bacterial enzyme with applications in cancer therapy. Structure 1997; 5:337-347.
114. Chevrier B, Schalk C, Dorchymont H et al. Crystal-structure of aeromonas-proteolytica aminopeptidase - A prototypical member of the cocatalytic zinc enzyme Family. Structure 1994; 2:283-291.
115. Dietmann S, Holm L. Identification of homology in protein structure classification. Nat Struct Biol 2001; 8:953-957.
116. Sedgewick R. Algorithms in C. MA: Addison-Wesley, Reading, 1990.
117. Havlin S, Benavraham D. Diffusion in disordered media. Adv Phys 1987; 36:695-798.
118. Stauffer D, Aharony A. Introduction to percolation theory. Philadelphia, 1994.
119. Bollobas B. Random graphs. London: Academic Press, 1985.
120. Albert R, Barabasi AL. Statistical mechanics of complex networks. Reviews of Modern Physics 2002; 74:47-97.
121. Finkelstein AV, Ptitsyn OB. Why do globular-proteins fit the limited set of folding patterns. Prog Biophys Mol Biol 1987; 50:171-190.
122. Vendruscolo M, Dokholyan NV, Paci E et al. Small-world view of the amino acids that play a key role in protein folding. Phys Rev E Stat Nonlin Soft Matter Phys 2002; 65:061910.
123. Deeds EJ, Dokholyan NV, Shakhnovich EI. Protein evolution within a structural space. Biophys J 2003; 85:2962-2972.
124. Deeds EJ, Shakhnovich B, Shakhnovich EI. Proteomic traces of speciation. J Mol Biol 2004; 336:695-706.
125. Aravind L, Koonin EV. Gleaning nontrivial structural, functional and evolutionary information about proteins by iterative database searches. J Mol Biol 1999; 287:1023-1040.
126. Jordan IK, Kondrashov F, Rogozin I et al. Constant relative rate of protein evolution and detection of functional diversification among bacterial, archaeal and eukaryotic proteins. Genome Biol 2001; 2:research0053.
127. Baker D, Sali A. Protein structure prediction and structural genomics. Science 2001; 294:93-96.
128. Li H, Tang C, Wingreen NS. Are protein folds atypical? Proc Natl Acad Sci USA 1998; 95:4987-4990.
129. Csete ME, Doyle JC. Reverse engineering of biological complexity. Science 2002; 295:1664-1669.
130. Todd AE, Orengo CA, Thornton JM. Evolution of function in protein superfamilies, from a structural perspective. J Mol Biol 2001; 307:1113-1143.
131. Brenner SA. Natural progression. Nature 2001; 409:459.
132. Ponting CP, Russell RB. Identification of distant homologues of fibroblast growth factors suggests a common ancestor for all beta-trefoil proteins. J Mol Biol 2000; 302:1041-1047.
133. Cooper VS, Schneider D, Blot M et al. Mechanisms causing rapid and parallel losses of ribose catabolism in evolving populations of Escherichia coli B. J Bacteriol 2001; 183:2834-2841.
134. Ashburner M, Ball CA, Blake JA et al. Gene ontology: Tool for the unification of biology. The Gene Ontology Consortium. Nat Genet 2000; 25:25-29.
135. Shakhnovich BE, Dokholyan NV, Delisi C et al. Functional fingerprints of folds: Evidence for correlated structure-function evolution. J Mol Biol 2003; 326:1-9.
136. Schug J, Diskin S, Mazzarelli J et al. Predicting gene ontology functions from ProDom and CDD protein domains. Genome Res 2002; 12:648-655.

Chapter 8

Gene Regulatory Networks

T. Gregory Dewey* and David J. Galas

Abstract

Two gene regulatory networks inferred from different types of data are considered in this chapter. **Gene expression networks** are networks inferred from microarray time series data and **transcription factor networks** are networks obtained from a new genome-wide technique that allows an identification of all of the DNA binding sites for each transcription factor (TF). While addressing the same underlying questions, these networks reflect different properties of gene regulation and provide different insights. The gene expression network is inferred from dynamic analysis of time series data of gene expression profiles. The TF networks, on the other hand, are a direct result of experimental observation of a physical association between a TF and a DNA binding site, which (except for experimental noise) is unique. While our knowledge of the transcription factor networks is limited, these networks provide insights into a regulatory core network of TFs that regulate each other, and drive all network interconnectivity. In both cases, the resulting networks show features that may be universal to biological systems. The global properties of such networks show the scale-free distributions of node connectivity indicative of a hierarchical network and also exhibit small world graph properties. We discuss a network growth model based on gene duplication that provides excellent agreement with the global network parameters derived from the analysis of experimental expression data. In addition to these global properties, the local properties of these gene expression networks can be used in data mining and classification.

Introduction

High throughput technologies allow a genome-wide interrogation of biological systems. These technologies permit the measurement of the many parameters and variables associated with life processes and reveal, in many cases, the inherent complexities of these processes. The current era of systems biology is marked by ongoing efforts to assimilate and integrate this avalanche of information into models of biological functions. To do this, the detailed information about molecular species cannot be considered in isolation but rather must be related to all of the other components of the system. These relationships are most easily represented by network structures or graphs. Thus, systems biology invariably means network analysis. To this end, systems-wide investigations have focused on specific functional network structures such as metabolic, signaling and gene regulatory networks. In this chapter we review progress in inferring and interpreting gene expression networks and transcription factor networks.

*Corresponding author: T. Gregory Dewey—Keck Graduate Institute of Applied Life Sciences, Claremont, California 91711, U.S.A. Email: greg_dewey@kgi.edu

Power Laws, Scale-Free Networks and Genome Biology, edited by Eugene V. Koonin, Yuri I. Wolf and Georgy P. Karev.

An emerging problem in bioinformatics is to identify the relationships between the various components of a system and infer how one component influences another. To understand the mechanism of gene expression, a detailed molecular picture of gene regulatory networks is required. Such a picture can be developed by probing the interactions between signaling pathways, transcription factors and the TF binding sites of the cis regulatory region of a gene. These interactions constitute the molecular circuitry that describes how external influences can trigger signal transduction pathways to activate transcription factors for a specific set of genes. The molecular circuitry not only reveals how the expression of individual genes are controlled but also how one effector or agent influences the entire network.

Two very different types of networks associated with gene expression are considered here. First, we consider **gene expression networks**. These are networks derived from microarray data through the measurement of time series of mRNA levels on a genome-wide scale. The second network to be discussed is the **transcription factor network**. These networks are derived from a genome-wide identification of all of the DNA binding sites for each transcription factor (TF). This network allows the direct interaction and control of each gene to be identified from the observation of TF binding at elements upstream from DNA coding regions. While these two networks address the same underlying questions, the regulation of gene expression, they are, by their nature, very different, and represent different manifestations of the underlying regulatory mechanism. The gene expression network, being inferred from dynamic analysis of time series data of gene expression profiles, must be considered phenomenological, reflecting dynamical observations from the data and an inherently incomplete modeling of this data. These networks describe how the mRNA level of one gene influences the level of another. These are not true, gene regulatory networks in the strict sense because they are correlative, and not necessarily causal, networks. The transcription factor networks, on the other hand, are a direct result of experimental observation of a physical association between a TF and a DNA binding site. No line of inference is required to generate these networks and they lead to direct mechanistic interpretation. However, these networks are silent as to the dynamics of the network, they are strictly structural and do not the indicate the extent of control exerted by the TFs. The two complementary approaches provide a dynamical, but incomplete, phenomenological model of the structure of the network and a precise structural model with unknown dynamical properties.

We discuss how gene regulatory networks are inferred from time series data using simple linear dynamical models. The resulting networks show features that may be universal to biological systems. The global network properties are discussed and it is seen that these inferred networks are scale-free and exhibit small world properties. A network growth model based on gene duplication is described that provides excellent agreement with the global network parameters derived from the experimental data. In addition to the global properties, the local properties of these networks provide powerful data mining tools. There is a limited literature on transcription factor networks thus far, but early results show intriguing network features for these as well. While these networks are of limited size, they also appear to show the scale-free behavior seen in the expression networks.We conclude with a discussion of how these two networks can be compared and used in concert to create more complete quantitative models of gene regulation.

Inferring Gene Expression Networks from Microarray Data

Often high throughput methods, such as gene expression arrays, focus on genome-wide profiles of individuals from a population. From a dynamic point of view, this represents a "snapshot in time" of a potentially heterogeneous population. The complexity of this snapshot arises from two influences. First, no two organisms are alike, even in the same species, and genetic variation will influence the expression profile. Second, each organism has its own history and this history can lead to a wide range of dynamic states of the system with very different expression profiles. To distinguish between intrinsic genetic variation and the dynamic variation is a challenging task.

To discriminate between effects of history versus generic variation requires that expression profiles be monitored over time. There are real advantages to determining and analyzing time series expression data. Just as chemical kinetics yields mechanistic information in a more straightforward fashion than chemical thermodynamics, expression time series data are more amenable to network modeling than expression data from a population. In time series data, the system is prepared in a given physiological state at the initial time point and changes in gene expression levels are measured as it moves to a new state. These experiments have some similarity to traditional perturbation-relaxation experiments in physics and chemistry. Time series profiles have been measured in a wide range of systems including responses to media growth conditions (diauxic shift in yeast[1]), cell cycle synchronization,[2] exposure to vaccines,[3] signaling responses to cytokines[4] and mechanical stimulation and insect feeding in Arabidopsis.[5]

Linear Models of Gene Expression

Perhaps the simplest model for analyzing expression time series is the linear response model.[6-15] Two forms of the model have been used in the literature to analyze microarray data: the differential form[8,15] and the difference or Markovian model.[6,7] The differential form follows a series of coupled equations given by:

$$\dot{a}_i(t) = \sum_{j=1}^{m} W_{ij} a_j(t) + b_i(t) + \xi_i(t) \tag{1}$$

where $a_i(t)$ is the expression level of the ith gene at time t after some exposure or treatment, the dot presents a time derivative, W_{ij} is a matrix of first order rate constants showing the influence of the jth gene on the production of the ith gene, $b_i(t)$ is an external forcing function and $\xi_i(t)$ is a noise term. The sum is over all m different genes that are measured.

Alternatively, a simple form of a linear finite difference model has also been employed:

$$a_i(t) = \sum_{j=1}^{m} \lambda_{ij} a_j(t-1) \tag{2}$$

The transition coefficients λ_{ij} are the respective elements of the mxm transition matrix (referred to as the Λ matrix). The matrix elements again represent the influence of the expression level of the jth gene on that of the ith gene. The Λ matrix is calculated from a time series data set using a generalized matrix inversion technique.[6,7] One could also add noise and forcing function terms to the difference equation, but such models have not appeared in the literature.

While these models may appear quite different, they can be directly related to each other. Equation 1 can be solved in closed form using standard methods. Such solutions can be substituted into Equation 2 and a complicated relationship between λ_{ij} and W_{ij} is obtained. In our work, we chose to deal with the finite difference form because it required no data manipulation such as calculation of time derivatives (cf. 10) and no assumptions on the nature of the noise or the driving forces. Under the current technology, the noise and driving forces are not experimentally accessible quantities.

Phenomenological networks of gene interactions are derived from the transition matrices.[5] The mxm transition matrix, Λ calculated from the linear model can be viewed as a weighted graph showing the influence of one expression level on another. This is the starting point for the description of the genetic circuitry. Rather than work with these weighted graphs, we consider a simpler approach in which Λ is converted into an adjacency matrix for digraphs, indicating the connectivity but not the strengths of the influence. We describe the operation (adj) as:

$$\Gamma(\varepsilon) = adj(\Lambda) \tag{3}$$

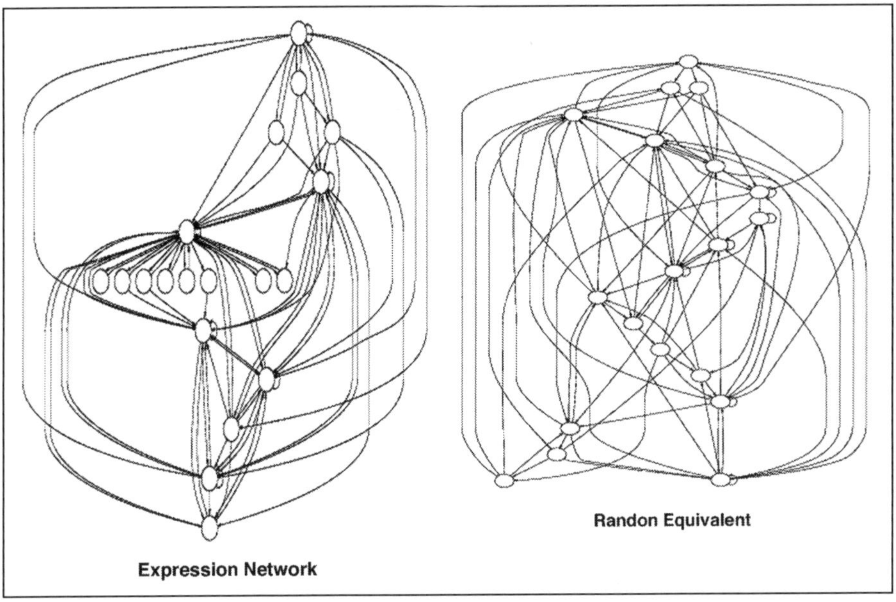

Figure 1. Comparison of gene expression network with random network. A small experimentally-derived gene expression network (left) shows more structure than the equivalent random network (right)

where the entries in Λ are set equal to 1 if the absolute values are above a certain threshold, ε, and are set equal to 0 below this threshold. For high values of the threshold, the resulting $\Gamma(\varepsilon)$ matrix will be a sparse adjacency matrix with a small network. As the value of ε is lowered, we can "grow" the network to include more nodes (genes). This threshold parameter is an adjustable parameter of the model. At this time we do not differentiate between positive and negative values for members in the transition matrix, as we are only interested in the underlying connectivity. Figure 1 gives an example of the type of network we obtain with this methodology. This network derived from the analysis of the diauxic shift in yeast shows a network characterized by central hubs connecting a large number of nodes of low connectivity. These networks are seen to be distinctly different from the equivalent random network (Fig. 1).

Validating Gene Expression Networks

A number of challenges are faced when inferring gene networks from linear models of microarray time series. First, the computation time to obtain the optimal coefficient matrix is prohibitive if the system gets larger than a few thousand genes. Secondly, the number of measurements in a gene expression time series is usually much smaller than the number of genes-the models are under determined. Both Dewey et al[6,7] and Yeung et al[15] took advantage of singular value decomposition (SVD) to solve the under determined problem. While this represents a least squares solution, it is not unique. Yeung et al[15] amend the SVD model by constructing a family of solutions and then use robust regression to identify solutions that create the sparsest network. In addition to the SVD solution, one has a class of solutions to Equation 3 that are governed by:

$$\Lambda' = \Lambda + CV^T \tag{4}$$

where Λ is the SVD solution obtained from Eq. 2 and V^T is the matrix of eigenvectors obtained from the SVD and C is a matrix to be determined from a constraint imposed on the problem. In the previous work,[15] sparseness of the network was assumed, $\Lambda' = 0$, and C was determined from the computationally intense problem of optimizing the zeros in solving: $CV^T = -\Lambda$. Simulations on model networks showed that this solution provided a more accurate reconstruction of the network. Our preliminary results show that networks generated using Equation 4 are essentially indistinguishable from those obtained with just the SVD. This suggests that the family of solutions all have a similar underlying network structure.

Currently, the networks inferred from gene expression time series cannot be considered quantitative predictive networks. The simplicity of the models, the quality of microarray data and the limited number of time measurements preclude a rigorous physico-chemical model. These shortcomings suggest that the inferred networks might best be used as data mining devices or as starting points for model development. Like other biological networks, such as protein-protein interaction maps, some expression networks doubtless suffer from a significant number of false positives and false negative connections. Thus, it is prudent to use these networks in an iterative loop with ongoing experimental efforts. It is important to identify constraints on the networks from independent experimental data. Simulations show that the addition of constraints can greatly enhance the accuracy of an inferred network.[15]

Networks as Classification Schemes

Currently the most popular way to analyze time series microarray data is cluster analysis. Based on the fundamental premise that genes having similar expression profiles may share similar functions, interesting genes and their functions can be inferred from clustering the relative expression profiles through the time course. Such simple clustering approaches are not easily extended to consider more complicated dynamics characteristic of the system. However, cluster analysis can be combined with dynamic modeling to show how dynamic characteristics of a biological system, such as the cell cycle, can be explored. By choosing a model parameter as a metric, one can extend the level of inference of the cluster analysis. Conversely, cluster displays provide a facile method for visualizing genome-wide parameters obtained from specific models.

Using the linear model parameter λ_{ij} as a metric, two-way clustering can be performed that shows how **influencing genes** affected the expression levels of **responding genes**. The application of this unsupervised method to the cell cycle data in yeast shows strikingly strong clustering of cell cycle regulated genes. Figure 2 shows a two-way clustering of λ_{ij} obtained from an analysis of yeast cell cycle data.[16] The two-way clustering is about j, the **influencing genes** and about i the **response genes**. The striking observation is that the blocks crossing clusters in columns (influencing genes) and rows (responding genes) can be used to infer the relationship between cell cycle phases. For example, if we look down the column representing S phase in Figure 2, we can see that S phase genes influence S and M phase genes positively (red color in image), but influence genes in M/G1 and G1 phase negatively (green color in image).

A schematic presentation of interaction between genes among different cell cycle phases, as well as alpha pheromone and heat shock activated genes, is shown in Figure 3. The influences are defined by the mean value of clustered blocks in Figure 2. We found that genes in one cell cycle phase activate genes in next phase (solid lines), and sometimes inhibit genes in the previous phase (dotted lines). Genes responding to alpha pheromone activate genes in S/G2 and G2/M, driving the cells into the cell cycle. Similar observations can be found with heat shock activated genes, which activate genes in M/G1 phase to drive the cells into the cell cycle. Interestingly, it is well known that alpha pheromone arrests cells in G1 phase while low temperature arrests cdc-15 strains in late mitosis.[2] Cells tend to reenter the cell cycle in the next phase beyond which they are arrested. The serial regulation of genes forms a connected regulatory network that is a cycle, as discovered by Simon et al.[17] Note that our result was obtained in an unsupervised fashion without any prior knowledge of chronological characteristics of the cell cycle. These results show how linear models can be used, not in a fully

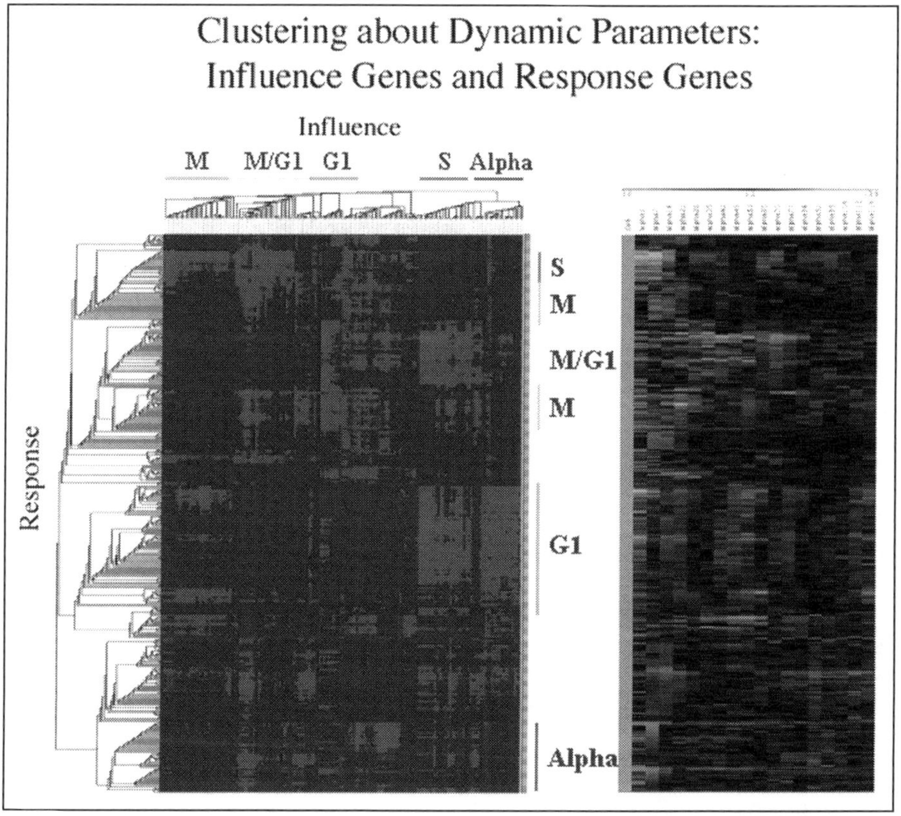

Figure 2. Hierarchical clustering of linear transition matrix. Clustering result of lambda matrix is shown on the left. Influencing genes are across the top and responding genes are along the side of the figure. Expression profile of genes in the same order as in clustering are shown on the right. Genes with similar expression profiles are grouped together, and they are labeled with the cell cycle phases. Alpha is alpha pheromone regulated genes. Reprinted with permission from Wu X, Dewey TG J Bioinf Comp Biol 2003;1:447-458. A color version of this figure is available online at http://www.Eurekah.com.

quantitative capacity, but rather in a qualitative, descriptive fashion. Considering the quality of the data and the phenomenological nature of the model, this is perhaps a more appropriate use of these models than as tools for quantitative prediction of expression levels.

Global Properties of Gene Expression Networks

There has been considerable recent interest in the network structure of a diverse range of systems, including the Internet, communities of actors, scholarly citations, metabolic networks and ecological systems, among others.[18-22] Three main categories of networks have been used to model these various systems. They are random networks,[23,24] small world networks[21,25] and growing random networks (GRNs).[18,19,26,27] Random graphs have been extensively studied and are constructed by randomly connecting a set of nodes. Small world graphs are generated from a regular starting lattice. Edges in this lattice are then randomly "rewired" to remote nodes. This provides strong local structure as well as global connectivity. Graphs can also be constructed from nonequilibrium growth models that start with a seed graph and add nodes and connections according to some prescribed set of preferences. Such models are referred to as

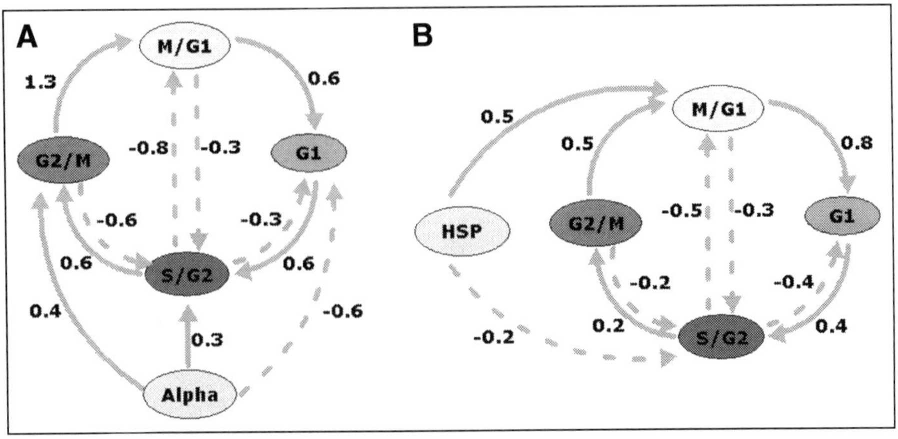

Figure 3. Interaction of genes among cell cycle phases. A) Alpha data set; B) cdc-15 data set. Cell cycle phases are shown in colored boxes, as well as alpha pheromone activated genes (Alpha) and heat shock activated genes (HSP). Solid lines indicate positive influence or activation, while dotted lines indicate negative influence or inactivation. Numbers along the edges are the mean value of transition matrix entries of each clustering blocks corresponding to the interaction strength between cell cycle phases. Reprinted with permission from Wu X, Dewey TG J Bioinf Comp Biol 2003;1:447-458.

growing random networks (GRN). Often a "rich get richer" set of preferences is used, where the newly added nodes are preferentially connected to nodes of high connectivity.

Global graph parameters can be used to characterize different types of networks. The cluster coefficient characterizes the extent to which vertices adjacent to any vertex are adjacent to each other. The cluster coefficient is calculated by averaging over all vertices, the fraction of vertices adjacent to a given vertex that are adjacent to each other. The cluster coefficient varies from 0 to 1 with 1 indicating that all the neighboring nodes are connected to one another. A second parameter, the characteristic path length is found by determining the number of edges on the shortest path connecting any two vertices and averaging this number over all pairs of vertices. This is a measure of the "connectedness" of the network. Finally, we will consider a third parameter, the scaling exponent for the node connectivity. Hierarchical networks often show scale-free or power law behavior between the number of nodes, $N(k)$, and the connectivity or degree per node, k. Such scale-free networks are hierarchical because a few nodes have many connection and many nodes have few connections.

Gene expression networks have a number of interesting properties. They have short mean pathlengths characteristic of highly connected networks and high clustering coefficients associated with very "clique-ish" graphs. Additionally, they show a scale-free distribution of connectivities with scaling exponents that are less than 2. This combination of graph traits is unique and is not observed in other real world networks analyzed to date. Studies of previous models for the growth of networks have elucidated the behavior of some properties of real networks such as the Internet, but they do not explain the biological networks represented by genetic regulatory networks. These models fail because they cannot yield exponents below 2 and because they often do not have either high cluster coefficient or low mean pathlengths.

Recently, the properties of a number of biological networks have been explored. Metabolic networks showing the connectivity of substrates show high cluster coefficient and a scaling exponent of 1.6.[28] Other studies of metabolic networks show a higher scaling exponent of 2.2.[18] The yeast protein-protein interaction map has been reported also to have high cluster coefficients and a higher exponent of 2.5. Our analysis of the protein-protein data however,

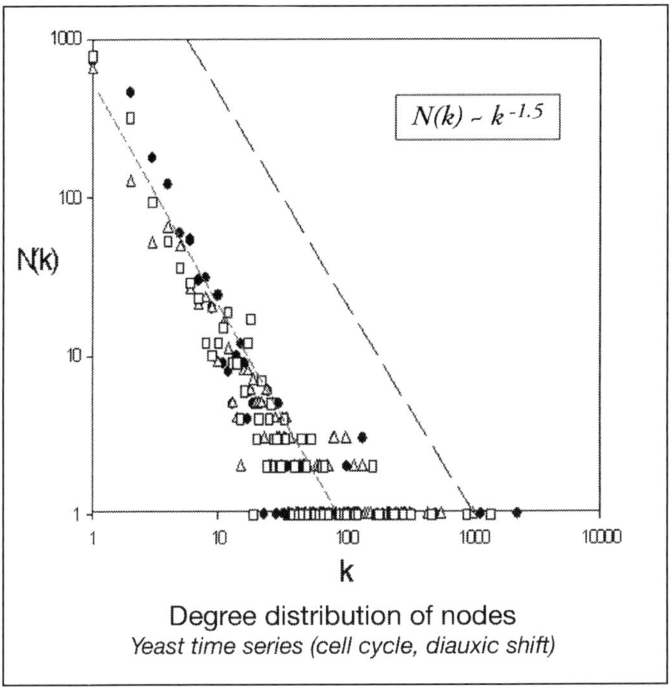

Figure 4. Plot of the distribution of number of nodes *N(k)*, plotted versus degree *k*. Networks for three different gene expression data sets were used: cell cycle data (cdc-28), •; cell cycle data (alpha), Δ; diauxic shift data, □. Dashed line is drawn with a slope of –3/2. Threshold parameter was set so that the adjacency matrix has 2×10^5 connections out of a possible 36×10^6 connections. Reprinted with permission from Bhan A, Galas DJ, Dewey TG Bioinformatics 2002;18:1486-1493.

using a composite of all the existing databases,[29,30] gives an exponent of 1.5. The results obtained here suggest that some biological networks show lower scaling than other observed networks and may obey a –3/2 power law (Fig. 4).

Gene Duplication Model of Expression Networks

Gene duplication provides a natural and compelling model for the growth of genetic regulatory networks. There is now abundant evidence from recent genome analysis from yeast[31] to human[32] that Ohno's original hypothesis that new genes are almost always created by duplication is largely valid. Gene duplication is now widely accepted as the single most important mechanism for generating new functions and processes.[33] This evolutionary mechanism must be at work in shaping the structure and function of interactions between genes and regulatory networks. We may be seeing evidence of this in the scaling law evident in the yeast expression data.

Specific duplication models can simulate the graph properties of the networks constructed from expression data. Figure 5 illustrates how a duplication event can affect a network. Duplication results in the creation of a new node that has inherited all the connectivity of the parent node, as would be true of a duplicated gene (including its *cis* regulatory elements). This results in an increase by one of the number of vertices with the degree of the parent. It also results in an increase of one in the degree of each of the neighbors. In a "pure" duplication model, this is the only event that occurs. This kind of growth model by itself has

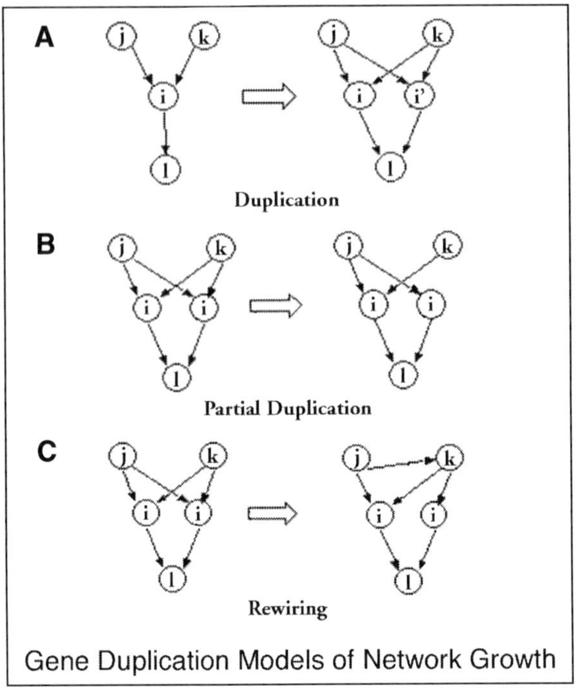

Figure 5. Schematic representation of network growth through gene duplication. A) shows pure gene duplication where a new node is created by duplicating the connectivity of the parent. This results in an increase in degree of the neighboring nodes. Node *i* is duplicated to give *i'*. Nodes *j*, *k*, *l* are neighbors. B) the partial duplication model where node *i* is duplicated to *i'* but not all the original connections are retained. C) shows a rewiring process where edge $j \rightarrow i$ is rewired to become $j \rightarrow k$. Reprinted with permission from Bhan A, Galas DJ, Dewey TG Bioinformatics 2002;18:1486-1493.

some interesting properties but it does not support a scale-free distribution of connectivities. We have, therefore, examined a number of "mixed" models that include gene duplication plus a second event. Features of two such models are illustrated in Figure 5. The "partial duplication" model (Fig. 5B) consists of duplication plus random removal of edges from the daughter node. A second model, "duplication plus preferential rewiring" (Fig. 5C) involves duplication followed by random rewiring of one of the edges in the network. In our preferential rewiring model, the new node to which the edge is rewired is chosen at random according to the same preference function in the previous GRN models,[18,19] i.e., the probability of connecting the edge to a node is proportional to the fraction of edges in the network that are incident at that node. These mixed models have formal similarity to a previous model used to describe the effect of gene duplication on protein-protein interaction networks.[34] Recently, a network growth model that yields scale-free networks has been described that involves gene duplication events.[35] This is a specific model involving domain shuffling and is distinctly different from the ones presented in this work. In all of these models, gene duplication is followed by a second event that breaks the parent-daughter symmetry inherent in a pure gene duplication model. This results in a broader range of node connectivities.

The results of the computer simulations of network growth are shown in Figure 6 for a variety of growth models and for the two different starting networks (network seeds). As can be

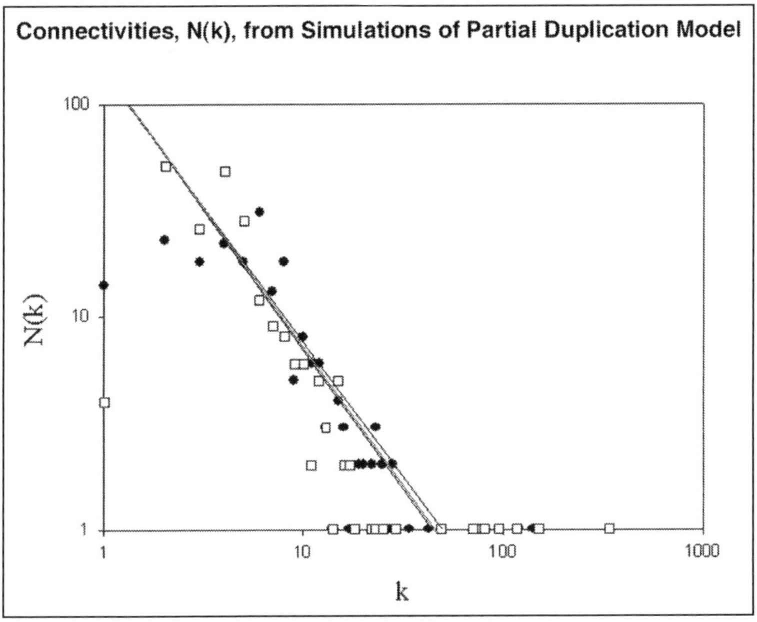

Figure 6. Plot of the distribution of number of nodes with degree *k*, plotted versus *k* for simulated networks. (See text for details.) Graphs correspond to three equally spaced time periods during network growth are shown. Leftmost graph is earliest time and rightmost is latest time. Top row are results from simulations with the duplication plus preferential rewiring model. Bottom row is from simulations with the partial duplication model.

seen, these networks reproduced a scaling exponent that is consistent with the experimental data (Fig. 4). Table 1 compares the clustering coefficient and the pathlength of the experimental data with the partial duplication model simulation. For comparison, we also show the results for the GRN model, originally introduced by Barabasi and coworkers.[19] As can be seen, the GRN model produces lower cluster coefficients and longer pathlengths than the experimental data. The simulation, on the other hand, faithfully reproduces the experimental data. Thus, it is seen that this biologically motivated, gene duplication model does account for the global network statistics of yeast gene expression networks.

Transcription Factor Networks

Recently, new array technologies have made it possible to determine where in the genome various TFs bind.[36] Since transcription factor binding to the cis regulatory region of the gene strongly influences the expression level of a given gene, this data provides linkages between the expression of TFs and other yeast genes and allows construction of a network. When 100 yeast TFs (out of an estimated 300) were examined in this fashion, it was found that there are many promoters that bind several factors. For example, there are about 100 promoters that bind four of these factors and about 40 that bind five, and several that bind even more. This statistic reveals the degree of complexity of the gene regulatory network in yeast, and the distribution of multiple binding sites on promoters (Fig. 7), also suggests a kind of hierarchy in the structure of the network. This hierarchy is implied by the distribution shown in Figure 7—that a minority of promoters bind a large number of regulatory factors, while a large number of promoters bind only a few factors.

Table 1. Statistical graph parameters for gene expression networks

	Λ_1		Λ_2*	
Data Set	**Cluster Coefficient**	**Average Pathlength**	**Cluster Coefficient**	**Average Pathlength**
Diauxic shift				
Original	0.58	3.0	0.67	2.3
Random	0.17	1.9	0.19	1.9
Cell cycle-alpha factor				
Original	0.66	2.6	0.46	3.5
Random	0.06	2.5	0.15	1.9
Cell cycle-cdc 28				
Original	0.88	2.2	0.71	2.4
Random	0.07	2.4	0.07	2.4
Gene duplication model	0.8	2.0		

Consider now only the network that is formed by the regulatory regions of transcription factor genes and the transcription factors themselves. This network can have mutual interactions, rather than the one way interactions between the transcription factor and other gene promoter links. This transcription factor network is at the heart of the regulatory processes of

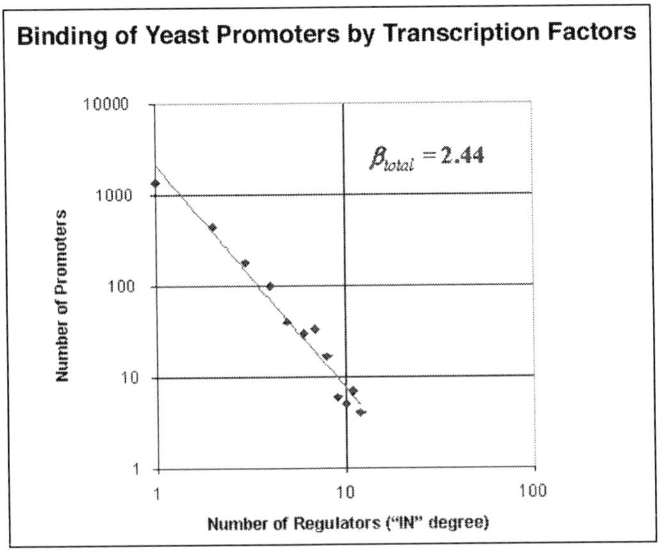

Figure 7. Distribution of number of promoters vs number of different TF binding sites in *Saccharomyces cerevisiae*. This is a log-log plot of data taken directly from Lee et al.[36] The red lines represents a least-squares fit to a power law. The resulting power-law exponent is 2.44 as indicated. A color version of this figure is available online at http://www.Eurekah.com.

the cell, because while other information is fed into this network by signaling from outside the cell, by coupling to the metabolic networks and by other proteins interacting with the transcription complex, the transcription factor network provides the sequence recognition mechanisms that processes all external information. The manner in which binding data implies a network is illustrated in Figure 8A. If we restrict our attention only to the transcription factors that form a single connected graph (no isolated nodes or pieces of two or three factors unconnected to the net) we can get a better view of the interlinking of the functional categories. This connected graph is shown in the figure. The directed graph representing this network depicts well the information derived from the experiments, but does not represent the network in detail. For example, no information about whether a given factor represses or activates a gene when it binds a given region, or the strength of such effects, is represented. Nonetheless, the basic structure of the graph, sometimes referred to as its "topological" structure, is well represented. Bear in mind that the graph in the figure is inherently a directed graph because the factors bind the regulatory regions of other's genes, giving a direction to each link.

Since this network includes a large number of transcription factors and is the most extensive such network known to date, we will examine it further, recognizing that it is incomplete in several different ways. The most important shortcoming is that genetic regulatory interactions can occur in ways other than just by transcription factor binding to cis-acting regulatory regions. The processing and translation of transcripts is often regulated, the binding of the TF proteins themselves may be regulated by modifications and protein cofactors that bind transcription factors, but do not bind DNA sites themselves also play important roles in regulation. The TF network by itself is only a part of the active regulatory network, albeit an important and well defined part. This network illustrates the strong degree of inter-linkage of different functional categories of TFs. While the Environmental Response category (green nodes) has some significant links within the category (see the tight coupling of Yap6, Rox1 and Cin5, for example) they are all directly linked to all other categories except the Development Processes one. The Cell Cycle category is coupled in multiple ways to all of the others. The Yap6 couplings just mentioned appear to be the only examples of mutual linkage between pairs of factor genes in this network. Each of the factors in such a pair (Yap6 - Cin5, and Yap6-Rox1) bind to the regulatory region of the other's gene, thereby creating a direct feedback loop of some kind. There are also several instances of self-regulation (Rap1, Rcs1, Nrg1, Yap6, Smp1 and Swi4). A glance at the graph in the figure reveals a major feature seen in most biological networks studied to date—the existence of a number of highly linked "hubs". For example, Sfl1, Abf1, Swi4, Swi5, Fkh2, Phd1 and Rap1, all have 5 or more links to other factors. The existence of hubs is a feature of a more general, mathematical property of many large networks as described in a previous section.

If we examine the distribution of connections of TF network, that is the number of nodes with k connections as a function of k, we must consider both the "in" connections as well as the "out" connections of these nodes. These form two distinct distributions, albeit with similar exponents. If we compare these distributions with the overall "in" distribution for the whole network (the nontranscription factor genes clearly have only "in" connections) we see that they are also power laws, but they have a different exponent (Fig. 8B). A smaller exponent indicates more highly connected nodes in the network than a higher exponent. One qualitative conclusion, then, is simply that the transcription factor network (the "core" network, as we are calling it) is more highly connected (in this statistical sense) than the other, peripheral, genes that it regulates. The conclusion has some intutitive appeal—the regulatory circuit, the computer, is more highly connected than the downstream output network of linkages to the "effectors" (Fig. 8C). It is also interesting that the "in" and "out" distributions of the network look to be the same.

Other genomes have now been sequenced and it is possible to make some preliminary comparisons between the yeast network and some others. While the transcription circuitry has

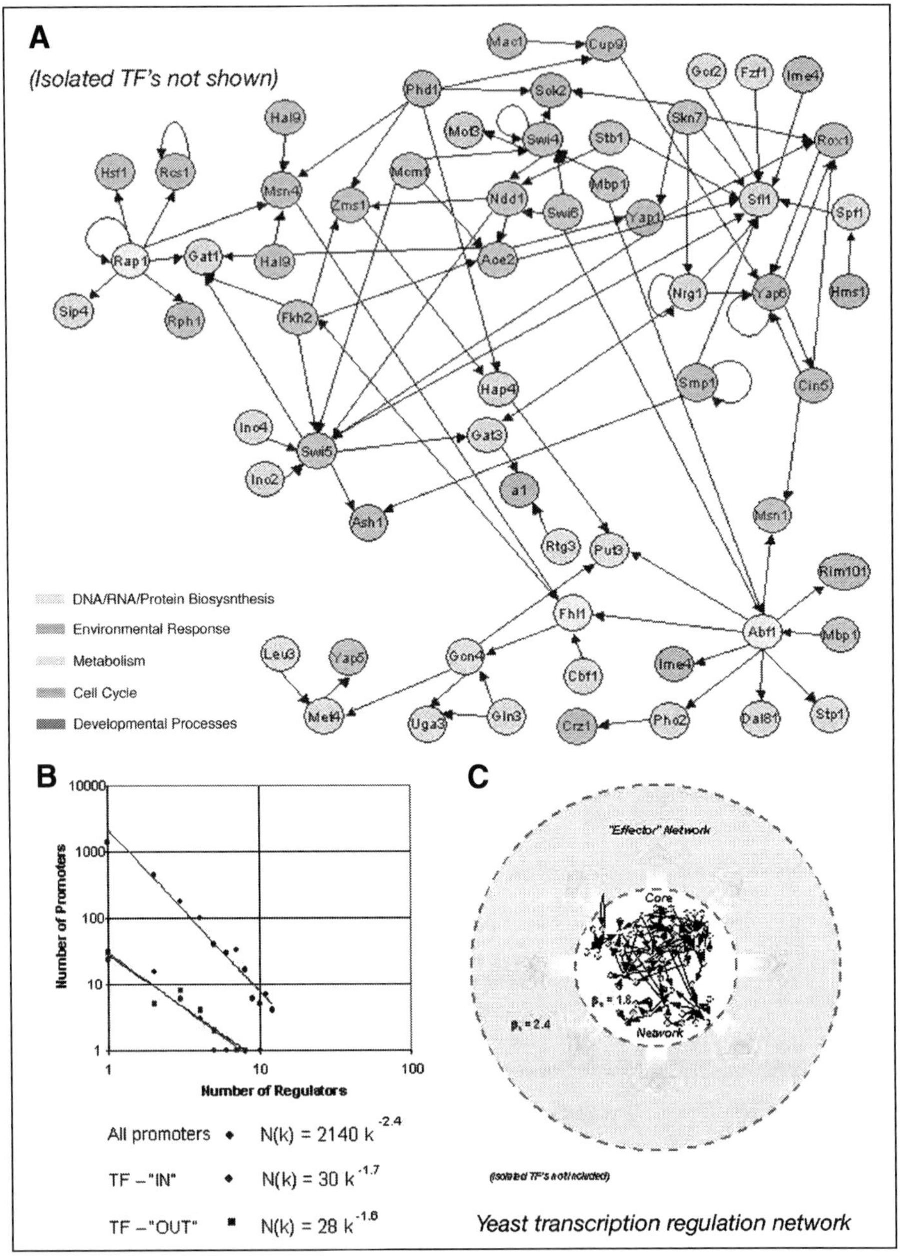

Figure 8. Transcription factor network of *Saccharomyces cerevisiae.* A) Diagram of the connections between 106 TF's. This graph is implied by the data reported in Lee et al.[36] The connections were inferred from the data using the thresholding criteria used in the original paper. Only the connected graph is shown. The characterization of the TFs into categories (color coding) is according to the paper. B) Power law fit of both the "core" network among TFs (described in the text and diagramed in A) for both "in" (round, pink-center points, pink line) and "out" (square points, blue line) networks. This is compared on the same graph to the overall promoter distribution (shown in A). C) Conceptual diagram showing how the "core" network and the "effector" network are related. A color version of this figure is available online at http://www.Eurekah.com.

not yet been fully elucidated for *E. coli* there is enough information to make a preliminary comparison. Although the data for this comparison is not obtained by the same methods, there is reasonable evidence for the overall structure of the regulatory network based on the integration of wide variety of approaches. Babu and Teichman have made such an integration in their presentation of the transcription network[37] based on comparative genomics of the transcription factors in *E. coli* and a detailed survey of the literature reporting regulatory relationships.[38] When we analyze this inferred network using the same statistical approach as for the yeast system we find (albeit with less statistical strength because of the smaller number of genes) that the picture is remarkably similar. We have represented the data in Figure 9A.

The transcription factor regulatory network comparison between yeast and *E. coli* then suggests an overall structure that has a central core network, consisting of transcription factors regulating each other's expression, and an external, "effector gene" network, regulated by the outputs from this core (Fig. 9B). The central cores are both more highly connected, as indicated by their lower exponent in the power-law fit, than the effector networks. What is more surprising than this, however, is that the exponents are very similar for the two organisms even though one is a prokaryote and the other a eukaryote.

This similarity is surprising, since yeast is much more complex in a variety of ways than *E. coli*. This raises the question of how an organism as complex as yeast is (relative to *E. coli* in any case) can have the same overall topological structure of its core regulatory network. The increased complexity may arise from several sources outside of the TF network. First, there are more transcription factors in yeast, and genes are regulated individually, rather than in operons, as in *E. coli*. Second, it is becoming clear, but is not yet fully understood, that there are a number of transcription cofactors (proteins that regulate gene transcription by binding to transcription factors, but not to DNA) that regulate genes by binding to multiple, bound TF's. This can lead to a large increase in regulatory complexity in yeast, and these kinds factors are not found (or are exceptional) in *E. coli*. Third, it is likely, judging from the number of RNA-binding proteins, that expression is regulated at the level of the RNA (post-transcriptional regulation) much more in yeast than in *E. coli*. Clearly the TF network comparison is only part of the story, yet it is significant that the core TF networks are very similar in structure.

Conclusions and Summary

We have discussed two substantially different methods and views of transcription networks, one based on correlative, "influence" analysis using time series analysis, and another based on direct TF binding analysis. While they are different and complementary in many ways, they are connected through the underlying mechanism of global gene regulation and control. Both methods provide insights into the structure of gene expression networks as well as powerful frameworks for data mining. All genes are regulated directly by TF's binding cis-regulatory regions. Thus, when a nontranscription factor gene is seen to "influence" another by time series analysis, we know that it does so through a "hidden" set of interactions that involve TFs, perhaps through regulated chemical modifications of TFs or transcription of TF genes. Most TF's are too weakly expressed, relative to the "effector genes" for the mRNA levels to be followed by array experiments. Thus the "central computer " we discussed as the core network plays the role, in some sense, as a set of "hidden variables" for the effector genes that are followed in a time series analysis.

Clearly a variety of methods must be used to elucidate fully the structure and function of gene expression networks, even in single celled organisms as "simple" as yeast and *E. coli*. The next stage of analysis of these networks should bring the details of the regulatory interactions, and a full picture of the network to the point where the dynamics, stable states and state transitions of the networks can be predicted and compared with experiment.

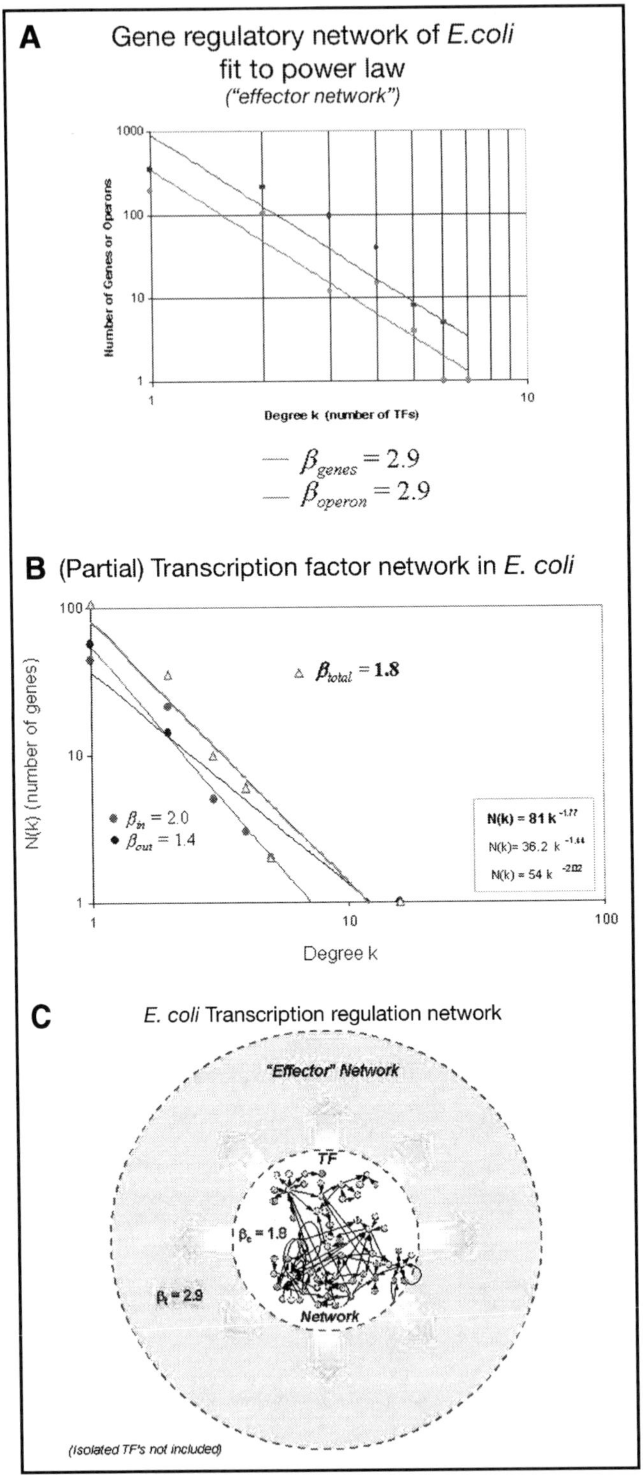

Figure 9. Transcription factor network for *E. coli.* A) Power-law plot of information reported in reference 37 for genes and operons (several, coregulated genes). The fit was done the same as in Figure 8. B) Power law fits for the core TF network, the "in" (red) and "out" (blue) networks. Combining the "in" and "out" networks to get a more statistically meaningful exponent gives a value of 1.8 (yellow). C) Conceptual diagram showing how the "core" network and the "effector" network are related. A color version of this figure is available online at http://www.Eurekah.com.

References

1. DeRisi J, Iyer A, Brown P. Exploring the metabolic and genetic control of gene expression on a genomic scale. Science 1997; 278:680-686.
2. Spellman PT, Sherlock G, Zhang MQ et al. Comprehensive identification of cell cycle-regulated genes of the yeast Saccharomyces cerevisiae by microarray hybridization. Mol Biol Cell 1998; 9:3273-3297.
3. Boldrick JC, Alizadeh AA, Diehn M et al. Stereotyped and specific gene expression programs in human innate immune responses to bacteria. Proc Natl Acad Sci USA 2002; 99:972-977.
4. Bechtel M, Wu X, Dewey TG. Analysis of time series expression data reveals cooperation of signaling pathways in chondrocytes. Submitted 2003.
5. Reymond P, Weber H, Damond M et al. Differential gene expression in response to mechanical wounding and insect feeding in Arabidopsis. Plant Cell 2000; 12:707-719.
6. Dewey TG, Galas DJ. Dynamic models of gene expression and classification. Func Integr Genomics 2001; 1:269-278.
7. Bhan A, Galas DJ, Dewey TG. A duplication growth model of gene expression networks. Bioinformatics 2002; 18:1486-1493.
8. Chen T, He HL, Church GE. Modeling gene expression with differential equations. Pacific Symp Biocomputing 1999; 4:29-40.
9. Holter NS, Maritan A, Cieplak M et al. Dynamic modeling of gene expression data. Proc Natl Acad Sci USA 2001; 98:1693-1698.
10. Heyer LJ, Kruglyak S, Yooseph S. Exploring expression data: Identification and analysis of coexpressed genes. Genome Res 1999; 9:1106-1115.
11. D'Haeseleer P, Wen X, Fuhrman S et al. Linear modeling of mRNA expression levels during CNS development and injury. Pac Symp Biocomput 1999; 41-52.
12. D'Haeseleer P, Liang S, Somogyi R. Genetic network inference: From coexpression clustering to reverse engineering. Bioinformatics 2000; 16:707-26.
13. Kim S, Imoto SS, Miyano S. Dynamic Bayesian network and nonparametric regression for nonlinear modeling of gene networks from time series gene expression data. In: International Workshop on Computational Methods in Systems Biology (CMSB2003), Lecture Notes in Computer Science. Springer-Verlag, 2003: 2602:104-113.
14. de Hoon MJ, Imoto S, Kobayashi K et al. Inferring gene regulatory networks from time-ordered gene expression data of Bacillus subtilis using differential equations. Pac Symp Biocomput 2003; 17-28.
15. Yeung MK, Tegner J, Collins JJ. Reverse engineering gene networks using singular value decomposition and robust regression. Proc Natl Acad Sci USA 2002; 99:6163-6168.
16. Wu X, Dewey TG. Cluster analysis of dynamic parameters of gene expression. J Bioinf Comp Biol 2003; 1:447-458.
17. Simon I, Barnett J, Hannett N et al. Serial regulation of transcriptional regulators in the yeast cell cycle. Cell 2001; 106:697-708.
18. Jeong H, Tombor B, Albert R et al. The large-scale organization of metabolic networks. Nature 2000; 407:651-654.
19. Barabási A-L, Albert R. Emergence of scaling in random networks. Science 1999; 286:509-512.
20. Albert R, Barabási A-L. Statistical mechanics of complex networks. Rev Mod Phys 2002; 74:47-97.
21. Strogatz S. Exploring complex networks. Nature 2001; 410:268-276.
22. Amaral LAN, Scala A, Barthélémy M et al. Classes of small-world networks. Proc Natl Acad Sci USA 2000; 97:11149-11152.
23. Cohen JE. Threshold phenomena in random structures. Discr Appl Math 1988; 19:113-128.
24. Kauffman S. Metabolic stability and epigenesis in randomly constructed genetic nets. J Theor Biol 1969; 22:437-467.
25. Watts DJ. Small worlds-the dynamics of networks between order and randomness. Princeton University Press, 1999.
26. Krapivsky PL, Redner S, Leyvraz F. Connectivity of growing random networks. Phys Rev Lett 2000; 85:4629-4632.
27. Dorogovtsev SN, Mendes JFF. Scaling properties of scale-free evolving networks: Continuous approach. Phys Rev E 2001; 63:056125-1-056125-19.

28. Wagner A, Fell D. The small world inside large metabolic networks. Proc Roy Soc London Ser B 2001; in press.
29. Uetz P, Giot L, Cagney G et al. A comprehensive analysis of protein-protein interactions in Saccharomyces cerevisiae. Nature 2000; 403:623-627.
30. Ito T, Chiba T, Ozawa R et al. A comprehensive two-hybrid analysis to explore the yeast protein interactome. Proc Natl Acad Sci USA 2001; 98:4569-4574.
31. Seioghe C, Wolfe K. Updated map of duplicated regions in the yeast genome. Gene 1999; 238:253-261.
32. Lander E et al. Initial sequencing and analysis of the human genome. Nature 2001; 409:860-921.
33. Ohno S. Evolution by Gene Duplication. Springer-Verlag, 1970.
34. Wagner A. The yeast protein interaction network evolves rapidly and contains few redundant duplicate genes. Mol Biol Evol 2001; 18:1283-1292.
35. Rzhetsky A, Gomez SM. Birth of scale-free molecular networks and the number of distinct DNA and protein domains per genome. Bioinformatics 2001; 17:988-996.
36. Lee TI, Rinaldi NJ, Robert F et al. Transcriptional regulatory networks in Saccharomyces cerivisiae. Science 2002; 298:799-804.
37. Babu MM, Teichman SA. Evolution of transcription factors and the gene regulatory network in Escherichia coli. Nucl Acids Res 2003; 31:1234-1244.
38. Shen-Orr SS, Milo R, Mangan S et al. Network motifs in the transcriptional regulation network of Escherichia coli. Nature Genet 2002; 31:64-68.

CHAPTER 9

Power Law Correlations in DNA Sequences

Sergey V. Buldyrev*

Introduction

A wide variety of natural phenomena is characterized by power law behavior of their parameters. This type of behavior is also called scaling. The first observation of scaling probably goes back to Kepler[1] who empirically discovered that squares of the periods of planet revolution around the Sun scale as cubes of their orbits radii. This empirical law allowed Newton to discover his famous inverse-square law of gravity.

In the nineteenth century, it was realized that many physical phenomena, for example diffusion, can be described by partial differential equations. In turn, the solutions of these equations give rise to universal scaling laws. For example, the root mean square displacement of a diffusing particle scales as the square root of time.

In the twentieth century, power laws were found to describe various systems in the vicinity of critical points. These include not only systems of interacting particles such as liquids and magnets[2] but also purely geometric systems, such as random networks.[3] Scaling is also found to hold for polymeric systems, including both linear and branched polymers.[4] Since then, the list of systems characterized by power laws has grown rapidly including models of rough surfaces,[5] turbulence and earthquakes. Empirical power laws are found to characterize also many physiological, ecological, and socio-economic systems. These facts give rise to the increasingly appreciated "fractal geometry of nature".[6-15]

A major puzzle concerning genomes of eukaryotic organisms, is that the large percent of their DNA is not used to code proteins or RNA. In human genome, this "junk" DNA constitutes 97% of the total genome length which is equal to 3 billion nucleotides also called base-pairs (bp). The role of non-coding DNA is poorly understood. It seems that it evolves by its own laws not restricted by a specific biological function. These laws are based on probabilities of various mutations and as such resemble the laws governing other complex systems listed above. In this chapter, I will review the degree to which power laws can characterize fluctuating nucleotide content of the DNA sequences, see also a critical review of W. Li.[16]

The term "long range correlations" is often misunderstood, implying some mystical long-range interactions or information propagation in space. Therefore, I will start with a brief introduction in the theory of critical phenomena, in which this concept has been developed. An impatient reader can jump directly to section "Correlation Analysis of DNA Sequences".

*Sergey V. Buldyrev—Department of Physics, Yeshiva University, 500 West 185th Street, New York, New York 10033, U.S.A. Email: buldyrev@yu.edu

Power Laws, Scale-Free Networks and Genome Biology, edited by Eugene V. Koonin, Yuri I. Wolf and Georgy P. Karev.

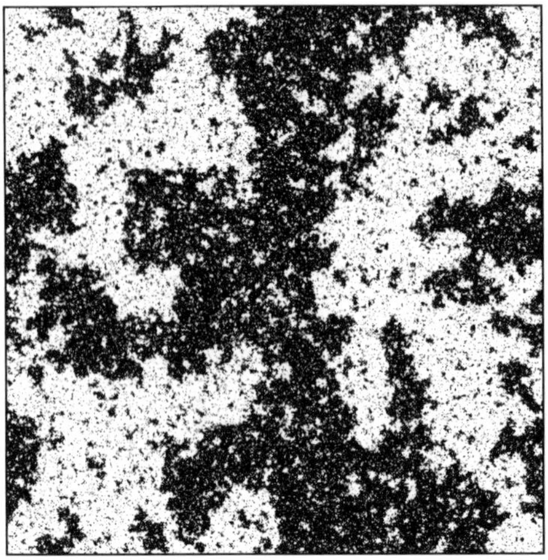

Figure 1. A snapshot of a two-dimensional system near its critical point. Black pixels represent gas particles. One can see density fluctuations of all different scales from a single particle to patches comparable with the entire system. This picture also represents an Ising magnetic near the Curie point, where black pixels are spins in positive orientation and white pixels are spins in negative orientation. The picture is obtained by computer simulations using the Metropolis algorithm at $T = T_c = 2.269185$.

Critical Phenomena and Long Range Correlations

One of the greatest advances in physics in the second half of twentieth century was the development of modern theory of critical phenomena.[2] The central paradigm of this theory is the importance of local fluctuations of the order parameter (Fig. 1). For a gas-liquid critical point, the order parameter is simply density. For the Curie point of a ferromagnetic, it is magnetization. Near the critical point T_c, the characteristic length scale ξ of the fluctuations, also known as the correlation length, grows according to a power law

$$\xi \sim |T - T_c|^{-\nu_c}. \tag{1}$$

The difference between the order parameters in the two phases (e.g., densities of gas and liquid) $\rho_l - \rho_g$ vanishes as the temperature approaches the critical point also according to a power law

$$\rho_l - \rho_g \sim (T_c - T)^{\beta_c}. \tag{2}$$

The positive quantities ν_c and β_c are called critical exponents. There are many other critical exponents α_c, γ_c, δ_c, η_c, etc., which characterize critical behavior of other parameters of the system.

The most spectacular manifestation of critical phenomena is critical opalescence. If one heats a closed transparent container filled by one third with water, the pressure inside it increases so that water and vapor remain at equilibrium: the water-vapor boundary is clearly visible and both phases are transparent. However, when the temperature approaches $T_c = 374°C$ within 1°C, the phase boundary disappears, and the substance in the container becomes milky: the density fluctuations scatter light because their average size becomes larger than the wave length of light which is about half a micron. Thus the correlation length becomes more than thousand times larger than the average distance between molecules which is about 0.3 nanometers.

Since the fluctuations near the critical point become extremely large, the details of the interaction potential which acts on much smaller scales become irrelevant and hence all liquids near the critical point have the same scaling behavior, i.e., they have exactly the same critical exponents, namely $\nu_c \approx 0.64$ and $\beta_c \approx 0.33$. Moreover, the theory predicts, that critical exponents are connected by several scaling relations, so that knowing any two exponents, for example ν_c and β_c one can predict the values of all the others. It turns out, that critical exponents depend only on dimensionality of space and some other major characteristics, such as dimensionality of spin orientations for magnetics. Thus, all variety of critical points can be classified by few universality classes so that all systems belonging to the same universality class have exactly the same values of critical exponents.

One of the simplest models for critical phenomena, the Ising model,[17] belongs to the same universality class as the liquid-gas critical point. We will discuss this model in greater detail, since it was first used by M. Ya. Azbel to describe possible correlations of nucleotides in the DNA.[18-20]

In the Ising model, atoms occupy sites on the *d*-dimensional lattice, for example on a square or a cubic lattice. In a one-dimensional system, atoms are placed equidistantly on a line. Each atom has a magnetic moment or spin, which may have only two orientations: up ($s = +1$) or down ($s = -1$). All pairs of spins occupying nearest neighboring sites interact with each other, so that they have a tendency to acquire the same orientation. The pair with the same orientations has negative potential energy $-\varepsilon$ while the pair with different orientations has positive potential energy $+\varepsilon$. Note that $\varepsilon < 0$ corresponds to the model of anti-ferromagnetic interactions. In addition, spins may interact with external magnetic field with energies $-h$ for positive spins and $+h$ for negative spins. It can be shown that this model is equivalent to the model of lattice gas, in which positive orientation of spins corresponds to the sites occupied by molecules, negative orientation indicates empty sites, two neighboring molecules attract with energy $-\varepsilon$, and the external field h corresponds to chemical potential which defines the average number of molecules in the system.

In 1973, M. Ya. Azbel[18] mapped a DNA sequence onto a one-dimensional Ising model by assigning positive spins $s = +1$ to strongly bonded pairs cytosine (C) and guanine (G) and negative spins $s = -1$ to weakly bonded pairs adenine (A) and thymine (T). (Complimentary base-pairs C and G located on the opposite strands of the DNA double helix are bonded by three hydrogen bonds, while A and T are boned only by two hydrogen bonds.)

One-Dimensional Ising Model

It is easy to solve the one-dimensional Ising model. According to the Boltzmann equation, the probability $p(U)$ to find a thermally equilibrated system in a state with certain potential energy is proportional to

$$p(U) \sim \exp(-U / k_B T), \tag{3}$$

where T is absolute temperature and k_B is Boltzmann constant. A striking simplicity of this equation is that it does not depend on any details of inter-atomic interaction and the details of motion of individual molecules. Once we know U and T, we can completely characterize our system in terms of the probability theory.

In the one-dimensional Ising model, a spin at position i can affect a spin at position $i + 1$ only through their direct interaction which is either $-\varepsilon$ if they orient the same way or $+\varepsilon$ if they orient in the opposite way. In the absence of magnetic field, the probabilities of these two orientations are proportional to $\exp(-U/k_BT)$, where $U = \pm\varepsilon$. Hence the probability of the same orientation is

$$p = \exp(\varepsilon / k_B T) / [\exp(\varepsilon / k_B T) + \exp(-\varepsilon / k_B T)] \equiv 1/(1+b), \tag{4}$$

where $b = \exp(-2\varepsilon/k_BT)$ and the probability of the opposite orientation is $q = 1 - p = b/(1 + b)$. Clearly, if T is small enough, b is also very small, and hence the probability for two neighboring spins to be in the same orientation is almost equal to one.

Do spins at a distant positions i and $(i + r)$ affect each other? To answer this question we must quantify this affect in mathematical terms. Two random variables $s(i)$ and $s(i + r)$ are called independent if the average of their product $\langle s(i)s(i + r)\rangle$ is equal to the product of their averages $\langle s(i)\rangle$ and $\langle s(i + r)\rangle$. Here and throughout the entire chapter $\langle\ldots\rangle$ denotes average taken over all possible positions i of the spins or nucleotide positions in a DNA sequence. The difference between these two quantities

$$\begin{aligned} C(r) &\equiv \langle s(i)s(i+r)\rangle - \langle s(i)\rangle\langle s(i+r)\rangle \\ &= \langle [s(i) - \langle s(i)\rangle][s(i+r) - \langle s(i+r)\rangle]\rangle \end{aligned} \tag{5}$$

characterizes the mutual dependence of two spins and is called correlation function. If $C(r) > 0$, the spins are correlated. If $C(r) < 0$, the spins are anti-correlated. Note that $C(0)$ coincides with the definition of variance of the variable $s(i)$. Note also that in general, for finite system of size L, $\langle s(i)\rangle \neq \langle s(i + r)\rangle$, because these two averages are taken over two different sets of positions $i = 1,2...L - r$ and $i + r = r + 1, r + 2,..., L$. When r is comparable to L, this difference becomes substantial.

It can be easily shown (see next section) that for a one-dimensional Ising model the correlations decay exponentially $C(r) \sim \exp(-r/\xi)$ at any temperature. The inverse speed of the exponential decay ξ is identical to the correlation length. In the one-dimensional model, correlation length can diverge only if temperature approaches absolute zero. Thus the critical point for the one-dimensional model is $T_c = 0$.

In the next section we will show this by making a mathematical excursion into the theory of Markovian processes, which is a very useful tool in bioinformatics. This chapter may be omitted by a reader who does not want to go deep into mathematical details, but is useful for those whose goal is to apply mathematics in biology.

Markovian Processes

In order to compute correlation function, we will represent a sequence of spins in the Ising model as a Markovian process. Markovian processes are very important in bioinformatics, thus we briefly summarize their definition and properties.

A Markovian process[21] is defined as a process obeying the following rules. (i) A system at any time step t, can be in n possible states $\mathbf{e}_1, \mathbf{e}_2, ...\mathbf{e}_n$. (ii) The probability to find a system in a certain state at any time step depends only on its state at the previous time step. Thus to fully characterize a Markovian process, we must define a $n \times n$ set of transition probabilities p_{ij} which are the probabilities to find a system in a state $\mathbf{e}_i$ at time $t + 1$ provided that a time t it was in a state $\mathbf{e}_j$. Obviously, $\sum_{i=1}^{n} p_{ij} \equiv 1$. (iii) It is assumed that p_{ij} do not depend on time.

It is convenient to describe the behavior of a Markovian process in terms of vector algebra, so that the probabilities $p_i(t)$ to find a system in any of its n states at time t is an n-dimensional vector-column $\mathbf{p}(t)$. The sum of its components $p_i(t)$ is equal to unity. Accordingly, it is natural to arrange the transition probabilities p_{ij} into a $n \times n$ matrix $\mathbf{P}$. The j-th column of this matrix is the set of transitional probabilities p_{ij}. Using the rule of matrix multiplication combined with the law of probability multiplication for independent events, we can find

$$\mathbf{p}(t + r) = \mathbf{P}^r \mathbf{p}(t), \tag{6}$$

where $\mathbf{P}^r$ is the r–th power of matrix $\mathbf{P}$, which can be easily found once we determine eigenvectors $\mathbf{a}_i$ and eigenvalues λ_i of matrix $\mathbf{P}$. By definition, eigenvectors and eigenvalues satisfy a homogeneous system of linear equations

$$\mathbf{P}\mathbf{a}_i = \lambda_i \mathbf{a}_i. \tag{7}$$

which has a nontrivial solution only if its determinant is equal to zero. Accordingly, the eigenvalues satisfy an algebraic equation of n-th power which turns the determinant of the matrix $\mathbf{P} - \lambda\mathrm{I}$, where I is the unity matrix, into zero.

Once we have determined the eigenvectors and eigenvalues, we can write

$$\mathbf{P}^r = \mathbf{A}\mathbf{\Lambda}^r\mathbf{A}^{-1}, \tag{8}$$

where Λ is the diagonal matrix formed by eigenvalues λ_i, and $\mathbf{A}$ is the matrix whose columns are eigenvectors $\mathbf{a}_i$.

Since the sum of elements in every column of matrix P is unity, the determinant of the matrix $\mathbf{P} - \mathrm{I}$ is equal to zero and one of the eigenvalues must be equal to unity: $\lambda_1 = 1$. The eigenvector $\mathbf{a}_1$, corresponding to this eigenvalue has a very special meaning. Its components yield the probabilities to find the system in each of its states for $r \to \infty$. We will show it in the following paragraph.

Except in some special degenerate cases, all the eigenvalues of a matrix are different. Assuming this, we can express the state of the system at time $t = r$ as a linear combination of the eigenvectors:

$$\mathbf{p}(t+r) = c_1\mathbf{a}_1 + c_2\lambda_2^r\mathbf{a}_2 + \ldots + c_n\lambda_n^r\mathbf{a}_n$$

where c_n are some coefficients, which can be obtained by multiplying the initial state of the system $\mathbf{p}(t)$ by matrix $\mathbf{A}^{-1}$. It can be easily seen from this equation that all eigenvalues must be less or equal to one: $|\lambda_i| \leq 1$. Indeed, if any $|\lambda_i| > 1$, the corresponding term in the above equation would diverge for $r \to \infty$, contradicting inequality $p_i(r) \leq 1$, which must be satisfied by the probabilities. Thus for all $i > 1$, $|\lambda_i| < 1$, and for any initial state of the system, we have $\lim_{r\to\infty}\mathbf{p}(r+t) = c_1\mathbf{a}_1$.

Thus, the average probability of finding the system in each of its states in a very long process is determined by the vector $c_1\mathbf{a}_1$, which can be readily found from the system of linear equations:

$$\mathbf{P}\mathbf{a}_1 = \mathbf{a}_1. \tag{9}$$

Since the determinant of this system is equal to zero, it has a nontrivial solution $c_1\mathbf{a}_1$, where c_1 is an arbitrary constant. Since the components of the vector $c_1\mathbf{a}_1$ have the meaning of the probabilities and, therefore, their sum must be equal to one, the coefficient c_1 must be the reciprocal of the sum of the elements of an arbitrary non-trivial solution $\mathbf{a}_1$ of Eq. (9).

The second-largest eigenvalue determines the decay of the correlations: $C(r) \sim \lambda_2^r = \exp(r \ln \lambda_2)$. By definition, the correlation length is the characteristic length of correlation decay which is determined by relation $C(r) \sim \exp(-r/\xi)$. Thus $\xi = 1/\ln(1/\lambda_2)$.

As an illustration of the Markovian formalism we can apply it to the one-dimensional Ising model. The matrix $\mathbf{P}$ in this case is simply

$$\mathbf{P} = \begin{pmatrix} p & 1-p \\ 1-p & p \end{pmatrix}, \tag{10}$$

where p is determined by Eq. (4). In order to find the eigenvalues, we must find the values of λ which turn the determinant of the matrix $\mathbf{P} - \lambda I$ into zero:

$$\begin{vmatrix} p-\lambda & 1-p \\ 1-p & p-\lambda \end{vmatrix} = 0. \tag{11}$$

This gives us a quadratic equation $(p-\lambda)^2-(1-p)^2=0$, with two roots $\lambda_1=1$, and $\lambda_2=2p-1$. The corresponding eigenvectors are

$$\mathbf{a}_1=\begin{pmatrix}\frac{1}{2}\\ \frac{1}{2}\end{pmatrix}\quad \mathbf{a}_2=\begin{pmatrix}-1\\ 1\end{pmatrix}. \tag{12}$$

Accordingly, we have

$$\mathbf{A}=\begin{pmatrix}\frac{1}{2} & -1\\ \frac{1}{2} & 1\end{pmatrix},\quad \mathbf{A}^{-1}=\begin{pmatrix}1 & 1\\ -\frac{1}{2} & \frac{1}{2}\end{pmatrix}. \tag{13}$$

and using Eq.(8),

$$\mathbf{P}^r=\begin{pmatrix}\frac{1+(2p-1)^r}{2} & \frac{1-(2p-1)^r}{2}\\ \frac{1-(2p-1)^r}{2} & \frac{1+(2p-1)^r}{2}\end{pmatrix}. \tag{14}$$

So we can see that the diagonal elements of this matrix, $p(r)=1/2+(2p-1)^r/2$ exponentially converge to $1/2$ for $r\to\infty$. The speed of convergence determines the correlation length:

$$\xi=-1/\ln(2p-1)=1/\ln\{[\exp(2\varepsilon/k_BT)+1]/[\exp(2\varepsilon/k_BT)-1]\}. \tag{15}$$

For $T\to 0$ the correlation length diverges $\xi\approx\exp(2\varepsilon/k_BT)\to\infty$, while for $T\to\infty$ the correlation length approaches zero: $\xi\approx 1/\ln(k_BT/\varepsilon)\to 0$. For finite temperature the correlation length is finite. Hence for one-dimensional Ising model, there is no critical point at positive temperatures, however the absolute zero $T=T_c=0$ can be treated as a critical point because in its vicinity the correlation length diverges faster than any power. So one can identify exponent ν_c as being infinite.

The eigenvector $\mathbf{a}_1$ gives us equal probabilities for a spin to be in positive and negative orientations, thus the spontaneous magnetization being determined as $\langle s(t)\rangle=a_{11}-a_{21}=1/2-1/2=0$ remains zero for all temperatures. In order to compute correlation function, we must compute the average product $\langle s(t)s(t+r)\rangle$. With probability a_{11} the value $s(t)=1$. Given $s(t)=1$, the probabilities of $s(t+r)=1$ and $s(t+r)=-1$ are equal to the elements of the first column of matrix $\mathbf{P}^r$. Analogously for $s(t)=-1$, which occurs with probability a_{21}, the probabilities of $s(t+r)=1$ and $s(t+r)=-1$ are given by the elements in the second column of matrix $\mathbf{P}^r$. So

$$\langle s(t)s(t+r)\rangle=a_{11}[p_{11}(r)-p_{21}(r)]-a_{21}[p_{21}(r)-p_{22}(r)]=(2p-1)^r \tag{16}$$

and, therefore,

$$C(r)=(2p-1)^r. \tag{17}$$

Exponential versus Power Law Correlations

In the previous section, we see that the one-dimensional Ising model in the absence of magnetic field is equivalent to a two-state Markovian process. In general, it is clear that any one-dimensional model with short range interaction is equivalent to a Markovian process with a finite number of states, and for such a process correlations must decay exponentially as λ_2^r, where $\lambda_2<1$. Thus the correlation length must be finite and can diverge only for $T\to 0$. Intuitively, we can imagine a one dimensional model as a row of dancing people each holding hands with two neighbors: one is on the left and one is on the right. Once they are holding hands, the correlation can pass from one neighbor to the next. No matter how strong they are holding hands, there is a finite chance q that they will separate, and the correlation will stop.

The probability that the correlation spreads distance r is proportional to $(1-q)^r \approx \exp(-qr)$, and hence the correlation length is finite and is inverse proportional to q.

In contrast, if the number of dimensions is larger than one, the interactions can propagate from point A to point B not only along a straight line from one neighbor to the next, but along an infinite number of possible paths connecting A and B. Accordingly, the correlation length can diverge for $T = T_c > 0$. Unfortunately, there are very few 2-dimensional models which can be solved exactly[17] and even those models have so complicated solutions that they are far beyond the scope of most physics textbooks. The most famous example of an exactly solvable 2-dimensional model is the Ising model, which was solved by Onsager in 1949. The solution is based on transfer matrices much more complicated than those we use in Section IV to solve the one dimensional model. It is much easier to simulate such a model on a computer and find an approximate numerical solution.

It can be shown that two-dimensional Ising model has a critical point at temperature $T_c = 2\varepsilon/\ln(1+\sqrt{2})/k_B = 2.269\varepsilon/k_B$. At the vicinity of this temperature, the correlation function acquires a non-exponential behavior

$$C(r) \sim r^{-\eta} \exp(-r/\xi) \tag{18}$$

where $\eta = 1/4$ is a new critical exponent proposed by M. Fisher in 1964. The correlation length ξ diverges as $|T-T_c|^{-1}$, which means that $\nu_c = 1$. The spontaneous magnetization for $T < T_c$ is not equal to zero, but, for any sample, it can be either positive or negative.

The absolute value of spontaneous magnetization approaches zero for $T \to T_c$ as $(T-T_c)^{1/8}$, so $\beta_c = 1/8$. If the temperature increases above T_c (also known as Curie point), the sample loses its magnetization. This phenomenon can be observed by everyone in a kitchen-style experiment: take an arm from a compass, place it into the fire of the burner and keep it there until it starts to glow red. The Curie point for Iron is 700°C. Cool it and place it back into the compass. It does not show North any longer!

Figure 1 shows the results of a computer simulation of a two-dimensional Ising model on the $L \times L = 1024 \times 1024$ square lattice. The program is very simple. At any time step, a computer attempts to "mutate" a spin at a randomly chosen lattice site. It first computes the energy change ΔU in such a would be mutation. If $\Delta U \leq 0$ the mutation always happens, if $\Delta U > 0$, it happens with probability $\exp(-\Delta U/k_BT)$. This algorithm invented by Metropolis in 1953,[22] leads to the Boltzmann distribution (3) of the probabilities to find a system in a state with total potential energy U. The proof of this fact is based on the theory of Markovian processes. Indeed, the set of Metropolis rules of flipping the spins can be represented as a transition matrix $\mathbf{P}$ with transition probabilities $p_{ij}\exp(-U_j/k_BT) = p_{ji}\exp(-U_i/k_BT)$, where U_i and U_j are the potential energies of the corresponding states. Obviously, vector $\mathbf{a}_1$ with components $a_{i1} = \exp(-U_i/k_BT)$ taken from the probability distribution (3) satisfies Eq. (9).

The system has periodic boundary conditions, so that pixels on the opposite edges of the system are in close proximity. In fact, the entire system can be viewed as a single line winded around the surface of a bagel. In such a system, site i has 4 neighbors $i+1$, $i-1$, $i+L$, and $i-L$, so the correlation can make really long jumps of length L and $-L$ along the line.

Black and white pixels show spins with positive and negative orientations respectively. One can see patches of irregular shapes and all possible sizes from very small, of one pixel size, to the giant one spanning the entire system. This scale-free property of patches is typical for systems with long range correlations with power law decay. Indeed, exponential decay of correlations $C(r) \sim \exp(-r/\xi)$ would imply a typical size ξ of patches so that the probability to find larger patches is exponentially small. The same picture can describe the behavior of gas particles near critical point. The molecules form clusters of all possible sizes which scatter light. Does this picture have anything to do with DNA?

It is well known that the DNA sequence has a mosaic structure[23] with patches of high concentration of strongly bonded CG base pairs alternating with patches of weakly bonded AT base pairs. These patches are called isochores and can span millions of base pairs. On a smaller scale of genes and exons, coding sequences have larger CG content than non-coding sequences. Finally there exist CpG islands of several hundred base pairs with high CG content.

May these patches have anything to do with Ising model? Of course DNA is not at thermal equilibrium and the concepts of temperature and potential energy cannot be applied to the study of its evolution. However, the evolution of DNA may be thought of as a Markovian process, similar to the Metropolis algorithm described above with mutation probabilities depending on the nature of neighboring nucleotides and on the pool of the surrounding nucleotides during replication process, which may be viewed as an external field or chemical potential.

There are several main objections to this idea:

1. **First objection**: the DNA chain is one dimensional. As we have seen above, long range correlations cannot exist in a one dimensional system.
   This objection can be easily overcome by the argument that the DNA molecule has an extremely complex three dimensional structure in which distant elements along the chain are in close geometrical proximity. Thus the correlation may propagate not only along the chain but may jump many steps ahead as in a toroidal Ising model shown in Figure 1. In 1993 Grosberg et al[24] proposed a model based on the distribution of loops in the polymer chain crumpled into a dense globular conformation. This simple model leads to the long range correlations decaying as a power law r^{γ}, where $\gamma \approx 2/3$.
2. **Second objection**. The long-range correlations emerge only in the narrow vicinity of the critical point. Why in the biological system such as DNA, the probabilities of mutations are such that they correspond to the vicinity of the critical point?
   This objection is more difficult to overcome. However there are examples of simple models which drive themselves to the critical behavior. The most relevant example is a polymer chain in the solvent,[11] in which the probability to find a monomer in a unit volume at distance r from a given monomer decays as $r^{1/\nu-3}$, where $\nu \approx 0.59$ is the correlation length exponent first determined by a Nobel prize winner P. Flory in 1949. In 1972, another Nobel prize winner P. G. de Gennes[4] mapped the problem of self-avoiding walks (which are believed to describe the behavior of polymers) to a model of a magnetic similar to an Ising model. He showed that the inverse polymer chain length $1/N$ is equivalent to the distance to the critical point $T - T_c$, and hence the correlation length ξ (which is equivalent to the radius of the polymer coil) grows as N^{ν}. A polymer chain has also a power law distribution of loops, determined by Des Cloiseaux.[25]
   In recent years, many models have been proposed that have a tendency of self organization (SOC) toward their critical points without any tuning of external parameters.[26,27] These models give rise to scaling, and produce sudden avalanche-like bursts of activity distributed according to a power law. Some SOC models are one-dimensional systems and have been applied to biological evolution.[28-30] Such models are of great interest and they might be relevant in studies of DNA sequences.
3. **Third objection**. Biological evolution is an extremely complex process which is governed by many different mechanisms acting at different length and time scales. The interplay of several characteristic length scales may lead to apparent power-law correlations, which thus lack universality of critical phenomena.[23]
   This objection is most probably correct. Indeed, the values of the correlation exponent are different for different species and change with distance r between the nucleotides (See section "Analysis of DNA Sequences"). Never the less, in the beginning of 1990s when the first long DNA sequences became publicly available, the idea to study them by correlation analysis attracted lot of attention.[31-41]

Correlation Analysis of DNA Sequences

Can correlation analysis be applied to DNA sequences? For a physicist or mathematician a DNA sequence looks like a text written in an unknown language, which is encoded in a 4-letter alphabet *A, C, G, T*. Each letter in this text corresponds to a DNA base pair. The first question one might ask is what is the overall fraction or frequency of each letter in this text. For example the frequency of letter "A", f_A is defined as $f_A = N_A/N$, where N_A is the number of letters "A" and N is the total length of the sequence.[42] This question is easy to answer, especially these days, when the total human genome is sequenced. In human genome, $f_A = f_T \approx 0.295$ and $f_G = f_C \approx 0.205$. Note that these numbers strongly depend on the organism under study. The second question one might ask: "Is there any apparent structure in this text, or it is indistinguishable from a text that would be typed by throwing a 4-sided dice?" (This dice can be made in the form of a Jewish toy, dreidel, with letters *A, C, G, T* on its sides which have slightly different surface areas, so that the probability of getting a letter on the top is equal to its frequency in the genome). For a text created by throwing such a dice, the events of getting any two letters at positions k and j are believed to be independent, so the probability of simultaneously getting letter X at position k and letter Y at position j is equal to the product $f_X f_Y$. If there is any structure in the text, the frequency $f_{XY}(r)$ of finding X at position k and Y at position $j = k + r$ will deviate significantly from the predicted value $f_X f_Y$.

To fully characterize all dependencies among four letters of the DNA alphabet one must compute 16 elements of dependence matrix $D_{XY}(r) = f_{XY}(r) - f_X f_Y$.[43] These dependence coefficients are equivalent to correlation functions used in the previous section to describe Ising model if the nucleotide sequence is replaced by a numerical sequence $s_x(k) = 1$ if nucleotide X is present at position k and $s_x(k) = 0$ otherwise:

$$D_{XY}(r) = \langle s_X(k) s_Y(r+k) \rangle - \langle s_X(k) \rangle \langle s_Y(k+r) \rangle, \qquad (19)$$

where $\langle \ldots \rangle$ indicates the average over all k.

All other measures of correlations including nonlinear measures such as mutual information[43-45] can be expressed via dependence coefficients. For example, one can introduce Purine-Pyrimidine (RY) correlation measure, in which any purine (A,G) is replaced by 1 and any pyrimidine (C,T) is replaced by -1. The numerical sequence for RY can be expressed as a linear combination of numerical sequences for each nucleotide $s_{RY} = s_A - s_C + s_G - s_T$. Accordingly,

$$\begin{aligned} C_{RY}(r) &= \langle s_{RY}(k)\, s_{RY}(k+r) \rangle - \langle s_{RY}(k) \rangle \langle s_{RY}(k+r) \rangle \\ &= D_{AA} + D_{CC} + D_{GG} + D_{TT} + 2(D_{AG} + D_{CT} - D_{GT} - D_{TT} - D_{AC} - D_{AT}). \end{aligned} \qquad (20)$$

Analogously, one can introduce C_{SW}, $(S = C,G;\ W = A,T)$ or C_{KM} $(K = A,C;\ M = G,T)$ or any other correlation function based on a linear combination of the elementary measures s_A, s_C, s_G and s_T.[32,46,47] The coefficients of this linear combination can be presented in the form of a vector $\mathbf{m} = (m_A, m_C, m_G, m_T)$ which we will call a mapping rule. For example, for *RY* mapping rule, we define $\mathbf{m} = (1,-1,1,-1)$, and for *C* mapping rule we define $\mathbf{m} = (0,1,0,0)$. Accordingly, any correlation measure could be expressed as a quadratic form $(\mathbf{m} \cdot \mathbf{Dm})$, where $\mathbf{D}$ is the dependence matrix.

Definitely, some of these correlation measures such as C_{SW} are not zero for at least the size of the isochore i.e., a chromosomal region with high or low $C + G$ content. Isochores have a typical size of about 10^5 base pairs, so the correlations would be non-zero for at least $r \approx 10^5$.

A physicist whose goal is to understand some general principles of DNA organization may attempt to fit the behavior $D_{SW}(r)$ by a power law function. A mathematical biologist[42,48-50] would rather try to characterize the size distribution of the isochores and their nucleotide content for various chromosomes and species and try to answer questions of biological relevance rather than to measure some power law exponent, which has an ambiguous biological meaning and characterize isochores in a very indirect fashion.

In general, DNA is known for its complex mosaic structure,[23,42] with structural elements such as isochores, intergenic sequences, CpG islands, LINE(long interspersed elements) and SINE (short interspersed elements) repeats, genes, exons, introns, and tandem repeats.[51,52,53] Each of these structural elements has its different size distribution, nucleotide frequencies, and laws of molecular evolution, so the correlations in the DNA sequence have very complex structure, are different for different spices and can not be characterized by a universal power-law exponent, in a way it is observed in critical phenomena. Correlation studies by their nature involve averaging over large portions of a sequence, so they have a tendency to gloss over particular details. This is the main reason why they are not very popular in bioinformatics whose main tool is the search for sequence similarities[54] analogous to finding in an unknown language some already known words or names, which may shed some light on the meaning of their neighbors.

Never the less, characterization of correlations in DNA sequences has some intellectual merit and even practical importance for a biologist whose goal is to understand molecular evolution of DNA sequences.[55] There are several reasonable models of DNA evolution in which exact power-law correlations emerge.[56-59] The values of the exponents of these power laws depend on the parameters of the model, such as mutation rates and thus can be used to test certain assumptions of the models. These models are discussed in the three sections starting with section "Mutation Duplication Model of DNA Evolution".

Another problem with correlation studies, is that they can be affected by many characteristics of the system, for example sequence length. In order to avoid many potential pitfalls it is very important to understand basic properties of correlation measures and fine-tune them on the well known systems which can serve as golden standards. In the next sections we will introduce various correlation measures and illustrate their usage, applying them to the Ising model, whose correlation properties are well known. Again, an impatient reader may proceed to "Mutation Duplication Model of DNA Evolution".

Correlation Function

In the next four sections we will describe several methods of correlation analysis. To develop some intuition on their advantages and disadvantages, we will apply them to the one-dimensional and two-dimensional Ising models, whose correlation properties are known theoretically.

The most straightforward analysis is the direct computation of the correlation function $C(r)$ defined in Eq. (5). Figure 2 shows the behavior of $\ln C(r)$ for the one-dimensional Ising model consisting of $L = 2^{16}$ spins for several values of T approaching zero. For small values of r, the graphs are straight lines with the slope equal to the inverse correlation length in complete agreement with Eq. (17). Figure 3 shows the behavior for the two-dimensional Ising model consisting of $L^2 = 2^8 \times 2^8$ spins above and below critical point. Figure 4 presents the corresponding snapshots of the system. The correlation length increases while temperature decreases toward $T_c \approx 2.27$ and then very quickly goes down again, as temperature continues to decrease.

This behavior may seem counterintuitive. Indeed, one can argue that correlations below T_c are so strong that the majority of spins acquire the same orientation. However, from a mathematical point of view, the majority of spins, say fraction p, has the same orientation. (In Fig. 4, $T = 2.17$, it is positive, but in other simulations, it may appear negative). White patches, indicating the negative orientation are small, isolated, and randomly distributed in the sample. These patches of the opposite orientation may be regarded as defects in the crystalline structure. Thus one can regard two spins at distant positions r and $r + k$ to be two independent random variables taking value 1 with probability p and value -1 with probability $1 - p$.

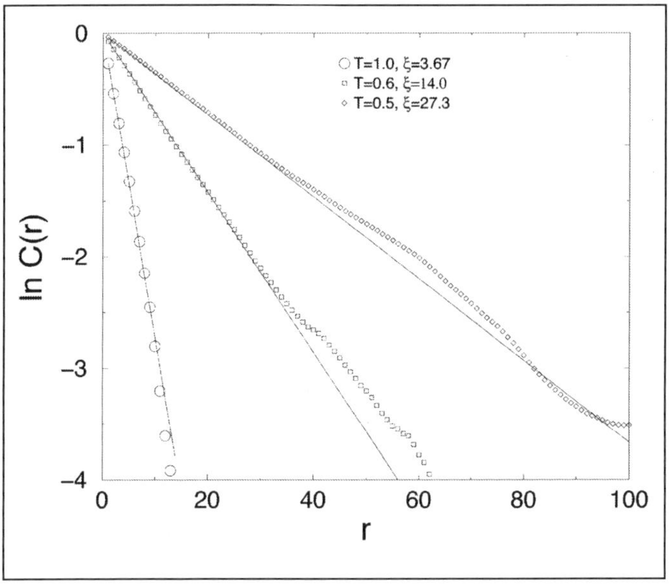

Figure 2. Logarithms of the correlation functions for the one-dimensional Ising model with $L = 2^{16}$ spins at three different temperatures $T = 1.0$ (○), $T = 0.6$ (□) and $T = 0.5$ (◊). The lines are drawn according to theoretical predictions of Eq. (17).

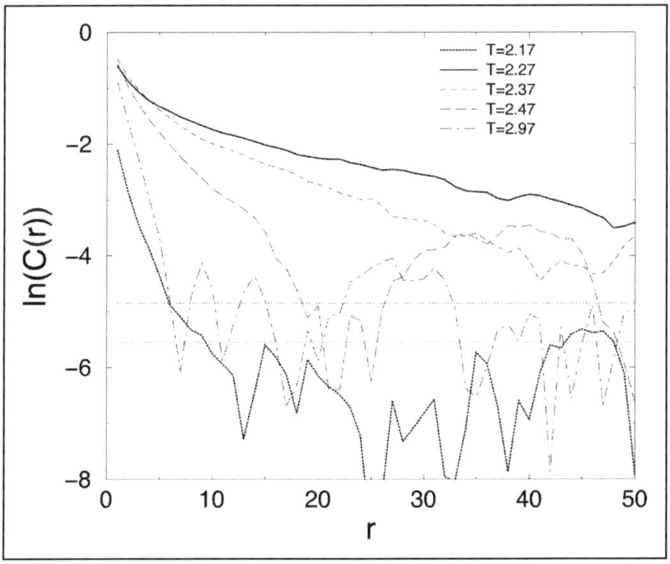

Figure 3. Logarithms of the correlation functions for the two-dimensional Ising model with $L = 2^8 \times 2^8$ spins at five different temperatures $T = 2.97$, $T = 2.47$, $T = 2.37$, $T = T_c = 2.27$, and $T = 2.17$. The straight horizontal lines show 68% and 95% confidence level for apparent correlations expected to be observed in an uncorrelated data of this length. Away from critical point, the behavior of correlations is well approximated by straight lines indicating exponential decay of correlations. The slopes of these lines are inverse proportional to the correlation length. Close to critical point, correlation length becomes extremely large.

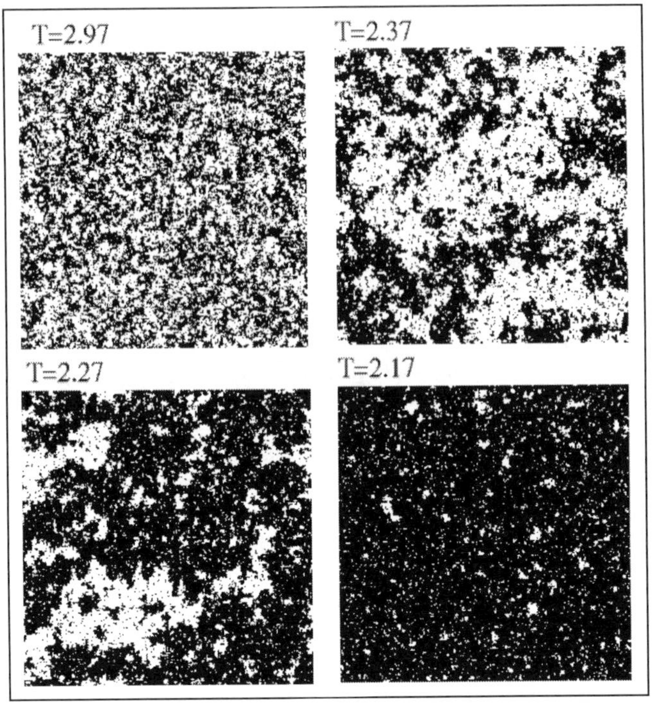

Figure 4. Snapshots of the Ising model far above the critical point $T = 2.97$, close to the critical point $T = 2.37$, at the critical point $T = 2.27$, and below the critical point $T = 2.17$. One can see that the patch sizes are the largest at the critical point.

Although $p >> 1 - p$, the average product $\langle s(k)s(r + k)\rangle$ of two independent variables $s(k)$ and $s(k + r)$ must be equal to the product of their averages $\langle s(k)\rangle\langle s(k + r)\rangle$, so the total correlation $C(r) = 0$. Note that $C(0) = 4p(1 - p)$, thus correlation function is small even for small r. Indeed, the graph corresponding to $T = 2.17$ starts at positions much below the graphs for $T \geq T_c$, for which $C(0) = 1$, since $p = 1/2$.

Note that calculations of $\ln C(r)$ become very inaccurate as $C(r)$ approaches zero. This is because the statistical error of calculating the correlation function becomes comparable with its value. Indeed, simple probabilistic analysis shows that for two independent variables x and y, the variable $(x - \langle x\rangle)(y - \langle y\rangle)$ has variance equal to the product of the variances of the variables x and y. When we compute correlation function, we average $\langle(x - \langle x\rangle)(y - \langle y\rangle)\rangle \equiv \langle s(k)s(r + k)\rangle - \langle s(k)\rangle\langle s(k + r)\rangle$ over $N = L \times L$ positions. In the best possible case, assuming all these measurements are independent, the standard error is the square root of variance divided by the square root of N. Since the variables $x \equiv s(k)$ and $y \equiv s(k + r)$ both have the variance $C(0) = 4p(1 - p)$, where p is the probability of a positive spin, we get this error $\sigma = 4p(1 - p)/\sqrt{N}$. Since for $T > T_c$ the probabilities of positive and negative spins are roughly equal, we have $\sigma = 1/256$. The horizontal lines indicate levels of σ, and 2σ corresponding to 68% and 95% confidence levels. Since in reality x and y are correlated, the number of independent measurements have to be divided by a factor proportional to ξ^d, where $d = 2$ is the dimensionality. The calculations of $C(r)$ become extremely inaccurate when we approach the critical point at which the correlation length diverges.

One can see that the values of the correlation function can be well approximated by the straight lines above the estimated standard error level, except for $T = 2.27$ and $T = 2.37$, when

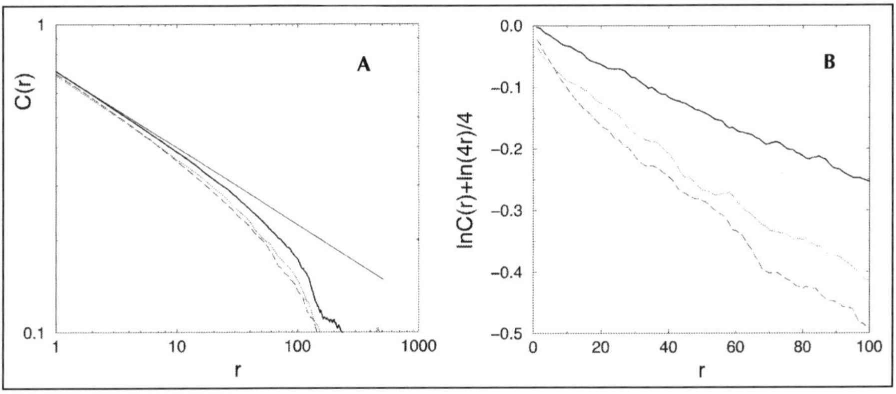

Figure 5. A) Double logarithmic plot of the correlation functions for the two-dimensional Ising model with $L = 2^{10} \times 2^{10}$ spins at critical temperature $T = T_c = 2.296$ for two realizations of the ferromagnetic model (dash and dotted lines) and the antiferromagnetic model (bold line). The straight line indicates theoretical fit $C(r) \sim r^{-1/4}/\sqrt{2}$. B) Logarithm of correlation function, multiplied by $r^{1/4}\sqrt{2}$. The linearity of the graph demonstrates exponential behavior of the corrected correlation function. The inverse slopes give values $\xi = 400$, $\xi = 270$, $\xi = 220$ comparable to half of the system size $L/2 = 512$.

the graphs start to bend upward as r decreases. This behavior may indicate a power law decay. To see this more clearly, we simulate a much larger system $L = 1024$ exactly at T_c. Figure 5A shows the behavior of correlation function on a double logarithmic plot. For small r the graph is approximately linear with slope −0.25 in agreement with the exact result for an infinite system $C(r) = r^{-1/4}/\sqrt{2}$.[17] However, one can see that the deviations rapidly increase with r and the agreement breaks down at about $r = 10$. Analyzing such a data, one can easily dismiss the possibility of power law correlations on the basis that their range is so small. In fact, this early deviation from the power law can be well explained by the finite size of the system $L = 1024$. Indeed, in a finite system, the correlation length cannot be larger than the radius of the system. In Figure 5B, we show that the correlation function can be well approximated by $C(r) \approx r^{-1/4} \exp(-r/\xi)/\sqrt{2}$, where ξ have different values comparable with the system radius ≈ 512. This example demonstrates difficulties associated with correct identification of power law correlations in a finite system.

It is illuminating to study also the anti-ferromagnetic Ising model, in which neighboring spins prefer to stay in the opposite direction, or be anti-correlated. At low temperatures, an anti-ferromagnetic system looks like a checker board. Mathematically, ferromagnetic and anti-ferromagnetic Ising models are identical, so that any configuration of the anti-ferromagnetic model corresponds to exactly one configuration of the ferromagnetic model which can be obtained by flipping all the spins according to a simple deterministic rule. Thus in both models, correlation length has the same finite value at any temperature, except at the critical point at which the correlation length in both models diverges. Nevertheless, the behaviors of correlation functions are totally different. In the anti-ferromagnetic case, correlation function is negative for all odd r and is positive for all even r (Fig. 6A.)

For $T > T_c$, the behavior of the absolute value of the correlation function is similar to that of the ferromagnetic model, both decaying exponentially with r, but below T_c in the anti-ferromagnetic case, the absolute values of correlations do not decay at all (Fig. 6B). However, if one average odd and even values of the correlation function, this averaged correlation function decays exponentially to zero as expected. This shows that the correlation length is finite and that there is no true long range correlations.

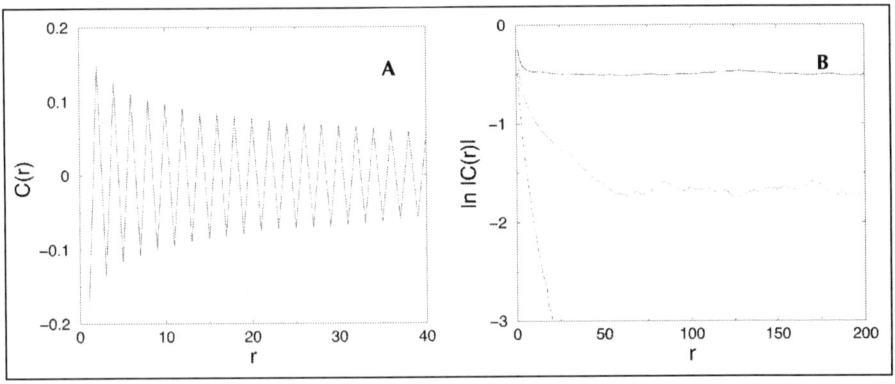

Figure 6. A) Correlation function for the two-dimensional anti-ferromagnetic Ising model. B) Absolute values of the correlation function below T_c (solid line), at T_c (dotted line) and above T_c, (dashed line).

The behavior of the anti-ferromagnetic model below critical temperature is similar to the behavior of coding sequences in DNA, which have a fixed reading frame[43] (see section "Models of Long Range Anti-Correlations'). For a totally uncorrelated sequence of codons in which codon frequencies are taken from real codon usage tables (i.e., no true long range correlations), certain correlation functions oscillate with period 3 with a fixed amplitude till the end of the reading frame. However, after averaging three successive values $C(r) + C(r + 1) + C(r + 2)$, the apparent correlations disappear.

Fourier Power Spectrum

In addition to large statistical errors in computation of $C(r)$, these calculations are also very slow, since the amount of operations is proportional to $r \times N$, where N is total number of points in the sample. An alternative way to study the correlations is to compute a power spectrum $S(f)$ which is the square of the absolute value of the Fourier transform of the function $s(k)$. This technique goes back to X-ray crystallography, in which the intensity of scattered X-rays at certain angle, appears to be a Fourier transform of the density correlation function in the sample under study.[60] It may also help to understand the Fourier transform technique in terms of a musical record. Imagine that $s(k)$ is a record of a melody. Now k is a continuum variable playing the role of time. Then $S(f)$ tells how much energy is carried by frequency (pitch) f. Unfortunately, applications of Fourier transform technique require substantial knowledge in mathematics involving complex numbers and trigonometry. In the following section, we give a brief review of the properties of Fourier transforms. Throughout this section we will use standard notations $i \equiv \sqrt{-1}$ for imaginary unity and $\pi = 3.14159...$. To simplify notations, we will also introduce an angular frequency $\omega = 2\pi f$.

Mathematically, the Fourier transform[61] of an infinitely long record is a result of an integral operator **F** acting on the function $s(x)$:

$$\tilde{s}(\omega) = \mathbf{F}s(\omega) = \int_{-\infty}^{\infty} e^{ix\omega} s(x)dx \equiv \int_{-\infty}^{\infty} \cos(\omega x)s(x)dx + i\int_{-\infty}^{\infty} \sin(\omega x)s(x)dx. \tag{21}$$

Since i is the imaginary unity, the result of a Fourier transform is a complex function $\tilde{s}(\omega) = a(f) + ib(\omega)$. The power spectrum $S(\omega)$ is defined as the square of the absolute value of the Fourier transform: $S(\omega) \equiv |\tilde{s}(\omega)|^2 \equiv \tilde{s}(\omega)\,\bar{\tilde{s}}(\omega)$, where $\bar{\tilde{s}}(\omega) = a(\omega) - ib(\omega)$ is a complex conjugate of $\tilde{s}(\omega)$. The signal $s(x)$ can be restored from $\tilde{s}(\omega)$ by the inverse Fourier transform

$$s(x) = \mathbf{F}^{-1}\tilde{s}(\omega) = \frac{1}{2\pi}\int_{-\infty}^{\infty} e^{-ix\omega}\tilde{s}(\omega)d\omega. \tag{22}$$

Fourier transforms have many interesting functional properties which make them a useful tool in data analysis. For example, $\mathbf{F}ds(x)/dx = -i\omega\tilde{s}(\omega)$ and $\mathbf{F}\int^{x} s(x)dx = i\tilde{s}(\omega)/\omega$. An important property of the Fourier transform is to turn a convolution of two functions into a product of their Fourier transforms:

$$\mathbf{F}\int_{-\infty}^{\infty} s_1(x)s_2(x+r)dx = \mathbf{F}s_1(\omega)\mathbf{F}s_2(-\omega). \tag{23}$$

Due to this property, the power spectrum of a function with zero average is equal to the Fourier transform of its autocorrelation function.

$$S(\omega) = \mathbf{F}C(r) \tag{24}$$

For example, if the correlations decay exponentially with correlation length ξ as for the one-dimensional Ising model or a one-step Markovian process, $C(r) = C(0)\exp(-r/\xi)$, we have

$$S(\omega) = 2C(0)\xi/\left(1+\omega^2\xi^2\right), \tag{25}$$

so the power spectrum is almost constant for low frequencies $\omega < 1/\xi$ and decays as $1/\omega^2$ for high frequencies $\omega >> 1/\xi$.

If the correlations decay as a power law (as at the critical point), $C(r) = |r|^{-\gamma}$, where $0 < \gamma < 1$, the power spectrum also decays as a power law $S(\omega) = c(\gamma)\omega^{-\beta}$, where

$$\beta = 1 - \gamma, \tag{26}$$

and $c(\gamma) = 2\cos[\frac{\pi}{2}(1-\gamma)]\Gamma(1-\gamma)$ does not depend on f. Here Γ is Euler's gamma-function.[61]

The case of approximately constant power spectrum is called white noise, since in this case all the frequencies carry the same energy (as in white light which is mixture of the colors of the rainbow corresponding to all different frequencies). The case $S(f) \sim 1/f^2$ is nicknamed "brown" noise since it describes Brownian motion and the case $S(f) \sim 1/f$ is called $1/f$-noise or "red" noise. The case $S(f) \sim 1/f^{\beta}$, with $0 < \beta < 1$corresponds to long range power-law correlations in the signal and is often called fractal noise. The power spectrum of the fractal noise looks like a straight line with slope $-\beta$ on a log-log plot.

In case of long range anti-correlations (as in the anti-ferromagnetic Ising model, Fig. 6) the correlation function oscillates with certain angular frequency ω_0. In this case, the behavior of the correlation function can be modeled as $C(r) \sim |r|^{-\gamma}\cos(\omega_0 r)$. Analogous calculations[61] lead to $S(\omega) = c(\gamma)(|\omega_0 - \omega|^{-\beta} + |\omega_0 + \omega|^{-\beta})/2$. This expression is analytical at $\omega = 0$, but it has power law singularities at $\omega = \pm\omega_0$. Thus in case of anti-correlations, the graph of power spectrum does not look like a straight line on a simple log-log plot. One must plot $\ln P(\omega)$ versus $\ln|\omega_0 - \omega|$ in order to see a straight line with the slope $-\beta$.

If the correlation function decays for $r \to \infty$ faster than r^{-1}, its Fourier transform must be a continuous function limited for $f \to \infty$ and, therefore, cannot have singularity at any f. The log-log graph of such a function plotted against $f - f_0$ has zero slope in the limit $\ln|f - f_0| \to \pm\infty$, so one can conclude that $\beta = 0$ if $\gamma > 1$. If $\gamma = 1$, the Fourier transform may have logarithmic singularities, which also corresponds to zero slope $\beta = 0$.

Discrete Fourier Transform

In reality, however, we never deal with infinitely long time series. Usually we have a system of N equidistant measurements. In this case, a sequence of N measurements $s(k)$, $k = 0,1,...N-1$, can be regarded as vector $\mathbf{s}$ of the N-dimensional space. Accordingly, one can define a discrete Fourier transform,[62,63] of this vector not as an integral but as a sum

$$\tilde{\mathbf{s}} \equiv \mathbf{F}\mathbf{s} = \sum_{k=0}^{N-1} s(k)e^{2\pi ikq/N}, \tag{27}$$

which can also be regarded as a vector in N-dimensional space with components $\tilde{s}(q)$, $q = 0,1,...N-1$. The fractional quantity $f = q/N$ plays the role of frequency. As one can see, the discrete Fourier transform can be expressed in a matrix form $\tilde{\mathbf{s}} = \mathbf{Fs}$, where $\mathbf{F}$ is the matrix with elements $f_{kq} = \exp(2\pi ikq/N)$. Analogously, vector $\mathbf{s}$ can be restored by applying an inverse Fourier transform:

$$\mathbf{s} \equiv \mathbf{F}^{-1}\tilde{\mathbf{s}} = \frac{1}{N}\sum_{q=0}^{N-1}\tilde{s}(q)e^{-2\pi ikq/N}. \tag{28}$$

If one assumes that the sequence $s(k)$ is periodic, i.e., $s(k+N) = s(k)$, then the square of the discrete Fourier transform is proportional to the discrete Fourier transform of the correlation function as in case of the continuum Fourier transform.[62,63] Indeed, $|\tilde{s}(f)|^2 = \mathbf{F}\sum_{k=0}^{N-1} s(k)s(k+r)$.

It is natural to define the discrete power spectrum $S(f)$ to be exactly equal to the Fourier transform of the correlation function. Since the correlation function is defined as $C(r) = 1/N\sum_{k=0}^{N-1} s(k)s(k+r) - \langle s\rangle^2$, which involves division by N and subtraction of the average value, $S(f) \equiv |\tilde{s}(f)|^2/N$ for $f > 0$ and $S(0) = 0$, because $\tilde{s}(0) \equiv N\langle s(k)\rangle$.

The correlation function can be thus obtained as an inverse discrete Fourier transform of a power spectrum. Since frequencies $-q/N$ and $1-q/N$ are equivalent (due to 2π– periodicity of sines and cosines) and, for real signal, $\tilde{s}(-f)$ and $\tilde{s}(f)$ are complex conjugates, the values $S(q/N)$ and $S(1-q/N)$ are equal to each other, so we can compute power spectra only up to the highest frequency $q/N = 1/2$.

If N is a natural power of two, $N = 2^n$, the discrete Fourier transform can be computed by a very efficient algorithm known as the Fast Fourier Transform (FFT).[62,63] The amount of operations in this algorithm grows linearly with N. This makes FFT a standard tool to analyze correlation properties of the time series.

Since the sequences we study are formed by random variables, the power spectra of such sequences are random variables themselves. Before proceeding further, it is important to calculate the power spectrum of a completely uncorrelated sequence of length N. As we have seen in section "Correlation Function", $C(0) > 0$ has the meaning of the average square amplitude (variance) of the original signal, while for $r > 0$, the values of $C(r)$ are Gaussian random variables with zero mean and standard deviation equal to $C(0)/\sqrt{N}$. Analogous conclusions can be made for $S(f)$. According to the central limit theorem,[21] the sum of N random uncorrelated variables $s(k)\exp(2i\pi kf)$ converges to a Gaussian distribution with mean equal to the sum of means and variance equal to the sum of variances of individual terms. Thus, we can conclude (after some algebra) that all $S(f)$ are identically distributed independent random variables with an exponential probability density $P(S(f)) = 1/[C(0)]\exp[-S(f)/C(0)]$. So the power spectrum of an uncorrected sequence has an extremely noisy graph. To reduce the noise one can average power spectra for many sequences, and the average value of the power spectrum will converge to a horizontal line $\langle S(f)\rangle = C(0)$ which is called the white noise level. An equivalent method is to average the values $S(f)$ for k neighboring frequencies $f, f+1/N, f+2/N,..., f+k/N$. Note that $\langle S(f)\rangle$ is equal to the Fourier transform of $\langle C(r)\rangle$, directly computed using Eq.(27), since as we see above, $\langle C(r)\rangle = 0$ for $r \neq 0$.

In the following, we will illustrate the usage of FFT computing power spectrum for a one- and two-dimensional Ising models near critical points.

Figure 7 shows the power spectrum for the one-dimensional Ising model consisting of $L = 2^{16}$ spins for $T = 0.5$ ($\xi = 27.3$), $T = 0.6$ ($\xi = 14.01$), $T = 1.0$ ($\xi = 3.67$). The power spectrum of the entire system for $N = L$ is very noisy so we show the running averages of the original data using window of 32 adjacent frequencies (gray fluctuating curves). The averages of 32 power spectra computed for 32 non-overlapping windows each of size $N = 2^{11}$ produce a very similar

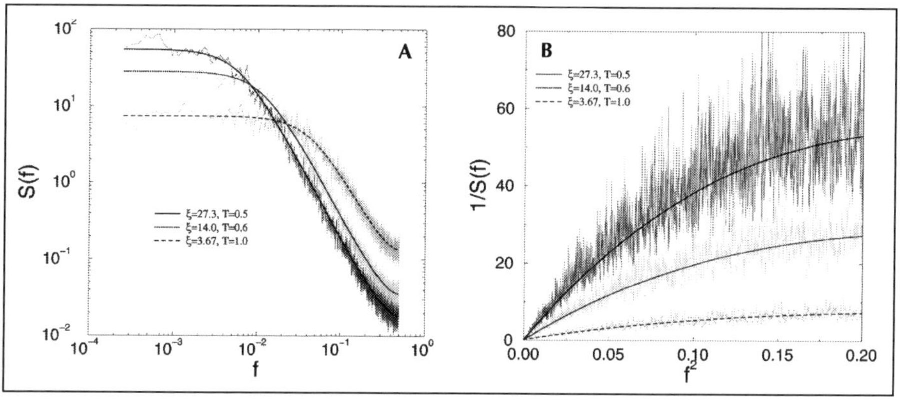

Figure 7. A) Power Spectrum of the one-dimensional Ising model with $L = 2^{16}$ spins for $T = 0.5$ ($\xi = 27.3$), $T = 0.6$ ($\xi = 14.0$), and $T = 1.0$ ($\xi = 3.67$). Smooth lines show analytical result Eq. (29). B) Inverse power spectrum of the same data plotted versus f^2. The slopes at $f^2 = 0$ are proportional to the values of the correlation length.

graph (not shown). The smooth bold lines represent exact discrete Fourier transform of the correlation function computed using Eqs. (4), (17) and (27)

$$S(f) = \frac{1-\lambda^2}{1+\lambda^2 - 2\lambda\cos(2\pi f)} \approx \frac{2\xi}{1+(2\pi\xi f)^2}, \tag{29}$$

where $\lambda = 2p - 1 = \exp(-1/\xi)$. These analytical results give excellent agreement with the numerical data. One way to estimate the correlation length is to measure a limit of $S(f)$ for $f \to 0$. This quantity can be applied to detect a characteristic patch size in the DNA sequence (see sections "Alternation of Nucleotide Frequencies" and "Models of Long Range Anti-Correlations"). Another, more accurate method[60] is to plot the inverse power spectrum $1/S(f)$ versus f^2 (Fig. 7B) and to measure the slope of this graph for $f^2 \to 0$. Indeed, according to Eq.(29), this slope is equal to $2\xi\pi^2$. These two methods give consistent results for exponentially decaying correlations, but technically speaking they measure two different properties of the power spectrum. In fact, the latter method gives the so called Debye persistence length $R^2 \sim \int_0^\infty C(r)r^2 dr$, which is not the same as correlation length ξ, but is proportional to ξ for exponentially decreasing correlations, $C(r) \sim \exp(-r/\xi)$.

Figure 8A shows the power spectrum for a two-dimensional Ising model on a $L \times L = 2^{10} \times 2^{10}$ square lattice computed averaging power spectra for L horizontal rows each consisting of $N = L = 2^{10}$ points. The figure shows a remarkable straight line indicating long range power law correlations. However, the slope of the line $\beta = 0.86$ corresponds to $\gamma = 0.14$ which is almost two times smaller than the theoretical exact value $\gamma = \eta = 0.25$. The discrepancy shows that the power spectrum analysis of a finite system may often give inaccurate values of the correlation exponents.

Figure 8B shows a log-log plot of the power spectrum for a two-dimensional anti-ferromagnetic Ising model, plotted versus $1/2 - f$. The analysis in the previous section shows that since $1/2$ is the frequency of the anti-ferromagnetic correlations, the power spectrum must have a power-law singularity in this point. Indeed, the graph gives an approximately straight line with slope $-\beta = -0.84$ similar to the case of ferromagnetic interactions.

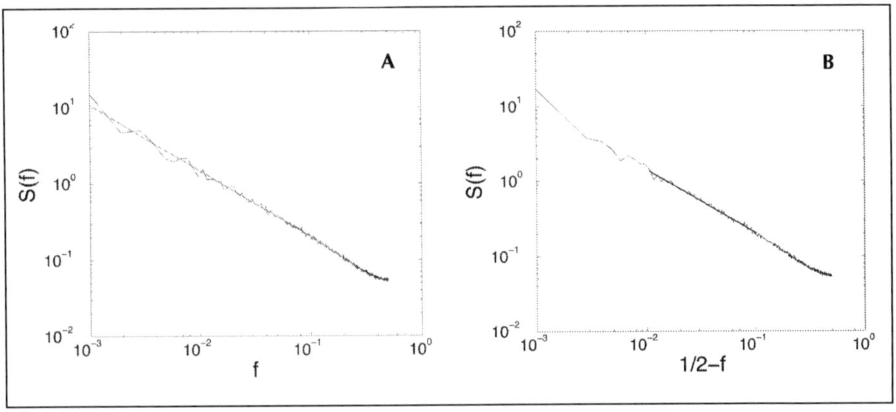

Figure 8. A) Power spectrum of the $2^{10} \times 2^{10}$ Ising model at the critical point. The slope of the straight line gives $\beta = 0.86$. B) Power spectrum of the $2^{10} \times 2^{10}$ anti-ferromagnetic Ising model at the critical point plotted versus $1/2 - f$. The slope of the straight line gives $\beta = 0.84$.

Detrended Fluctuation Analysis (DFA)

A somewhat more intuitive way to study correlations was proposed in the studies of the fluctuations of environmental records by Hurst in 1964.[7] This method is especially useful for short records. The idea is based on comparison of the behavior of the standard deviation of the record averaged over increasing periods of time with the analogous behavior for an uncorrelated record. According to the law of large numbers, the standard deviation of the averaged uncorrelated time series must decrease as the square root of the number of measurements in the averaging interval. This method naturally emerges when the goal is to determine an average value of a quantity (e.g., magnetization in the Ising model, or concentration of a certain nucleotide type in a DNA sequence) obtained in many successive measurements and to asses an error bar of this averaged value. Since the average is equal to the sum divided by the number of measurements, the same analysis can be performed in terms of the sum. In addition to its analytical merits, this method provides a useful graphical description of a time series which otherwise is difficult to see due to high frequency fluctuations.

A variant of Hurst analysis was developed in reference 64 under the name of detrended fluctuation analysis (DFA). The DFA method comprises the following steps:

1. For a numerical sequence $s(k)$, $k = 1,2,\ldots L$ compute a running sum:

$$y(n) \equiv \sum_{k=1}^{n} s(k), \tag{30}$$

which can be represented graphically as a one dimensional landscape, (see Fig. 9A).

2. For any sliding observation box of length r which includes $r + 1$ values $y(k), y(k+1), \ldots y(k+r)$ define a linear function $y_k(x) = a_k + b_k x$ which provides the least square fit for these values, i.e., a_k and b_k are such that the sum of $r + 1$ squares

$$F_k^2(r) = \sum_{n=k}^{k+r} \left[y(n) - y_k(n) \right]^2 \tag{31}$$

has a minimal possible value $F_{k,\min}^2(r)$. Note that b_k has the meaning of the average value $\langle s(k) \rangle$ for this observation box, which is the local trend of the values $y(k)$. For a non-stationary sequence, the local average values $\langle s(k) \rangle$ can change with time. Since these trends are subtracted in each observation box, this analysis is called detrended. Note that $F_{k,\min}^2(1) \equiv 0$, so it is a trivial value which can be excluded from the analysis.

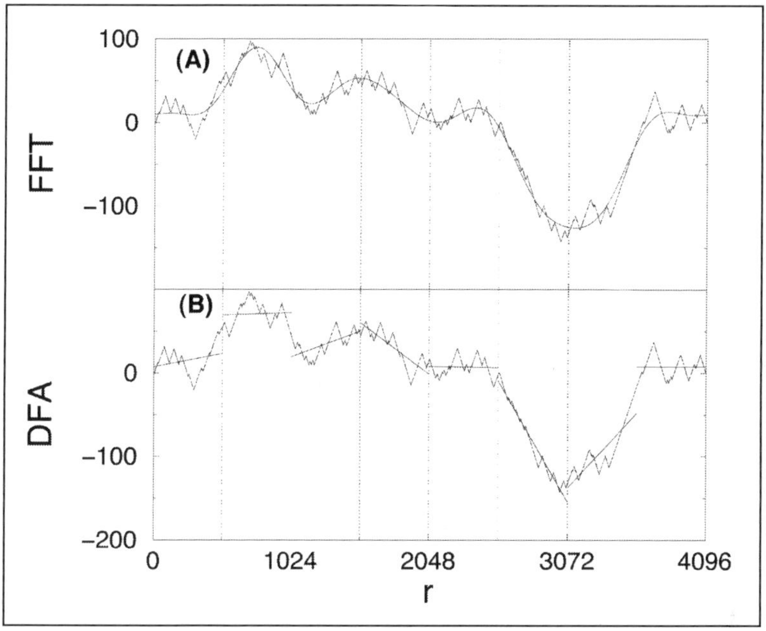

Figure 9. A) Low frequency Fourier approximation of the detrended landscape $y_D(n)$ for the one-dimensional Ising model. All the frequencies $f > 1/r$ are removed. Fourier DFA computes the average square deviation of this approximation from the landscape. B) Linear DFA of the same landscape. Straight lines show least square linear fits obtained for different windows of size $r = 512$. Linear DFA computes the average square deviation of these fits from the landscape.

3. For $r > 1$, compute the average value of $F^2_{k,\min}(r)$ from $k = 1$ to $k = L - r$ and define the deternded fluctuation function as

$$F_D^2(r) \equiv \frac{1}{(L-r+1)(r-1)} \sum_{k=1}^{L-r} F_{k,\min}(r). \tag{32}$$

It can be shown, that for a long enough sequence $L \to \infty$ of uncorrelated values $s(k)$ (i.e., $C(r) = 0$ for $r > 1$) with finite mean and variance $C(0)$, we must have $F_D^2(r) \to (r+3)C(0)/15$. Thus the graph of $F_D(r)$ for such a sequence on a log-log plot is a straight line with slope the $\alpha = 1/2$ if plotted versus $r + 3$. Any deviation from the straight line behavior indicates the presence of correlations or anti-correlations. It can be also shown that for a sequence with long range power law correlations $C(r) \sim r^{-\gamma}$ for $0 < \gamma < 1$, the detrended fluctuation also grows as a power law $F_D(r) \sim r^{\alpha}$ as $r \to \infty$, where

$$\alpha = 1 - \gamma/2 > 1/2, \tag{33}$$

is called the Hurst exponent of the time series.

A Relation between DFA and Power Spectrum

There are many different ways to subtract local trends in Eq. (31).[65] One can subtract polynomials of various powers or linear combinations of sines and cosines of certain frequency instead of linear functions. All these different types of DFA have certain advantages and disadvantages. One way to subtract local trends is first to subtract a global trend and plot a sequence

$y_D(k) \equiv y(k) - ky(L)/L$. Next, compute a discrete Fourier transform with $N = L$ of this function $\tilde{y}(f) = \mathbf{F}y_D$ and subtract from the function $y_D(k)$ a low frequency approximation

$$y_r(k) = 1/L\sum_{|f|<1/r}\tilde{y}(f)\exp(-2\pi ifk),$$

(see Fig. 9B). A visual comparison of Figures 9A and 9B, suggests that these two procedures of subtracting local trends are equivalent. Thus we can define a Fourier detrended fluctuation as

$$F_{DF}^2(r) \equiv \frac{1}{L}\sum_{k=1}^{L}\left[y_D(k) - y_r(k)\right]^2. \tag{34}$$

According to Eq. (28), the residuals in the right hand side of Eq. (34) are equal to the high frequency part of the inverse Fourier transform:

$$y_D(k) - y_r(k) = 1/L\sum_{|f|\geq 1/r}\tilde{y}(f)\exp(-2\pi ifk).$$

The Fourier basis vectors are mutually orthogonal, i.e., $\sum_{k=1}^{L}\exp(2\pi iqk/L)\exp(-2\pi iqk/L) = L\delta_{pq}$, where $\delta_{pq} = 1$ if $p = q$ and $\delta_{pq} = 0$, otherwise. Thus, according to the L-dimensional analogy of the Pythagorean theorem, the square of the vector $\mathbf{y}_D(k) - \mathbf{y}_r(k)$ is equal to the sum of the squares of its orthogonal components and therefore,

$$F_{DF}^2(r) = 1/L^2\sum_{|f|\geq 1/r}^{1/2}\left|\tilde{y}(f)\right|^2 = 1/L\sum_{|f|\geq 1/r}^{1/2}S_y(f). \tag{35}$$

The latter sum is nothing but the sum of all the high frequency components of the power spectrum $S_y(f)$ of the integrated signal.

Equation (35) allows us to derive the relation (33) between the exponents α and γ. Indeed, in continuum limit, this sum corresponds to the integral $\int_{f=1/r}^{\infty}S_y(f)df \sim \int_{f=1/r}^{\infty}S(f)f^{-2}df$, where $S(f)$ is the power spectrum of the original, non-integrated sequence $s(x)$ and the factor f^{-2} comes from the fact that the Fourier transform of the integrated sequence is proportional to the Fourier transform of the original sequence divided by f. As we see above (26), in case of power law correlations with exponent γ, we have $S(f) \sim f^{\gamma-1}$. Thus

$$F_{DF}^2(r) \sim \int_{f=1/r}^{\infty}S_y(f)df \sim \int_{f=1/r}^{\infty}S(f)f^{-2}df \sim (1/r)^{\gamma-1-2+1} = r^{2-\gamma}$$

If we assume that $F_{DF}(r) \sim F_D(r) = r^{\alpha}$ as visual inspection of Figure 9 suggests, we have $\alpha = 1 - \gamma/2$.

Figure 10A shows linear DFA and Fourier DFA for a one-dimensional Ising model on a double logarithmic plot. These two methods are graphically introduced in Figure 9. One can see a sharp transition from the correlated behavior for $r \approx \xi$ with slope $\alpha(r) > 1$ to an uncorrelated behavior for $r >> \xi$ with slope $\alpha(r) \approx 1/2$. The change of the slope can be also studied by plotting the local slope $\alpha(r)$ versus r (Fig. 10B). This graph shows that Fourier DFA can detect the correlation length more accurately than the linear DFA.

Figure 11 shows analogous plots for the two-dimensional Ising model with long range correlations $\gamma = 1/4$. One can see again that the Fourier DFA is more accurate in finding the correct value of the exponent $\alpha = 1 - \gamma/2 = 0.875$ than linear DFA.

In summary, we introduce three methods to study correlations: autocorrelation function $C(r)$, power spectrum $S(f)$, and DFA or Hurst analysis $F_D(r)$. For a signal with long range power law correlations $\gamma < 1$, all three quantities behave as power law:

$$\begin{aligned} C(r) &\sim r^{-\gamma} && r \to \infty \\ S(f) &\sim f^{-\beta} && f \to 0 \\ F_D(r) &\sim r^{\alpha} && r \to \infty \end{aligned} \tag{36}$$

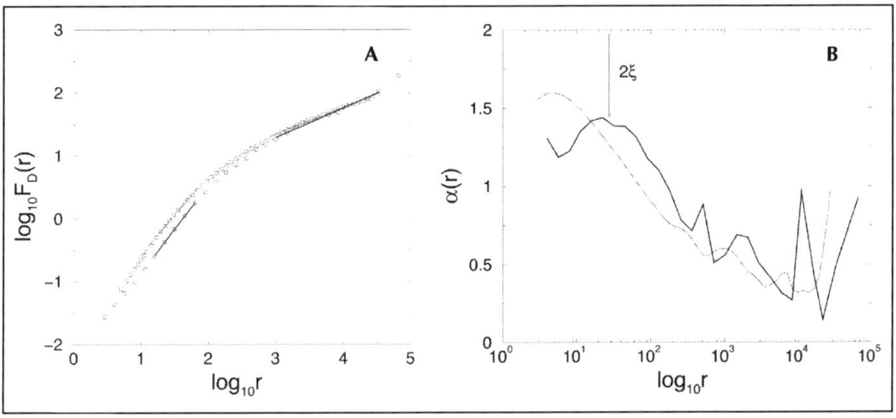

Figure 10. A) Linear detrended fluctuation (O) and Fourier detrended fluctuation (□) of the one dimensional Ising model for ($T = 0.6$, $L = 2^{16}$). The slopes of linear fits give local values of $\alpha = 1.24$ (thin line) and $\alpha = 1.38$ (bold line) for small $r \approx \xi = 14$ and $\alpha = 0.42$ (thin line), $\alpha = 0.47$ (bold line) for an uncorrelated regime $r >> \xi$. B) The slope $\alpha(r)$ of the detrended fluctuations as function of r. Note that Fourier DFA gives a strong maximum at $r = 2\xi$ while linear DFA shows monotonic decay of α.

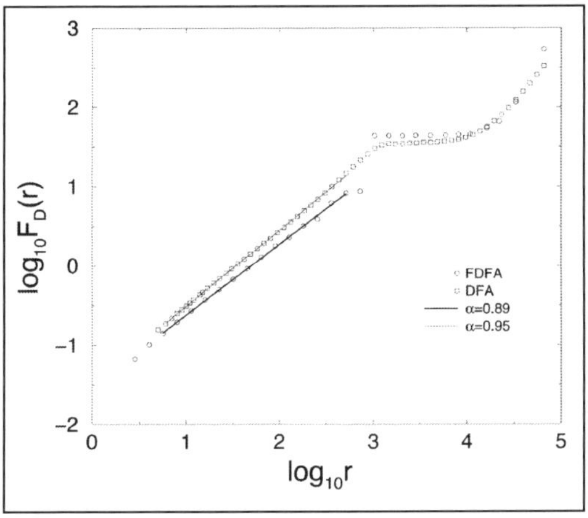

Figure 11. Linear detrended fluctuation (O) and Fourier detrended fluctuation (□) of the two dimensional Ising model for ($T = T_c$. $L = 2^{10}$). The slopes of linear fits give local values of $\alpha = 0.95$ (thin line) and $\alpha = 0.88$ (bold line) for small $r < L$. The steep jump in Fourier DFA at $L = 2^{10}$, indicates quasi-periodicity with period $L = 2^{10}$ due to the "bagel" geometry of the model.

where the exponents α, β, and γ are related via the following linear relations:

$$\begin{aligned} \beta &= 1 - \gamma \\ \alpha &= 1 - \gamma/2 \\ \alpha &= (\beta + 1)/2. \end{aligned} \tag{37}$$

If $\gamma > 1$, the exponents $\beta = 0$, $\alpha = 1/2$ are the same as for a short range correlated sequence with finite correlation length ξ.

Duplication-Mutation Model of DNA Evolution

In 1991, W. Li proposed a duplication-mutation model of DNA evolution which predicted long-range power law correlations among nucleotides.[56] As we see above, in a one dimensional system with finite range interactions, correlations must decay exponentially with distance. So in order to produce a power law decay of correlations, one must assume long-range interactions among nucleotides. In the model of W. Li, such interactions are provided by the fact that the time axes serves as an additional spatial dimension which connects distant segments of DNA developed from a single ancestor. The model is based on two assumptions both of which are well biologically motivated:

1. Every nucleotide can mutate with certain probability.
2. Every nucleotide can be duplicated or deleted with certain probability.

First phenomenon is known as point mutation which can be caused by random chemical reactions such as methylation.[51] Second phenomenon often happens in the process of cell division (mitosis and meiosis) when pairs of sister chromosomes exchange segments of their DNA (genetic crossover). If the exchanging segments are of identical length the duplication does not happen. However, if two segments differ in length by n nucleotides, the chromosome that acquires larger segment obtains an extra sequence of length n which is identical to its neighbor, while another chromosome loses this sequence. In many cases, duplications can be more evolutionary advantageous than deletions. This process leads to creation of large families of genes developed from the same ancestor. For simplicity, we will start with a model similar to the original model of Li[56] which neglects deletions and deals only with duplication of a single nucleotide ($n = 1$). Next, we will discuss the implications of deletions. Schematically, this model can be illustrated by Figure 12. For simplicity, we assume only two types of nucleotides X and Y (say purine vs. pyrimidine or A vs. not A). Each level of the tree-like structure represents one step of the evolutionary process during which every nucleotide duplicates, a nucleotide X can mutate with probability p_Y into Y, and a nucleotide Y can mutate with probability p_X into X. This model can be illustrated by a "family" tree in which every nucleotide is connected to its parent in the previous generation and eventually to a single ancestor at the root of the tree.

After k duplication steps, this process will lead to a sequence of total 2^k nucleotides. The frequencies of nucleotides X and Y in this sequence can be computed using the theory of Markovian processes. Indeed, the sequence of mutations along any brunch of the tree connect-

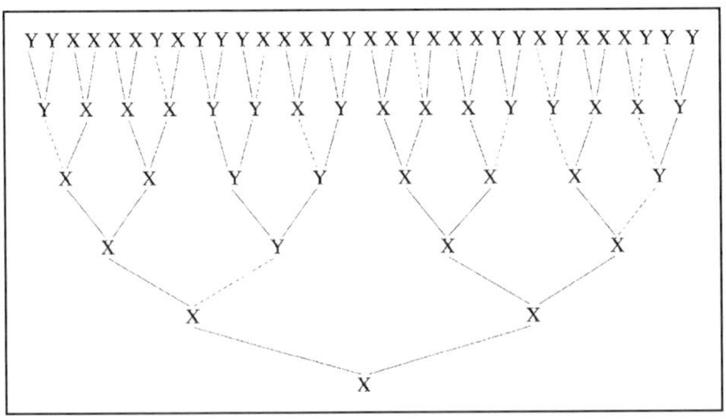

Figure 12. Mutation duplication model of W. Li.[56] At each time step, nucleotides or genes X and Y duplicate and may mutate with probability $P_X + P_Y \approx 1/12$. Mutations are indicated by dashed lines. The correlations can spread along solid lines. Thus nucleotides that are far away along the chain are still closely correlated since they descend from the same ancestor. The above values of mutation probabilities correspond to the long range power-law correlations with $\gamma = 0.25$.

ing a nucleotide to a single ancestor can be regarded as a one-step Markovian process with a matrix of transition probabilities

$$\mathbf{P} = \begin{pmatrix} 1-p_Y & p_X \\ p_Y & 1-p_X \end{pmatrix}. \tag{38}$$

Simple calculations of the eigenvector corresponding to the largest eigenvalue $\lambda = 1$ described in section "Markovian Processes" gives the frequencies of nucleotides X and Y after many steps: $f_X = p_X/(p_X + p_Y)$ and $f_Y = p_Y/(p_X + p_Y)$. In addition, Markovian analysis predicts that all dependence coefficients along any branch of the tree decay as λ_2^k, where k is number of generations, and $\lambda_2 = 1 - p_X - p_Y$ is the second largest eigenvalue.

Let us compute the dependence coefficients between two nucleotides which are at distance r from each other in the resulting sequence. The reason of why the correlations are now long-range is obvious. Indeed, the nucleotides which are $r = 2^{k'}$ apart from each other in space are only $2k'$ apart from each other in time, since they are both descendants of one common ancestor $k' = \log_2 r$ generations before. As we see above, the correlations decay exponentially with k' and hence as a power law with r. After some elementary algebra, we get that all dependence coefficients D_{XX}, D_{XY}, D_{YX}, and D_{YY} decay as power law

$$D(r) \sim r^{-\gamma} \tag{39}$$

where

$$\gamma = -\frac{2\ln|p_X + p_Y - 1|}{\ln 2}. \tag{40}$$

If the deletions may occur with some probability $P_d < 1/2$, the number of descendants of one common ancestor grows as $z^{k'}$ where $z = 2(1 - P_d)$ and k' is the number of generations. Thus, replacing ln2 by ln$z$ in the denominator of the expression for (40), we get

$$\gamma = -\frac{2\ln|p_X + p_Y - 1|}{\ln 2(1 - p_d)}. \tag{41}$$

The true long range correlations correspond to the case $\gamma < 1$, or $(p_X + p_Y - 1)^2(1 - p_d) > 1/2$, which means that the mutation rates must be very small: $p_X + p_Y \approx 0$ or alternatively very large: $p_X + p_Y \approx 2$, while the deletion rate must be small. This simple example shows that the exponent of the power law crucially depends on the parameters of the model.

In real DNA sequences, the duplication unit is rather a gene or a part of a gene coding for a protein domain. One can generalize this model assuming that coding sequences X and Y can duplicate, and with some probability jump from place to place effectively mimicking mutations X to Y and Y to X in the above scheme. One can also introduce various point mutation rates for nucleotides in the sequences X and Y. These alternations may change the formula for γ, but the model will still produce power law decaying correlations $D(r) \sim (r/\langle n\rangle)^{-\gamma}$, where $\langle n\rangle$ is the average length of sequences X and Y. The problem with the application of this model to a real situation is that the model has many parameters, describing point mutations, duplications and deletions, while resulting in a single observable parameter γ.

Alternation of Nucleotide Frequencies

Let us assume that a nucleotide sequence consists of two types of patches,[57] in one of which the frequency of nucleotide X is f_{X1} while in the other it is f_{X2}. The patches can alternate at random, so that after a patch of type 1 a patch of type 2 can follow with probability 1/2 and

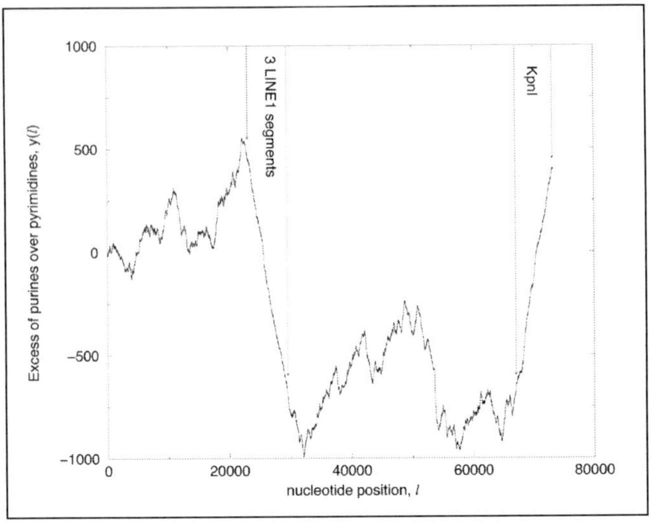

Figure 13. Purine-pyrimidine landscape representation (excess of purines over pyrimidines) of the human beta globin chromosomal region (GenBank accession HUMHBB) of the total length $L = 73{,}308$. The overall frequency of purines (50.27%) is almost equal to the frequency of pyrimidines (49.73%). HUMHBB contains a KpnI repeat from position 67071 to position 73195. This region is very purine rich with 58.57% of purines. KpnI repeat belongs to the LINE1 family of repetitive elements. A region from 23,137 to 29,515 is very purine poor (41.43%). It contains 3 segments of LINE1 repetitive elements inserted into the opposite DNA strand, so that all purines are exchanged with pyrimidines.

vice versa. Let us assume that the lengths of these patches l are distributed according to the same probability distribution $P(l)$.

The motivation for this model could be the insertion of transposable elements,[53,66] e.g., LINEs and SINEs into the opposite strands of the DNA molecule. It is known that LINE-1 sequence has 59% purines (A,G) and 41% pyrimidines (T,C).[66-68] Obviously, due to A-T and C-G complementarity, if LINE-1 is inserted into the opposite strand, it will have 41% purines and 59% pyrimidines (see Fig. 13).

Of course, much more complex models with many parameters can be introduced. These types of models are similar to hidden Markov processes.[16,69] However we will study only the above simple case in order to understand under which conditions this model can lead to power-law correlations.

Let us compute the correlation function for this model. Obviously, the average frequency of a nucleotide in the entire sequence is $f_X = (f_{X1} + f_{X2})/2$, so if both nucleotides k and $k + r$ belong to the same patch their correlation $D_{XX}(r)$ will be $f_{X1}^2 - f_X^2$ if it is a patch of type 1 or $f_{X2}^2 - f_X^2$ otherwise. Since both events have the same probability 1/2, the overall correlation $D_{xx}(r) = \left(f_{X1}^2/2 + f_{X2}^2/2\right)\Pi(r) = \Pi(r)\left(f_{X1} - f_{X2}\right)^2/4$, where

$$\Pi(r) = \sum_{l=1}^{\infty} P(r+l)l / \sum_{l=1}^{\infty} P(l)l \tag{42}$$

is the probability that a randomly chosen pair of nucleotides at distance r belongs to the same patch.

If the distribution of patch sizes is exponential $P(l) = \lambda^{l-1}(1 - \lambda)$, the overall correlation is easy to compute using summation of geometric series $D_{XX}(r) = (f_{X1} - f_{X2})^2\lambda^r/4$, which decays exponentially with r. However, this correlation can be extremely small comparatively to the "white noise level" $D_{XX}(0) = f_X(1 - f_X)$, and thus can be very difficult to detect. For example, if $f_{X1} = 0.3$ and $f_{X1} = 0.2$ $D_{XX}(0) = 3/16$, while $D_{XX}(1) = \lambda/400$, which is almost 100 times smaller even for very large $\lambda \to 1$.

If we have the distribution of patch sizes decaying for $l \to \infty$ as a power law

$$P(l) \sim l^{-\mu}, \tag{43}$$

where $\mu > 2$, one can show that $\prod(r) \sim r^{2-\mu}$. This can be easily seen if one approximates summation by integration in the expression (42) for $\prod(r)$. Thus, in this case the correlations are indeed power law with

$$\gamma = \mu - 2. \tag{44}$$

For $\mu > 3$, we have $\gamma > 1$ and the power spectrum of the model is finite for $f \to 0$, which means $\beta = 0$, $\alpha = 1/2$. The value $\lim_{f\to 0} S(f) \sim \sum_{l=1}^{\infty} l^2 P(l) / \sum_{l=1}^{\infty} l P(l)$ has the meaning of the weighted average patch length, i.e., the average length of the patch containing a randomly selected base pair.

The case $2 < \mu < 3$ is equivalent to the behavior of the displacement in the so called Lévy walks,[70,71] i.e., walks in which distribution of step lengths are taken from a power law with exponent μ. In this case, $\beta = 3 - \mu$, $\alpha = 2 - \mu/2$.

If $\mu \leq 2$, the sums in (42) do not converge, this means that summation in Eq. (42) must be taken up to the largest $l \approx L$, where L is the total sequence length. Thus $\prod(r) \sim (L - r)/L = 1 - r/L$ and we can assume $\gamma = 0$, $\beta = 1$, $\alpha = 1$.

Figure 14A shows the behavior of the correlation function of a sequence for which $p_{X1} = 0.3$, $p_{X2} = 0.2$ and $P(l) = l^{-3/2} - (l+1)^{-3/2}$, corresponding to $\mu = 2.5$. In this case $\prod(r) = \sum_{l=r+1}^{\infty} l^{-3/2} / \sum_{l=1}^{\infty} l^{-3/2} \sim r^{-0.5}$. We present the results of correlation analysis for a very long sequence of $L = 2^{23} \approx 8 \cdot 10^6$. One can see good agreement with Eq. (44). For a short sequence, $L = 2^{13} = 8192$, there is no agreement: the correlations sink below random fluctuations, whose amplitude is equal to $C(0)/\sqrt{L}$. This means that the sequence must be very long so that the long range correlations can be seen on top of random noise.

Figure 14B shows the power spectrum for the case of $N = L = 2^{23}$ obtained by averaging power spectra for 2048 non-overlapping windows of size $N = 4096$. The power spectrum is almost flat corresponding to the white noise level $C(0) = 3/16$. If the white noise level is subtracted, the long-range correlations become apparent (Fig. 14C). Indeed the graph of $|S(f) - C(0)|$ on a log-log scale is a perfect straight line with slope -0.57 in a good agreement with the theoretical prediction. The DFA method gives exponent $\alpha(r)$ monotonically increasing from an uncorrelated value 0.5 for small r to $\alpha = 1 - \gamma/2 = 0.75$ for large r. Similar situation is observed in coding DNA, in which the long range correlations may exist but are weak comparatively to the white noise level. These correlations are limited to the third nucleotide in each codon[72] and can be detected if the white noise level is subtracted.

If the length of the largest patch is comparable with the length of the entire sequence as in case $\mu \leq 2$, $\beta = 1$, the global average frequency f_X of a nucleotide cannot be accurately determined no matter how large is the entire sequence length. The average frequency we obtain will be always the frequency of the largest patch. This behavior known as non-stationarity is observed in many natural systems in which different parts are formed under different conditions. Non-stationarity makes the correct subtraction of the white noise level problematic, since its calculation involves estimation of $C(0) \sim f_X(1 - f_X)$, which depends on f_X.

Applying subtraction of the white noise level procedure, Richard Voss[34,35] found that both coding and noncoding DNA sequences from any organism, have exponent $\beta \approx 1$, corresponding to the $1/f$ noise. Note that $\beta = 1$ is exactly the case when this procedure is not quite reliable. Earlier[73] he applied the same type of analysis to the music of different composers from J.-S. Bach to the Beatles and showed that all their music is just $1/f$ noise! No matter how

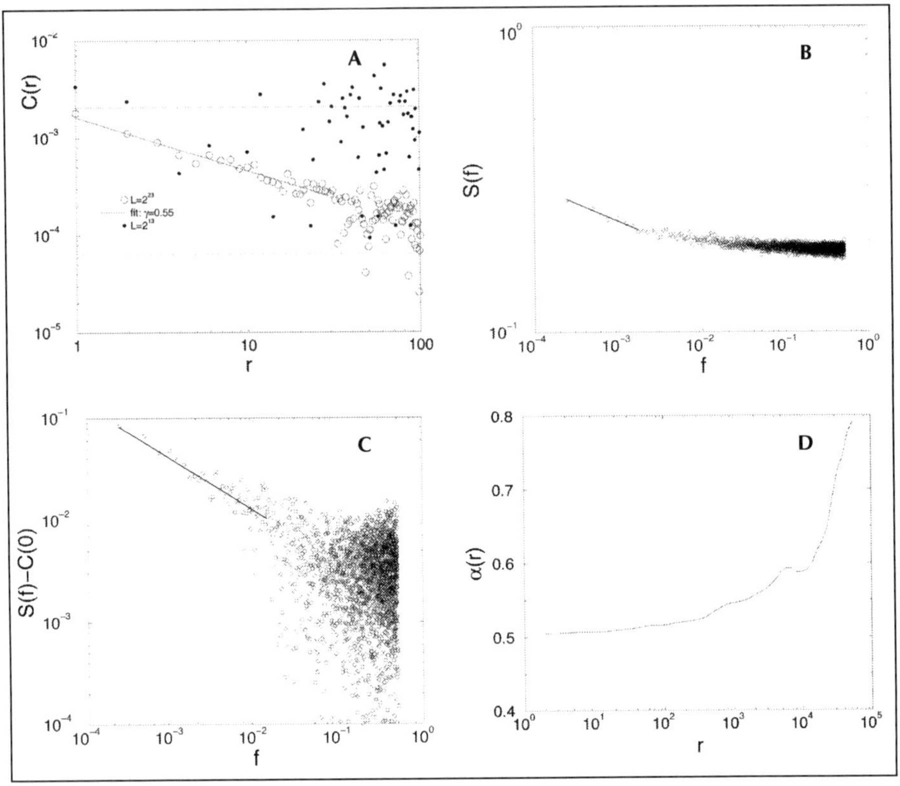

Figure 14. A) Correlation functions for the Lévy walk model for a long $L = 2^{23}$ and a short $L = 2^{13}$ sequences. The lower and upper horizontal lines show random noise levels for the long and short sequences, respectively. B) The power spectrum for the long sequence. The spectrum is almost flat indicating that long-range correlations are small comparatively to the white noise level. Effective exponent $\beta = 0.12$ is very small. C) After the white noise level is subtracted, the power spectrum shows long-range correlations with exponent $\beta = 0.57$. D) The effective exponent $\alpha(r)$ obtained by the DFA method.

intriguing this observation might seem, the explanation is somewhat trivial. The case $\mu < 2$, ($\beta = 1$) i.e., the case when the length of the largest patch is comparable with the entire sequence length is indeed likely to be true for music as well as for DNA. In music, fast pieces follow slow pieces, while in DNA, CG rich isochores follow CG poor ones.

It is interesting to note that similar long range correlations with exponent $\alpha = 0.57$ have been found in human writings.[74,75] These correlations can be explained by the changes in local frequenies of letters caused by changes in the narrative which excessively uses the names of currently active characters.

In DNA, these patches may represent different structural elements of 3D chromosome organization, e.g., the DNA double helix with period 10.5 bp,[76] nucleosomes about 200 bp long,[76] 30 nm fiber, looped domains of about 10^5 bp, and chromatin bands or isochores[72,77] that may consist of several million nucleotides. Such hierarchical structure of several length-scales may produce effective long-range power law correlations. In fact,[78,79] it is enough to have three discrete sizes $r = 100$, $r = 1000$ and $r = 10000$ of these patches in the distribution $P(r)$ in order to get a sufficiently straight double logarithmic plot of the power spectrum over three decades in the frequency range.

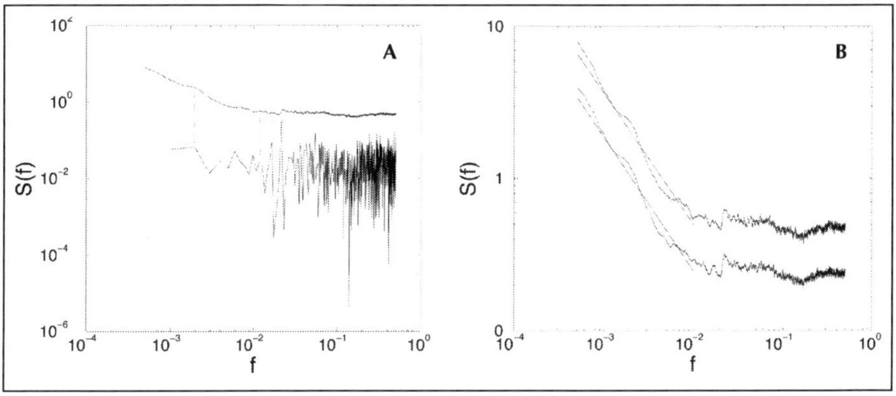

Figure 15. A) Power spectrum of the "chromosome" of length $L = 2^{20}$ (upper curve) in comparison with the power spectrum of the inserted "transposon" $\ell = 2^{10}$ (lower curve) in the insertion-deletion model. Dotted lines indicate peaks present in both sequences. B) Power spectra of the "chromosome" after 1024 iterations (lower curve) and after 16384 iterations (upper curve) showing that the model reaches steady state after $L/\ell = 2^{10}$ iterations.

An interesting model can reproduce some feature of the human genome, namely the abundance of interspersed repeats or retroposons,[68] virus-like sequences that can insert themselves into different places of the chromosomes by reverse transcriptase. An example of such a sequence is LINE-1, which we discussed earlier in this section.

Suppose, we have an initially uncorrelated "chromosome" consisting of L base-pairs with equal concentration of purines and pyrimidines and a "transposon" of length $\ell << L$ with strong strand bias (60% purines) and no correlations (Fig. 15A). Let us assume that at every simulation step our "transposon" can be inserted at random places into one of the two opposite strands of the "chromosome" with equal probabilities. In order to keep the length of the chromosome constant, let us delete exactly ℓ nucleotides selected at random after each insertion. After approximately L/ℓ insertions, the power spectrum of the "chromosome" reaches a steady state shown in Figure 15B. In this example, we use $L = 2^{20}$, $\ell = 2^{10}$. Note the presence of strong peaks in the flat spectral part for $f > 0.01$ and a steep slope with average slope $\beta \approx 0.8$ for $0.0005 < f < 0.01$. One can easily see (Fig. 15A) that the power spectrum of the "transposon" which is kept unchanged during the entire simulation has strong peaks coinciding with the peaks of the resulting "chromosome". This example shows that the presence of many copies of interspersed repeats (some of which have partially degraded) can lead to the characteristic peaks at high frequencies larger than the inverse length of the retroposons and strong power-law like correlations at low frequencies comparable with the inverse length of the retroposons.

Models of Long Range Anti-Correlations

Another interesting situation may exist in coding DNA which preserves the reading frame. The reading frame is a non-interrupted sequence of codons each consisting of three nucleotides. One of the most fundamental discoveries of all time, is the discovery of the universal genetic code, i.e., that in all leaving organisms, with very few exceptions, each of the twenty amino acids is encoded by the same combinations of three nucleotides or codons. Since there are $4^3 = 64$ different codons and only 20 amino acids, some amino acids are encoded by several codons. In the different codons used for coding the same amino acid, the first letter is usually preserved. Since the amino acid usage is non-uniform, the same is true for the codon usage, particularly for the frequency of the first letter in the codon. It is known[80] that for all coding sequences in the GenBank, there is a preference for purine in the first position in the codon (32% G and 28% A) and for weakly bonded pair in the second position (31% A and 28% T). This preference exists for any organism in the entire phylogenetic spectrum and is the basis for the species independence of mutual information.[44]

Accordingly, let us generate many patches of different length l in which the frequencies of a certain nucleotide at positions $3k + 1 + c$, $3k + 2 + c$, and $3k + c$ are f_1, f_2 and f_3. Here c is a random offset which is constant within each patch and can take values 0,1,2 with equal probabilities. Following Herzel and Grosse,[43] we will call this construction a random exon model.

All the correlation properties of the random exon model can be computed analytically. But even without lengthy algebra, it is clear that the correlation function will oscillate with period three being positive at positions $r = 3k$ and negative at positions $r = 3k + 1$ and $r = 3k + 2$. The envelope of these oscillations will decay, either exponentially if the patch length is distributed exponentially or as a power law if the distribution of patches is a power law $P(l) \sim l^{-\mu}$. Accordingly, in the power spectrum, there will be either a finite strong peak at frequency $f = 1/3$ with intensity proportional to the weighted average patch length or a power law singularity $|f - 1/3|^{\mu-3}$ if $2 < \mu < 3$. If $\mu \leq 2$, it will be $1/f$-singularity $|f - 1/3|^{-1}$.

Figure 16A,B shows the correlation function for the random exon model with $f_1 = 0.29$, $f_2 = f_3 = 0.2$ and a power law distribution of reading frame lengths with $\mu = 2.5$. Figure 16C

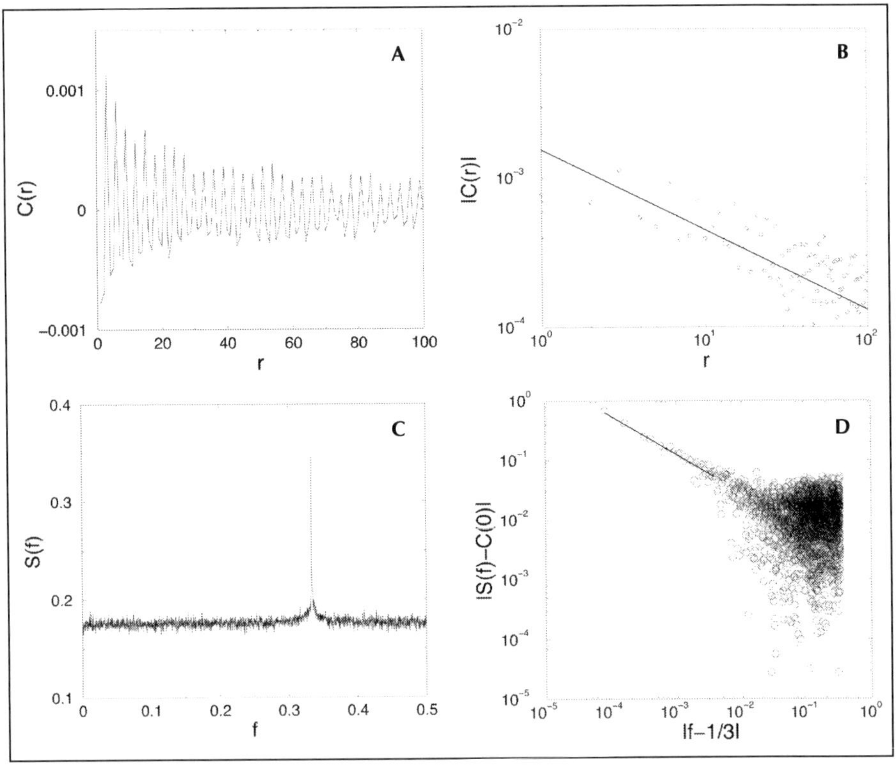

Figure 16. A) Correlation function for the random exon model with power-law distribution of reading frame lengths $P(l) \sim l^{-2.5}$ oscillates with period three. B) The log-log plot of the absolute value of the correlation function for the same sequence. The power-law correlations with exponent $\gamma = 0.57$ are clearly seen. C) The power spectrum of the same sequence. It is almost flat with a strong peak at $f = 1/3$. D) The log-log plot of the power spectrum with the subtracted white noise level. The power law correlations with $\beta = 0.64$ are clearly seen.

shows the power spectrum of this sequence and finally Figure 16D shows the log-log plot of $|P(f) - C(0)|$ versus $|f - 1/3|$. One can see approximate straight line behavior with the slope 0.6. The DFA analysis fails to show the presence of any power law correlations except for a small bump at $r = 3$ (not shown).

Analysis of DNA Sequences

In this section, we finally present the analysis of real DNA sequences. The examples of the previous sections show us that among different methods of analysis, the power spectrum usually gives the best results. In contrast to $C(r)$ method, it provides natural averaging of the long range correlations from a broad interval of large distances $[r, r + \Delta r]$ adding them up into a narrow range of low frequencies $[1/r - \Delta r/r^2, 1/r]$. Thus, the power spectrum restores useful information which cannot be seen from $C(r)$ quickly sinking below the white noise level for large r. On the other hand, the power spectrum does not smooth out the details on the short length scales corresponding to high frequencies as DFA does. Also it is much less computationally intensive than the two other methods. Once the intuition on how to use the power spectrum analysis is developed, it can be applied to DNA sequences with the same success as in X-ray crystallography, especially, today when the length of the available DNA sequences becomes comparable with the number of atoms in the nano-scale experimental systems. Not surprisingly, power spectra of the DNA from different organisms have distinct characteristic peaks,[81] similarly to the X-ray diffraction patterns of different substances. Accordingly, in this section, we will use only the power spectrum analysis.

In the beginning of 1990, when the first long DNA sequences became available, an important practical question was to find coding regions in the "sea" of noncoding DNA which constitutes 97% of human genome. The problem was not only to determine genes, i.e., the regions which are transcribed in the process of RNA transcription, but also the exons, the smaller segments of genes which remain in the messenger RNA after the noncoding introns are spliced out. Only the information from exons is translated into proteins by the ribosomes.[51,52] That is why, the claim of reference 31 that the non-coding DNA sequences have stronger power law correlations than the coding ones attracted much attention and caused a lot of controversy.[34] The results of reference 31 were based on the studies of a small subset of sequences using DNA landscape technique (see Fig. 13). Later these results were confirmed by the DFA method, the wavelet,[55,72,82] the power spectrum[80] and modified standard deviation analyses.[83] However, the difference between coding and noncoding DNA appeared to be not as dramatic as it was originally proposed. In Figure 17 we present the results[80] of the analysis of 33301 coding and 29453 noncoding sequences of the eukaryotic organisms. These were all the genomic DNA sequences published in the GenBank release of August 15th, 1994 whose length was at least 512 nucleotides. The power spectrum is obtained by averaging power spectra calculated by FFT of all non-overlapping intervals of length $N = 2^9 = 512$ contained in the analyzed sequences. The conclusions hold not only for the average power spectrum of all eukaryotes but also for the average power spectra of each organism analyzed separately.

Unlike the graphs for Ising model, the log-log graphs for coding and non-coding DNA are not straight but have three distinct regimes for high (H) ($f > 0.09$), medium (M) $0.012 < f < 0.09$ and low (L) $f < 0.012$ frequencies. The slopes β_M in the region of medium frequencies can be obtained by the least square linear fit. For RY mapping rule (see section "Correlation Analysis of DNA Sequences") presented in Figure 17 for coding DNA, we see $\beta_M = 0.03$ which corresponds to the white noise, while for non-coding DNA we see weak power-law correlations with $\beta_M = 0.21$. Reference 80 contains the tables of the exponents β_M obtained for various eukaryotic organisms for seven different mapping rules (RY, SW, KM, A, C, G, T). For all the rules and all the organisms, the exponent β_M for the averaged power spectra of non-coding regions is always larger than β_M for coding regions. For some rules, such as SW, the exponent β_M is negative for coding DNA and

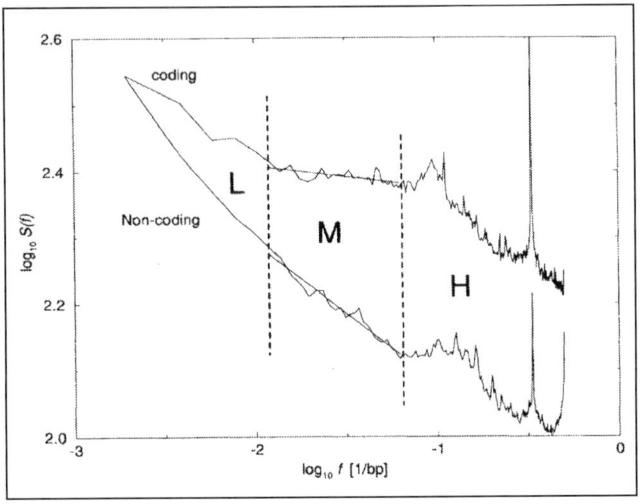

Figure 17. The RY power spectrum obtained by averaging power spectra of all eukaryotic sequences longer than 512 bp, obtained by FFT with window size 512. Upper curve is average over 29,453 coding sequences; lower curve is average over 33,301 noncoding sequences. For clarity, the power spectra are shifted vertically by arbitrary quantities. The straight lines are least squares fits for second decade (Region M). The values of β_M for coding and noncoding DNA obtained from the slopes of the fits are 0.03 and 0.21, respectively. (From ref. 80.)

is close to zero for non-coding DNA. But the algebraic values of the exponents for non-coding DNA is always larger than for coding DNA. The histogram of values of β_M computed for individual 512-bp sequences has a roughly Gaussian shape with standard deviation $\sigma = 0.3$ which is several times larger than the difference between mean values of β_M for coding and non-coding DNA. This makes the use of fractal exponent β_M impractical for finding coding regions.[84]

A much more important characteristic of the power spectrum is the height of the peak at the codon frequency $f = 1/3$, which was included in the standard gene finding tool boxes.[85,86] Figure 17 shows that the peak for coding regions is several times higher than for non-coding ones. The presence of the weak peak in the noncoding regions can be attributed to the non-identified genes or to pseudo-genes which have recently (on the evolutionary time scale) become inactive (like olfactory genes for humans). The presence of the peak can be explained by the non-uniform codon usage, (see section "Models of Long Range Anti-Correlations", Fig. 16).

Another interesting and distinctive feature of non-coding DNA is the presence of the peak at $f = 1/2$ as in the anti-ferromagnetic Ising model. This peak can be attributed to the presence of long tandem repeats ...CACACA... and ...TGTGTG... which are prolific in non-coding DNA but very rare in the coding (see next section).

Presently, when several complete or almost complete genomes are just a mouse-click away, it is easy to test if the true power-law long-range correlations do exist in the chromosomes of different species. Figure 18A,B shows power spectra of the 88 million base-pair contig of the human chromosome XIV computed according to the seven mapping rules described in section "Correlation Analysis of DNA Sequences". A very interesting feature of the human genome is the presence of the strong peaks at high frequencies. These peaks are much stronger than the peak at $f = 1/3$ for coding DNA. It is plausible that these peaks are due to the hundreds of thousands almost identical copies of the SINE and LINE repeats,[87] which constitute a major portion of human genome.[68] If one compares the peaks in the power spectrum of the chromosome, with

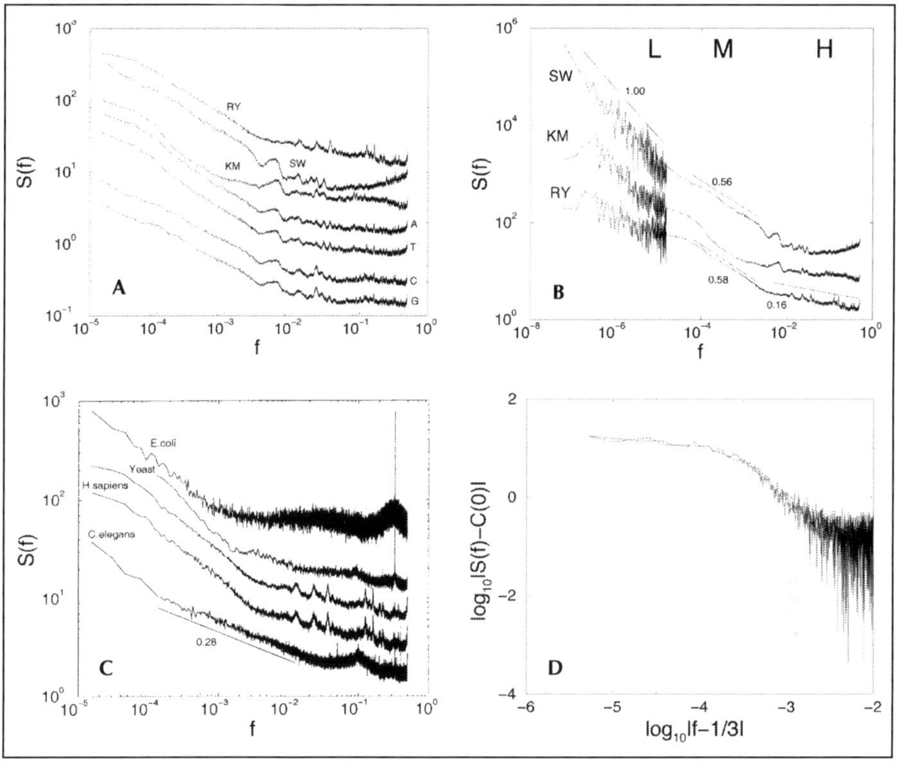

Figure 18. A) Power spectra for seven different mapping rules computed for the Homo sapiens chromosome XIV, genomic contig NT_026437. The result is obtained by averaging 1330 power spectra computed by FFT for non-overlapping segments of length $N = 2^{16} = 65536$. B) Power spectra for SW, RY, and KM mapping rules for the same contig extended to the low frequency region characterizing extremely long range correlations. The extension is obtained by extracting low frequencies from the power spectra computed by FFT with $N = 2^{24} = 16M$ base pairs. Three distinct correlation regimes can be identified. High frequency regime ($f < 0.003$) is characterized by small sharp peaks. Medium frequency regime ($0.5{\cdot}10^{-5} < f < 0.003$) is characterized by approximate power-law behavior for RY and SW mapping rules with exponent $\beta_M = 0.57$. Low frequency regime ($f < 0.5{\cdot}10^{-5}$) is characterized by $\beta = 1.00$ for SW rule. The high frequency regime for RY rule can be approximated by $\beta_H = 0.16$ in agreement with the data of Figure 17. C) RY Power spectra for the entire genome of *E. coli* (bacteria), *S. cerevisae* (yeast) chromosome IV, *H. sapiens* (human) chromosome XIV and the largest contig (NT_032977.6) on the chromosome I; and *C.elegans* (worm) chromosome X. It can be clearly seen that the high frequency peaks for the two different human chromosomes are exactly the same, while they are totally different from the high frequency peaks for other organisms. One can also notice the presence of enormous peaks for $f = 1/3$ in *E. coli* and yeast, indicating that their genomes do not have introns, so that the lengths of coding segments are very large. The *C. elegans* data can be very well approximated by power law correlations $S(f) \sim f^{-0.28}$ for $10^{-4} < f < 10^{-2}$. D) Log-log plot of the RY power spectrum for *E. coli* with subtracted white noise level versus $|f - 1/3|$. It shows a typical behavior for a signal with finite correlation length, indicating that the distribution of the coding segments in *E. coli* has finite average square length.

the peaks in the power spectra of various SINE and LINE sequences, one can find that some of these peaks coincide as in the model of insertion-deletion discussed in section "Alternation of Nucleotide Frequencies". The absence of these peaks in the genomes of primitive organisms (see Fig. 18C) is in agreement with the fact that these organisms lack interspersed repeats.

It is clear that the long-range correlations lack universality, i.e., they are different for different species and strongly depend on the mapping rule. The slopes of the power spectra change with frequency and undergo sharp crossovers which do not coincide for different organisms. The strongest correlations with the spectral exponent $\beta = 1$ are present for *SW* rule at low frequencies, indicating the presence of the isochores. The middle frequency regime which can be particularly well approximated by power law correlations in *C. elegans* can be explained by the generalized duplication-mutation model of W. Li in which duplications and mutations occur on the level genes, consisting of several hundred base pairs. The high frequency correlations, sometimes characterized by small positive slopes of the power spectra can be attributed to the presence of simple sequence repeats (see next section). In contrast, the high frequency spectrum of the bacterium *E. coli* is almost flat with the exception of the huge peak at $f = 1/3$. Bacterial DNA practically does not have noncoding regions, thus (in agreement with refs. 31,72,80,82) it does not have long range correlations on the length scales smaller than the length of a typical gene. Large peaks at $|f - 1/3|$ in the power spectra of *E. coli* and yeast are consistent with fact that these primitive organisms do not have introns and therefore their open reading frames are very long. The spectrum of *E. coli* printed versus $|f - 1/3|$ shows a horizontal line for $f \to 1/3$ on a double logarithmic plot indicating that the length distribution of the open reading frames has finite second moment.

Distribution of Simple Repeats

The origin, evolution, and biological role of tandem repeats in DNA, also known as microsatellites or simple sequence repeats (SSR), are presently one of the most intriguing puzzles of molecular biology. The expansion of such SSR has recently become of great interest due to their role in genome organization and evolutionary processes.[88-100] It is known that SSR constitute a large fraction of noncoding DNA and are relatively rare in protein coding sequences.

SSR are of considerable practical and theoretical interest due to their high polymorphism.[97] The formation of a hairpin structure during replication is believed to be the cause of the *CAG* and *CTG* repeat expansions, which are associated with a broad variety of genetic diseases. Among such diseases are fragile X syndrome,[101] myotonic dystrophy, and Huntington's disease[94,102] SSR of the type $(CA)_\ell$ are also known to expand due to slippage in the replication process. These errors are usually eliminated by the mismatch-repair enzyme MSH2. However, a mutation in the MSH2 gene leads to an uncontrolled expansion of repeats—a common cause of ovarian cancers.[103] Similar mechanisms are attributable for other types of cancer.[85,92,93] Telomeric SSR, which control DNA sequence size during replication, illustrate another crucial role of tandem repeats.[51]

Specifically, let us consider the distribution of the most simple case of SSR—repeats of identical dimers *XYXY...XY* ("dimeric tandem repeats"). Here *X* and *Y* denotes one of the four nucleotides: adenine (A), cytosine (C), guanine (G), and thymine (T). Dimeric tandem repeats are so abundant in noncoding DNA that their presence can even be observed by global statistical methods such as the power spectrum. For example, Figures 17 and 18A-C show presence of a peak at $(1/2)\text{bp}^{-1}$ in the power spectrum of **noncoding** DNA (corresponding to repetition of dimers) and the absence of this peak in **coding** DNA. The abundance of dimeric tandem repeats in noncoding DNA suggests that these repeats may play a special role in the organization and evolution of noncoding DNA.

First, let us compute the number of repeats in an uncorrelated sequence. Suppose that we have a random uncorrelated sequence of length $2L$ which is a mixture of all 16 possible types of dimers *XY*, each with a given frequency f_{XY}. The probability that a randomly selected dimer belongs to a dimeric tandem repeat $(XY)_\ell$ of length ℓ can be written as

$$P_{XY}(\ell) = f_{XY}^{\ell} \cdot (1 - f_{XY})^2 \cdot \ell, \tag{45}$$

where $(1 - f_{XY})$ is the terminating factor responsible for not producing an additional unit XY at the beginning (or end) of the repeating sequences and the factor ℓ takes into account ℓ possible positions of a dimer XY in a repeat $(XY)_\ell$. Since the total number of dimers in our sequence is L, the number of dimers in the repeats $(XY)_\ell$ is $LP_{XY}(\ell) = \ell N_{XY}(\ell)$, where $N_{XY}(\ell)$ is the total number of repeats $(XY)_\ell$ in a sequence of length $2L$. Finally,

$$N_{XY}(\ell) = f_{XY}^{\ell} \cdot (1 - f_{XY})^2 \cdot L \sim e^{-\ell \cdot |\ln f_{XY}|}, \tag{46}$$

which decreases exponentially with the length of the tandem repeat. Thus, a semi-logarithmic plot of $N_{XY}(\ell)$ versus ℓ must be a straight line with the slope

$$-k_{\text{unc}} = \ln(f_{XY}). \tag{47}$$

In order to compare the prediction of this simple model with real DNA data, we estimate f_{XY} for the real DNA as follows: (i) divide the DNA sequence into L non-overlapping dimers, (ii) count n_{XY}, the total number of occurrences of a dimer XY in this sequence, and calculate

$$f_{XY} \equiv \frac{n_{XY}}{L}. \tag{48}$$

Indeed, most dimeric tandem repeats in **coding** DNA produce linear semilogarithmic plots, (Fig. 19A) but with slopes significantly different from those predicted by (47). The deviation of the slopes from prediction (47) can be explained by the short order Markov correlations.[106,107]

On the other hand, semilogarithmic plots of the length distributions of dimeric repeats for noncoding parts (Fig. 19C) are usually not straight, but display negative slope with constantly decreasing absolute value which indicates that their probability decays less rapidly than exponentially. Indeed, these distributions can be better fit by straight lines on a double logarithmic plot (Fig. 19D)

$$N_{XY}(\ell) \sim \ell^{-\mu}. \tag{49}$$

A simple model to explain the power law behavior (49) was presented in references 106 and 108. The mechanism proposed in references 106 and 108 is based on random multiplicative processes, which can reproduce the observed non-exponential distribution of repeats. The increase or decrease of repeat length can occur due to unequal crossover[51,109] or slippage during replication.[92,100,110,111] It is reasonable to assume (see ref. 110 and refs. therein) that in these types of mutations, the new length ℓ' of the repeat is not a stepwise increase or decrease of the old length ℓ, but is defined as a product $\ell' = \ell r$, where r is some random variable.

For simplicity, we neglect point mutations and assume that with conditional probability $\pi(r, \ell)$ in a single mutation, a repeat of length ℓ can expand or shrink to a repeat of length $r\ell$, where the function $\pi(r, \ell)$ is normalized:

$$\int_0^\infty \pi(r, \ell) dr = 1. \tag{50}$$

After t steps of evolution the length of the repeat is given by

$$\ell_t = \prod_{i=1}^{t} r_i \ell_0, \tag{51}$$

where r_i is a random variable taken from a distribution with probability density $\pi(r, \ell)$. Such a process is called a random multiplicative process and, in many cases, leads in the long time limit ($t \to \infty$) to a stable distribution of repeat length $P(\ell)$. According to Eq. (51), repeats may fluctuate in length and even disappear. Thus, to prevent the extinction of repeats, one can either set a non-zero probability for a repeat to reappear, or set $\pi(r, \ell) = 0$ when $r\ell \leq 1$. Both ways are mathematically equivalent and might be biologically controlled by point mutations.

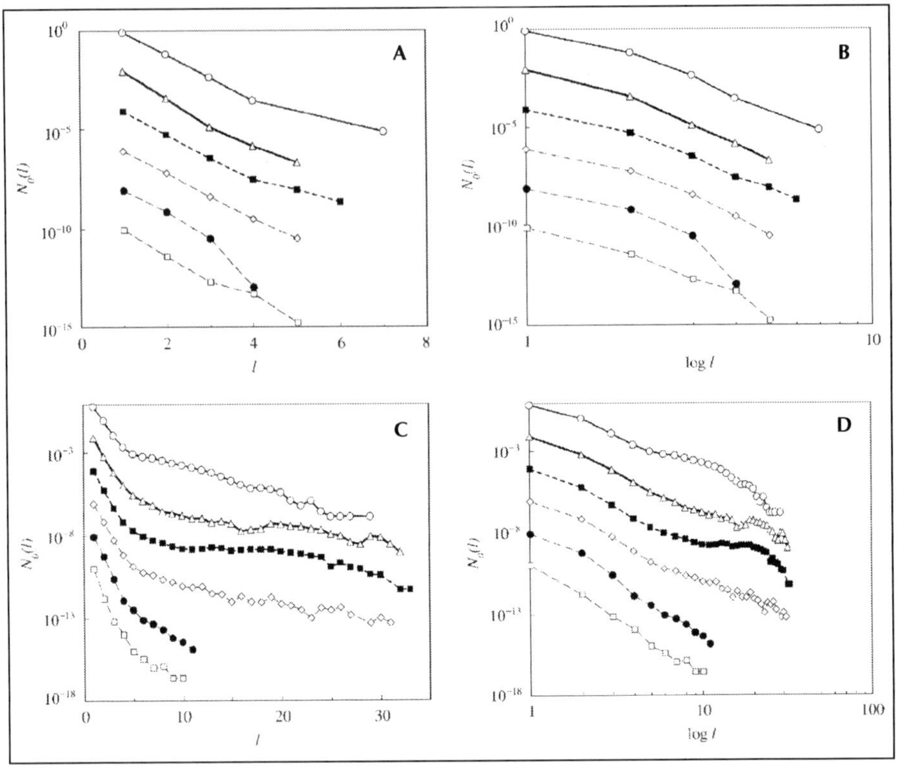

Figure 19. The combined plot of the normalized number $N_0(\ell) \equiv N_{XY}(\ell)/N_{XY}(1)$ of repeats for six groups of dimeric tandem repeats in human genome averaged over analogous repeats in each group: $(AA)\,\ell$ and $(TT)\,\ell$ (○); $(TA)\,\ell$ and $(AT)\,\ell$ (Δ); $(CA)\,\ell$, $(AC)\,\ell$, $(TG)\,\ell$ and $(GT)\,\ell$ (■); $(GA)\,\ell$, $(AG)\,\ell$, $(TC)\,\ell$ and $(CT)\,\ell$ (◊); $(CC)\,\ell$ and $(GG)\,\ell$ (●); $(GC)\,\ell$ and $(CG)\,\ell$ (□). Semi-logarithmic plot for coding DNA (A), double-logarithmic plot for coding DNA (B), semi-logarithmic plot for noncoding DNA (C), and double-logarithmic plot for noncoding DNA (D). For clarity, we separate plots for these six groups by shifting them by a factor of 100 on the ordinate. The values of μ for six groups of repeats in (D) are 3.6, 3.3, 3.2, 4.1, 6.7, and 5.4 from top to bottom, fitting range is $\ell > 5$. The values of μ for strongly bonded repeats GC, CG and CC, GG are significantly larger than for other repeats. (From ref. 107.)

If we take the logarithm of both parts of Eq. (51) and change variables to $z \equiv \ln \ell$, the process becomes a random diffusion process in semi-infinite space $z > 0$ in which a particle makes steps $v_i = \ln r_i$. The distribution of steps $\tilde{\pi}(v,z)$ can be related to the original distribution of growth-rates, $\pi(r, \ell)$. Indeed, in the continuum limit $\tilde{\pi}(v,z)dv = \pi(r,\ell)dr$, or $\tilde{\pi}(v,z) = \pi(e^v,e^z)e^v$.

A classical example of such a process is Brownian motion in a potential field $U(z)$, which leads for $t \to \infty$ to a Boltzmann probability distribution (3). The strength and the direction of the potential force $f(z) = -dU/dz$ depends on the probability distribution $\tilde{\pi}(v,z)$. If probability to go up is larger than the probability to go down, the force acts upward, so the particle travels upward indefinitely and no stable probability distribution is observed. (This situation corresponds to the uncontrollable expansion of repeats as in some types of cancers.) If the distribution $\tilde{\pi}(v,z)$ does not depend on z, the force is constant. If the force is constant and acting down as the gravitational force on the Earth, the final probability distribution decays exponentially with z as the density of Earth's atmosphere

$$P(z) \sim e^{-kz}, \tag{52}$$

where k is a positive constant which depends on the distribution $\tilde{\pi}(\nu,z) = \tilde{\pi}(\nu)$.

Using the theory of Markovian processes (Section III), one can show that the final probability distribution $P(z)$ must satisfy an equation analogous to (9) in which $P(z)$ plays the role of eigenvector $\mathbf{a}_1$ and $\tilde{\pi}(\nu,z)$ plays the role of transition matrix $\mathbf{P}$. In the continuum, limit we have $P(z) = \int_{-\infty}^{\infty} P(z-\nu)\,\tilde{\pi}(\nu,z-\nu)d\nu$, which in case $\tilde{\pi}(\nu,z) = \tilde{\pi}(\nu)$ has solution (52) and $k > 0$ must satisfy equation

$$\int_{-\infty}^{\infty} \exp(k\nu)\tilde{\pi}(\nu)d\nu = 1. \tag{53}$$

After transforming back to our original variables, the solution (52) can be rewritten in the form of a power law,

$$P(\ell) = \ell^{-\mu} \tag{54}$$

where $\mu = k + 1 > 1$. Accordingly (53) must be rewritten in the form

$$\int_0^{\infty} r^{\mu-1} \cdot \pi(r)dr \tag{55}$$

Equation (55) always has a trivial solution $\mu = 1$ (due to the normalization (50)). However, Eq. (55) may also have additional roots, $\mu > 1$. If it does not have such roots then the final distribution does not exist. This case corresponds to the uncontrollable expansion of repeats.

Let us discuss two examples, in which Eq. (55) has simple solutions. For example, if $\pi(r)$ is a step-function

$$\pi(r) = \begin{cases} 1/2, & 0 \le r \le 2 \\ 0, & r < 1, r > 2, \end{cases} \tag{56}$$

equation (55) becomes

$$\frac{1}{2} \cdot \frac{2^{\mu}}{\mu} = 1. \tag{57}$$

Eq. (57) has a solution $\mu = 2$. The above case can serve as the simplest model of unequal crossover,[108] after which a repeat of length ℓ becomes of length $\ell \cdot (1 + r)$ in the first allele and of length $\ell \cdot (1 - r)$ in the second allele. If both alleles have equal probability of becoming fixed in the population, we arrive to Eq. (56).

In another simple example we take

$$\pi(r) = \pi_1 \cdot \delta(r - 1/2) + \pi_2 \cdot \delta(r - 2), \tag{58}$$

where $\pi_1 + \pi_2 = 1$ and $\delta(r)$ is the Dirac delta-function, i.e., with probability π_1 the repeat can shrink by factor of two and with probability π_2 it can grow by factor of two. In this case, Eq. (55) can be written as

$$\pi_1 \cdot \left(\frac{1}{2}\right)^{\mu-1} + \pi_2 \cdot 2^{\mu-1} = 1, \tag{59}$$

which has a root $\mu = 1 + \log_2(\pi_1/\pi_2)$. If probability to grow is larger than probability to shrink, $\pi_2 > \pi_1$, we have $\mu < 1$, which, as we see above, leads to an uncontrollable expansion of repeats as in some diseases. These simple examples show that our multiplicative model is capable to explain the power law distribution of simple repeats with any exponent $\mu > 1$.

In the general case of discrete multiplicative processes, one cannot obtain analytical solutions. However, numerical simulations[106] show that Eq. (54) still provides a good approximation for large ℓ. The deviation of the actual distributions (Fig. 19) from an exact power law can

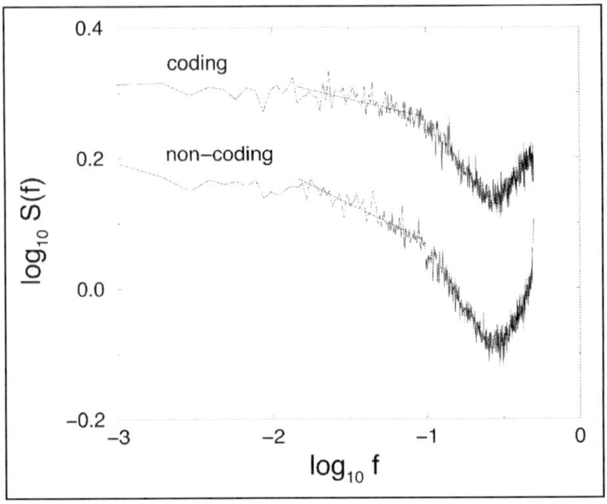

Figure 20. RY power spectra of the random dimeric tandem repeat model for coding and noncoding DNA for the mammalian sequences.

be explained if one takes into account that the distribution of growth rates $\pi(r, \ell)$ may depend on the length of the repeat ℓ.[107,112] This is especially plausible for the slippage during replication mechanism, since the ability for a DNA molecule to form hairpins clearly depends on the length of a segment involved in the slippage and on its biophysical properties and thus must depend on the type of repeat. Therefore, it is not surprising that different types of repeats have different length distribution. For example, the distribution of $(AC)_\ell$ and $(TG)_\ell$ repeats in vertebrates have plateaus in the range $10 < \ell < 30$. In contrast, the distributions of $(CC)_\ell$, $(CG)_\ell$ and $(GG)_\ell$, and repeats decay much faster than other repeats which include weakly bonded base pairs.

A different model proposed by reference 113 can also reproduce long tails in the repeat length distribution. This model assumes the stepwise change in repeat length with the mutation rate proportional to the repeat length. It is possible to map this model to a random multiplicative process with a specific form of distribution $\pi(r, \ell)$, where $r = \ell'/\ell$, ℓ is the original length and ℓ' is a repeat length after a time interval during which several stepwise mutations can occur.

From the analysis in section "Alternation of Nucleotide Frequencies", it follows that simple tandem repeats randomly distributed along the sequence can produce long-range power-law correlations if, and only if, $\mu < 3$. However, in almost all real DNA sequences $\mu > 3$, which means that simple tandem repeats alone cannot explain long-range correlations. On the other hand, simple tandem repeats may be the primary source of the difference in correlation properties of coding and noncoding sequences at relatively short length scales of $\ell \approx 100$ bp.[78,79] In order to test such a possibility, we construct a random dimeric repeat model by randomly selecting all possible repeats $(XY)_\ell$ from the empirically observed distribution $N_{XY}(\ell)$ and concatenating them into an artificially constructed sequence of nucleotides. Figure 20 shows the power spectra of two sequences produced by random concatenations of various dimeric repeats taken from the noncoding and coding mammalian DNA. These power spectra show a slight difference in the spectral exponent β_M in the region of medium frequencies analogous to Figure 17. Note that the difference in the spectral exponents in the random repeat model is smaller than in real sequences. However, here we consider only dimeric tandem repeats, thus

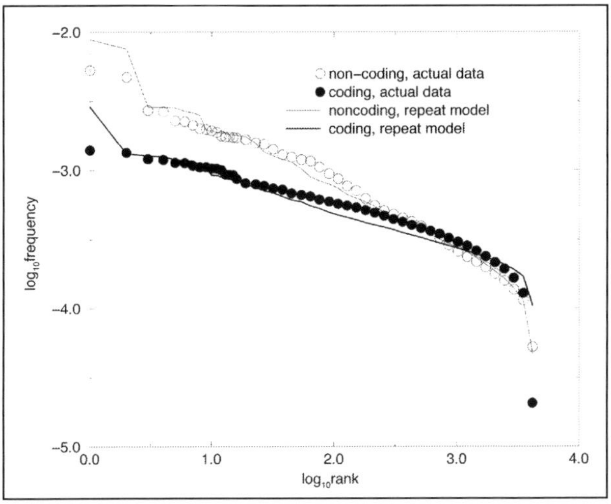

Figure 21. Log-log plot of the frequency of 6-letter words (hexamers) versus their rank for invertebrate coding and non-coding sequences in comparison with the same graphs produced by the random dimeric repeat model.

neglecting the repeats of other types. We also neglect the possibility of imperfect repeats interrupted by several point mutations.

Finally, dimeric tandem repeats can explain the difference observed in the distribution of *n*-letter words in coding and non-coding DNA (see Fig. 21). As an example, we show the rank-frequency of the 6-letter words (hexamers) for invertebrate coding and noncoding sequences in the form of the so called Zipf plots.[114] For natural languages, Zipf graphs show that the frequency of a word in a text is inverse proportional to its rank. For example, in an English text, the most frequent word is "the" (rank 1), the second most frequent word is "of" (rank 2), the third most frequent word is "a" (rank 3) and so on. Accordingly, the frequency of word "of" is roughly two times smaller than the frequency of word "the" and the frequency of word "a" is roughly three times smaller than the frequency of "the". Thus on the log-log scale, the Zipf graph is a straight line with the slope -1. In a DNA sequence, there is no precise definition of the "word", so one can define "word" as any string of the fixed number of consecutive nucleotides that can be found in the sequence. One can notice that the Zipf graph for non-coding DNA is approximately straight but with a slope smaller than 1, while for coding DNA, the graph is more curvy and is less steep. This observation led Mantegna et al [115,116] to conclude that noncoding DNA have some properties of natural languages, namely redundancy. Accordingly, noncoding DNA may contain some "hidden language". However, this conjecture was strongly opposed by the bioinformatics community.[117] Indeed, Zipf graphs of coding and noncoding DNA can be trivially explained by the presence of dimeric tandem repeats (Fig. 21).

To conclude, noncoding DNA may not contain any hidden "language" but it definitely has lot of hidden biological information. For example, it contains transcription regulatory information which is very difficult to extract. Application of correlation analysis may help to solve this problem.[118]

Conclusion

Long range correlations of different length scales may develop due to different mutational mechanisms. The longest correlations, on the length scales of isochores may originate due to base-substitution mutations during replication (see ref. 77). Indeed, it is known that different parts of chromosomes replicate at different stages of cell division. The regions rich in C+G replicate earlier than those rich in A+T. On the other hand, the concentration of C+G precursors in the cell depletes during replication. Thus the probability of substituting A/T for C/G is higher in those parts of the chromosome that replicate earlier. These unequal mutation rates may lead to the formation of isochors.[77] Correlations on the intermediate length scale of thousands of nucleotides may originate due to DNA shuffling by insertion or deletion[57,58] of transposable elements such as LINES and SINES[66,68,119] or due to a mutation-duplication process proposed by W. Li[56] (see also ref. 120).

Finally, the correlations on the length scale of several hundreds of nucleotides may evolve due to simple repeat expansion[106,108] As we have seen in the previous section, the distributions of simple repeats are dramatically different in coding and noncoding DNA. In coding DNA they have an exponential distribution; in noncoding DNA they have long tails that in many cases may be fit by a power law function. The power law distribution of simple repeats can be explained if one assumes a random multiplicative process for the mutation of the repeat length, i.e., each mutation leads to a change of repeat length by a random factor with a certain distribution (see ref. 106). Such a process may take place due to errors in replication[110] or unequal crossing over (see ref. 108 and refs. therein). Simple repeat expansion in the coding regions would lead to a loss of protein functionality (as, e.g., in Huntington's disease[110]) and to the extinction of the organism.

Thus the weakness of long-range correlations in coding DNA is probably related to the coding DNA's conservation during biological evolution. Indeed, the proteins of bacteria and humans have many common templates, while the noncoding regions can be totally different even for closely related species. The conservation of protein coding sequences and the weakness of correlations in the amino acid sequences[121] are probably related to the problem of protein folding. Monte-Carlo simulations of protein folding on the cubic lattice suggest that the statistical properties of the sequences that fold into a native state resemble those of random sequences.[122]

The higher tolerance of noncoding regions to various mutations, especially to mutations involving the growth of DNA length—e.g., duplication, insertion of transposable elements, and simple repeat expansion—lead to strong long-range correlations in the noncoding DNA. Such tolerance is a necessary condition for biological evolution, since its main pathway is believed to be gene duplication by chromosomal rearrangements, which does not affect coding regions.[123] However, the payoff for this tolerance is the growth of highly correlated junk DNA.

Acknowledgements

I am grateful to many individuals, including H.E. Stanley, S. Havlin, C.-K. Peng, A.L. Goldberger, R. Mantegna, M.E. Matsa, S.M. Ossadnik, F. Sciortino, G.M. Viswanathan, N.V. Dokholyan, I. Grosse, H. Herzel, D. Holste, and M. Simons for major contributions to those results reviewed here that represent collaborative research efforts. Financial support was provided by the National Science Foundation and National Institutes of Health (Human Genome Project).

References

1. Stauffer D, Stanley HE. From Newton to Mandelbrot: A Primer in Theoretical Physics. Heidelberg, New York: Springer-Verlag, 1990.
2. Stanley HE. Introduction to Phase Transitions and Critical Phenomena. London: Oxford University Press, 1971.
3. Stauffer D, Aharony A. Introduction to Percolation Theory. Philadelphia: Taylor & Francis, 1992.
4. de Gennes PG. Scaling Concepts in Polymer Physics. Ithaca: Cornell University Press, 1979.

5. Barabási AL, Stanley HE. Fractal Concepts in Surface Growth, Cambridge: Cambridge University Press, 1995.
6. Mandelbrot BB. The Fractal Geometry of Nature. San Francisco: WH Freeman, 1982.
7. Feder J. Fractals. New York: Plenum, 1988.
8. Bunde A, Havlin S, eds. Fractals and Disordered Systems. Berlin: Springer-Verlag, 1991.
9. Bunde A, Havlin S, eds. Fractals in Science. Berlin: Springer-Verlag, 1994.
10. Garcia-Ruiz JM, Louis E, Meakin P et al, eds. Growth Patterns in Physical Sciences and Biology. New York: Plenum, 1993.
11. Grosberg AY, Khokhlov AR. Statistical Physics of Macromolecules, New York: AIP Press, 1994; Grosberg AY, Khokhlov AR. Giant Molecules. London: Academic Press, 1997.
12. Bassingthwaighte JB, Liebovitch LS, West BJ. Fractal Physiology. New York: Oxford University Press, 1994.
13. Vicsek T. Fractal Growth Phenomena. Singapore: World Scientific, 1992.
14. Vicsek T, Shlesinger M, Matsushita M, eds. Fractals in Natural Sciences. Singapore: World Scientific, 1994.
15. Guyon E, Stanley HE. Fractal Formes. Amsterdam: Elsevier, 1991.
16. Li W. The study of correlation structures of DNA sequences: a critical review. Computers Chem 1997; 21:257-271.
17. Baxter RJ. Exactly Solvable Models in Statistical Mechanics. London: Academic Press, 1982.
18. Azbel MY. Random two-component, one-dimensional Ising model for heteropolymer melting. Phys Rev Lett 1973; 31:589-593.
19. Azbel MY, Kantor Y, Verkh L et al. Statistical Analysis of DNA Sequences. Biopolymers 1982: 21:1687-1690.
20. Azbel MY. Universality in a DNA statistical structure. Phys Rev Lett 1995; 75:168-171.
21. Feller W. An introduction to ptobability theory and its applications. Vols. 1-2. New York: Jhon Wiley & Sons, 1970.
22. Binder K, ed. Monte Carlo Methods in Statistical Physics. Berlin: Springer-Verlag, 1979.
23. Karlin S, Brendel V. Patchiness and correlations in DNA sequences. Science 1993; 259:677-680.
24. Grosberg AY, Rabin Y, Havlin S et al. Crumpled globule model of the 3-dimensional structure of DNA. Europhys Lett 1993; 23:373-378.
25. des Cloizeaux, J. Short range correlation between elements of a long polymer in a good solvent. J Physique 1980; 41:223-238.
26. Bak P. How Nature Works. New York: Springer 1996.
27. Bak P, Tang C, Wiesenfeld K. Self-organised criticality: an explanation of 1/f noise. Phys Rev Lett 1987; 59:381-384.
28. Bak P, Sneppen, K. Punctuated equilibrium and criticality in a simple model of evolution. Phys Rev Lett 1993; 71:4083-4086.
29. Paczuski M, Maslov S, Bak, P. Avalanche dynamics in evolution, growth and depinning models. Phys Rev E 1996; 53:414-443.
30. Jovanovic B, Buldyrev SV, Havlin S et al. Punctuated equilibrium and history-dependent percolation. Phys Rev E 1994; 50, R2403-2406.
31. Peng C-K, Buldyrev SV, Goldberger AL et al. Nature 1992; 356:168.
32. Li W, Kaneko K. Long-range correlations and partial 1/f a spectrum in a noncoding DNA sequence. Europhys Lett 1992; 17:655.
33. Nee S. Uncorrelated DNA walks. Nature 1992; 357:450-450.
34. Voss R. Evolution of long-range fractal correlations and 1/f noise in DNA base sequences. Phys Rev Lett 1992; 68:3805-3808.
35. Voss R. Long-Range Fractal Correlations in DNA Introns and Exons. Fractals 1994; 2:1-6.
36. Maddox J. Long-range correlations within DNA. Nature 1992; 358:103-103.
37. Munson PJ, Taylor RC, Michaels GS. DNA correlations. Nature 1992; 360:636-636.
38. Amato I. Mathematical biology-DNA shows unexplained patterns writ large. Science 1992; 257:747-747.
39. Prabhu VV, Claverie J-M.Correlations in intronless DNA. Nature 1992; 359:782-782.
40. Chatzidimitriou-Dreismann CA, Larhammar D. Long-range correlations in DNA. Nature 1993; 361:212-213.

41. Li W, Kaneko K. DNA correlations, Nature 1992; 360:635-636.
42. Karlin S, Cardon LR. Computational DNA sequence analysis. Annu Rev Microbiol 1994; 48:619-54.
43. Herzel H, Grosse I. Correlations in DNA sequences: The role of protein coding segments. Phys Rev E 1997; 55:800-810.
44. Grosse I, Herzel H, Buldyrev SV et al. Species independence of mutual information in coding and noncoding DNA. Phys Rev E 2000; 61:5624-5629.
45. Holste D, Grosse I, Herzel H et al. Optimization of coding potentials using positional dependence of nucleotide frequencies. J Theor Biol 206:525-537.
46. Berthelsen CL, Glazier JA, Skolnick MH. Global fractal dimension of human DNA sequences treated as pseudorandom walks. Phys Rev A 1992; 45:8902-8913.
47. Borovik AS, Grosberg AY, Frank-Kamenetski MD. Fractality of DNA texts. J Biomolec Struct Dyn 1994; 12:655-669.
48. Li WT. Are isochore sequences homogeneous? Gene 2002; 300:129-139.
49. Bernaola-Galvan P, Carpena P, Roman-Roldan R et al. Study of statistical correlations in DNA sequences. Gene 2002; 300:105-115.
50. Oliver JL, Carpena P, Roman-Roldan R et al. Isochore chromosome maps of the human genome. Gene 2002; 300:117-127.
51. Alberts B, Bray D, Lewis J et al. Molecular Biology of the Cell. New York: Garland Publishing, 1994.
52. Watson JD, Gilman M, Witkowski J et al. Recombinant DNA. New York: Scientific American Books, 1992.
53. Chen CF, Gentles AJ, Jurka J et al. Genes, pseudogenes, and Alu sequence organization across human chromosomes 21 and 22. Proc Natl Acad Sci USA 2002; 99:2930-2935.
54. Altschul SF, Madden TL, Schaffer AA et al. Gapped BLAST and PSI-BLAST: a new generation of protein database search programs, Nucl Acids Res 1997; 25:3389-3402.
55. Audit B, Vaillant C, Arneodo A et al. Wavelet analysis of DNA bending profiles reveals structural constraints on the evolution of genomic sequences. J Biol Phys 2004; 30:33-81.
56. Li WH. Expansion-modification systems: A model for spatial 1/f spectra. Phys Rev A 1991; 43:5240-5260.
57. Buldyrev SV, Goldberger AL, Havlin S et al. Generalized Levy Walk Model for DNA Nucleotide Sequences. Phys Rev E 1993; 47:4514-4523.
58. Buldyrev SV, Goldberger AL, Havlin S et al. Fractal Landscapes and Molecular Evolution: Modeling the Myosin Heavy Chain Gene Family. Biophys J 1993; 65:2673-2681.
59. Vieira MD, Herrmann HJ. A growth model for DNA evolution. Europhys Lett 1996; 33:409-414.
60. Hansen JP, McDonald IR. Theory of Simple Liquids. London: Academic Press, 1976.
61. Abramowitz M, Stegun IA, eds. Handbook of Mathematical Functions. New York: Dover, 1965
62. Press WH, Flannery BP, Teukolsky SA et al. Numerical Recipes. Cambridge: Cambridge Univ Press, 1989.
63. Burrus CS, Parks TW. DFT/FFT and Convolution Algorithms. New York: John Wiley and Sons, Inc. 1985.
64. Peng CK, Buldyrev SV, Havlin S et al. Mosaic Organization of DNA Sequences. Phys Rev E 1994; 49:1685-1689.
65. Chen Z, Ivanov PC, Hu K et al. Effect of nonstationarities on detrended fluctuation analysis. Phys Rev E 2002; 65:041107.
66. Jurka J, Walichiewicz T, Milosavljevic A. Prototypic sequences for human repetitive DNA. J Mol Evol 1992; 35:286-291.
67. Hattori M, Hidaka S, Sakaki Y. Sequence analysis of a KpnI family member near the 3' end of human beta-globin gene. Nucleic Acids Res 1985; 13:7813-7827.
68. Hwu RH, Roberts JW, Davidson EH et al. Insertion and/or deletion of many repeated DNA sequences in human and higher apes evolution. Proc Natl Acad Sci USA 1986; 83:3875-3879.
69. Churchill GA. Hidden Markov chains and the analysis of genome structure. Computers Chem 1992; 16:107-116.
70. Zolotarev VM, Uchaikin VM. Chance and Stability: Stable Distributions and their Applications. Utrecht: VSP BV, 1999.

71. Shlesinger MF, Zaslavsky GM, Frisch U, eds. Lévy Flights and Related Topics in Physics. Berlin: Springer-Verlag, 1995.
72. Arneodo A, D'Aubenton-Carafa Y, Audit B et al. What can we learn with wavelets about DNA sequences? Physica A 1998; 249:439-448.
73. Voss RF, Clarke J. 1/f noise in music: music from 1/f noise. J Acoust Soc Amer 1978; 63:258-263.
74. Schenkel A, Zhang J, Zhang, YC. Long Range Correlation in Human Writings. Fractals 1993; 1:47-57.
75. Amit M, Shmerler Y, Eisenberg E et al. Language and codification dependence of long-range correlations in texts. Fractals 1994; 2:7-13.
76. Trifonov EN. 3-, 10.5-, 200- and 400-base periodicities in genome sequences. Physica A 1998; 249:511-516.
77. Gu X. Li WH. A model for the correlation of mutation-rate with gc content and the origin of gc-rich isochores. J Mol Evol 1994; 38:468-475.
78. Viswanathan GM, Buldyrev SV, Havlin S et al. Quantification of DNA patchiness using correlation measures. Biophys J 1997; 72:866-875.
79. Viswanathan GM, Buldyrev SV, Havlin S et al. Long-range correlation measures for quantifying patchiness: Deviations from uniform power-law scaling in genomic DNA. Physica A 1998; 249:581-586.
80. Buldyrev SV, Goldberger AL, Havlin S et al. Long-range correlation properties of coding and noncoding DNA sequences: GenBank analysis. Phys Rev E 1995; 51:5084-5091.
81. Nyeo SL, Yang IC, and Wu CH. Spectral classification of archaeal and bacterial genomes. J Biol Syst 2002; 10:233-241.
82. Arneodo A, Bacry E, Graves PV et al. Characterizing long-range correlations in dna-sequences from wavelet analysis. Phys Rev Lett 1995; 74:3293-3296.
83. Nikolaou C, Almirantis Y. A study of the middle-scale nucleotide clustering in DNA sequences of various origin and functionality, by means of a method based on a modified standard deviation. J Theor Biol 2002; 217:479-492.
84. Ossadnik SM, Buldyrev SV, Goldberger AL et al. Correlation approach to identify coding regions in DNA sequences. Biophys J 1994; 67:64-70.
85. Uberbacher EC, Mural RJLocating protein-coding regions in human dna-sequences by a multiple sensor neural network approach. Proc Natl Acad Sci USA 1991; 88:11261-11265.
86. Fickett JW, Tung CS. Assessment of protein coding measures. Nucleic Acids Research 1992; 20:6441-6450.
87. Holste D, Grosse I, Beirer S et al. Repeats and correlations in human DNA sequences. Phys Rev E 2003; 67:061913.
88. Bell GI. Roles of repetitive sequences. Comput Chem, 1992; 16:135-143
89. Bell GI. Repetitive DNA sequences: some considerations for simple sequence repeats. Comput Chem 1993; 17:185-190.
90. Bell GI. Evolution of simple sequence repeats. Comput Chem 1996; 20:41-48.
91. Bell GI. and Jurka J. The length distribution of perfect dimer repetitive DNA is consistent with its evolution by an unbiased single step mutation process. J Mol Evol 1997; 44:414-421.
92. Richards RI, Sutherland GR. Simple repeat DNA is not replicated simply. Nature Genetic 1994; 6:114-116.
93. Richards RI, Sutherland GR. Simple tandem DNA repeats and human genetic disease. Proc Natl Acad Sci USA 1995; 92:3636-3641.
94. Chen X, Mariappan SV, Catasti P et al. Hairpins are formed by the single DNA strands of the fragile X triplet repeats: structure and biological implications. Proc Natl Acad Sci USA 1995; 92:5199-5203.
95. Gacy AM, Goellner G, Juramic N et al. Trinucleotide repeats that expand in human disease form hairpin structures in vitro. Cell 1995; 81:533-540.
96. Orth K, Hung J, Gazdar A et al. Genetic instability in human ovarian cancer cell lines. Proc Natl Acad Sci USA 1994; 91:9495-9499.
97. Bowcock AM, Ruiz-Linares A, Tomfohrde J et al. High resolution of human evolutionary trees with polymorphic microsatellites. Nature 1994; 368:455-457.

98. Olaisen B, Bekkemoen M, Hoff-Olsen P et al. VNTR mutation and sex. In: Pena SDJ, Chakraborty, R, Epplen JT et al, eds. DNA Fingerprinting: State of the Science. Basel: Springer-Verlag, 1993.
99. Jurka J. Pethiyagoda G. Simple repetitive DNA sequences from primates: compilation and analysis. J Mol Evol 1995; 40:120-126.
100. Li YC, Korol AB, Fahima T et al. Microsatellites: genomic distribution, putative functions and mutational mechanisms: a review. Mol Ecol 2002; 11:2453-2465.
101. Kremer E, Pritchard M, Lynch M et al. Mapping of DNA instability at the fragile X to a trinucleotide repeat sequence p(CCG)n. Science 1991; 252:1711-1714.
102. Huntington's Disease Collaborative Research Group. A novel gene containing a trinucleotide repeat that is expanded and unstable on Huntington's disease chromosomes. Cell 1993; 72:971-983.
103. Ionov Y, Peinado MA, Malkhosyan S et al. Ubiquitous somatic mutations in simple repeated sequences reveal a new mechanism for clonic carcinogenesis. Nature 1993; 363:558-561.
104. Kunkel TA. Slippery DNA and diseases. Nature 1993; 365:207-208.
105. Wooster R, Cleton-Jansen AM, Collins N et al. Instability of short tandem repeats (microsatellites) in human cancers. Nat Genet 1994; 6:152-156.
106. Dokholyan NV, Buldyrev SV, Havlin S et al. Distribution of base pair repeats in coding and noncoding DNA sequences. Phys Rev Lett 1997; 79:5182-5185.
107. Dokholyan NV, Buldyrev SV, Havlin S et al. Distributions of dimeric tandem repeats in noncoding and coding DNA sequences. J Theor Biol 2000; 202:273-282.
108. Dokholyan NV, Buldyrev SV, Havlin S et al. Model of unequal chromosomal crossing over in DNA sequences. Physica A 1998; 249:594-599.
109. Charlesworth B, Sniegowski P, Stephan W. The evolutionary dynamics of repetative DNA in eukaryotes. Nature 1994; 371:215-220.
110. Wells RD. Molecular basis of genetic instability of triplet repeats. J Biol Chem 1996; 271:2875-2878.
111. Levinson G, Gutman GA. Slipped-strand mispairing: a major mechanism for DNA sequence evolution. Mol Biol Evol 1987; 4:203-221.
112. Buldyrev SV, Dokholyan NV, Havlin S et al. Expansion of tandem repeats and oligomer clustering in coding and noncoding DNA sequences. Physica A 1999; 273:19-32.
113. Kruglyak S, Durrett RT, Schug MD et al. Equilibrium distributions of microsatellite repeat length resulting from a balance between slippage events and point mutations. Proc Natl Acad Sci USA 1998; 95:10774-10778.
114. Zipf KG. Human Behavior and the Principle of Least Effort. Redwood City: Addison-Wesley 1949.
115. Mantegna RN, Buldyrev SV, Goldberger AL et al. Linguistic features of noncoding DNA sequences. Phys Rev Lett 1994; 73:3169-3172.
116. Mantegna RN, Buldyrev SV, Goldberger AL et al. Phys Rev E 1995; 2939.
117. Bonhoeffer S, Herz AVM, Boerlijst MC et al. Explaining "linguistic features" of noncoding DNA. Science 1996; 271:14-15.
118. Makeev VJ, Lifanov AP, Nazina AG et al. Distance preferences in the arrangement of binding motifs and hierarchical levels in organization of transcription regulatory information. Nucl Acids Res 2003; 31:6016-6026.
119. Jurka J, Kohany O, Pavlicek A et al. Duplication, coclustering, and selection of human Alu retrotransposons. Proc Natl Acad Sci USA 2004; 101:1268-1272.
120. Stanley HE, Afanasyev V, Amaral L AN et al. Anomalous fluctuations in the dynamics of complex systems: From DNA and physiology to econophysics. Physica A 1996; 224 302-321.
121. Pande V, Gosberg A Ya, Tanaka T. Nonrandomness in protein sequences - evidence for a physically driven stage of evolution, Proc Natl Acad Sci USA 1994; 91:12972-12975.
122. Shakhnovich EI, Gutin AM. Implications of thermodynamics of protein folding for evolution of primary sequences. Nature 1990; 346:773-775.
123. Li W-H, Marr TG, Kaneko K. Understanding long-range correlations in DNA sequences. Physica D 1994; 7:392-416.

CHAPTER 10

Analytical Evolutionary Model for Protein Fold Occurrence in Genomes, Accounting for the Effects of Gene Duplication, Deletion, Acquisition and Selective Pressure

Michael Kamal, Nicholas M. Luscombe, Jiang Qian and Mark Gerstein*

Abstract

Motivation

Global surveys of protein folds in genomes measure the usage of essential molecular parts in different organisms. In a recent survey, we showed that the occurrence of protein folds in 20 completely sequenced genomes follow a power-law distribution; i.e., the number of folds (F) with a given genomic occurrence (V) decays as $F(V) = aV^{-b}$, with a few occurring many times and most occurring infrequently. Clearly, such a distribution results from the way in which genomes have evolved into their current states.

Results

Here we develop and discuss a minimal, analytically tractable model to explain these observations. In particular, we demonstrate that (i) stochastic gene duplication and (ii) overall acquisition of new folds are sufficient to accurately replicate the power-law distributions. Furthermore by optimizing the model using genomic data, we gain a quantitative insight into otherwise unattainable data. In particular, as the rate at which genomes acquire new folds is directly related to the power-law exponent-b, we can easily estimate this rate by measuring the gradient of the distribution on a log-log graph. In addition, extensions to the model suggest that gene deletion and selective pressure are important to the fate of individual genes, but do not significantly affect the final power-law distribution. That is, although gene deletion and selective pressure will affect the choice of the most common fold type in an organism, it will not change the overall power-law distribution found across different genomes. Finally, we gain an indication of the initial sizes of genomes, from the starting states of the simulations. We find that the power-law dependence of the fold distribution is independent of the composition of the starting genome.

*Corresponding author: Mark Gerstein—Department of Molecular Biophysics and Biochemistry, and Department of Computer Science, Yale University, 266 Whitney Avenue, P.O. Box 208114, New Haven, Connecticut 06520-8114, U.S.A. Email: mark.gerstein@yale.edu

Power Laws, Scale-Free Networks and Genome Biology, edited by Eugene V. Koonin, Yuri I. Wolf and Georgy P. Karev.

Availability

Additional data pertaining to this work is found at http://www.partslist.org/powerlaw.

Introduction

The power-law behavior is frequently found in many different population distributions. Also referred to as Zipf's law, a well-documented example is the usage of words in text documents.[1] By grouping words that have similar occurrences, it was noted that a small selection such as "the" and "of" are used many times, while most occur infrequently. When the size of each group is plot against its usage, the distribution is described by a power-law function: the number of words (F) with a given occurrence (V) decays with the equation $F = a/V^{-b}$. The distribution is linear when plotted on log-log axes, where-b describes the slope. Such distributions are also found for the relative sizes of cities, income levels and the number of papers published by scientists in a field of research.[1]

Significantly, the power-law behavior is also prevalent in many aspects of genomic biology.[2] It is found in the usage of short nucleotide sequences,[3-7] the populations of gene families,[8,9] the occurrence of protein superfamilies and folds in genomes[10,11] and several biological networks.[12-14] The distribution extends even further to the number of distinct protein functions associated with a particular fold, the number of protein-protein interactions that are made by each fold type, and the variations in expression levels between genes present in the yeast genome. These observations have been made in at least 20 prokaryotic and eukaryotic genomes, and so are likely to be universal to most other genomes that are yet to be analyzed. Given the prevalence of this behavior, we suggest that all of these biological distributions arise because of a common mechanism for genomic evolution, primarily by duplicating existing genes to increase the presence of particular types of proteins.[11]

The current study focuses on the distribution of protein folds in different organisms (Fig. 1A). Most proteins encoded in a genome have a defined three-dimensional structure that can be classified into distinct protein folds. Although these folds are defined by the topology of the peptide chain, it is possible to determine whether two proteins adopt the same fold by sequence comparison. So even if structures are unavailable for all the genes, we can classify them into equivalent folds by sequence similarity. Using these classifications, one way of representing the contents of a genome is to count the number of times different folds occur and then group together those with similar occurrences (Fig. 1B). Like word usage, the number of folds (F) with a certain genomic occurrence (V) decays according to the power-law function; we display the distribution for the *E. coli* genome in Figure 1A, and plots for 19 further organisms is available from our supplementary website.

There have been several efforts to understand this nonuniform distribution of protein families. A number of models suggested that the observation of non-uniform population distributions of protein families depends on the "designability" of the protein structure; that is, the relative size of a family depends on the fraction of all sequences that could successfully fold into any particular protein fold.[15,16] Others have modelled the occurrence of non-uniform distributions by simulating the evolution of genomes. In the model of Huynen and van Nimwegen,[9] families expand or shrink in size through successive multiplications by a random factor, which represents duplication or deletion events depending on its value. More recently, Yanai et al[17] introduced a model in which a genome evolves from a set of precursor genes to a mature size by iterative gene duplications and gradual accumulation of modifications through point mutations. When an individual family member acquires enough random mutations, it breaks away to form a new family.

We recently presented an equally minimal, but more biologically realistic model.[11] Here genomes evolve through stochastic gene duplications and steady acquisition of new protein

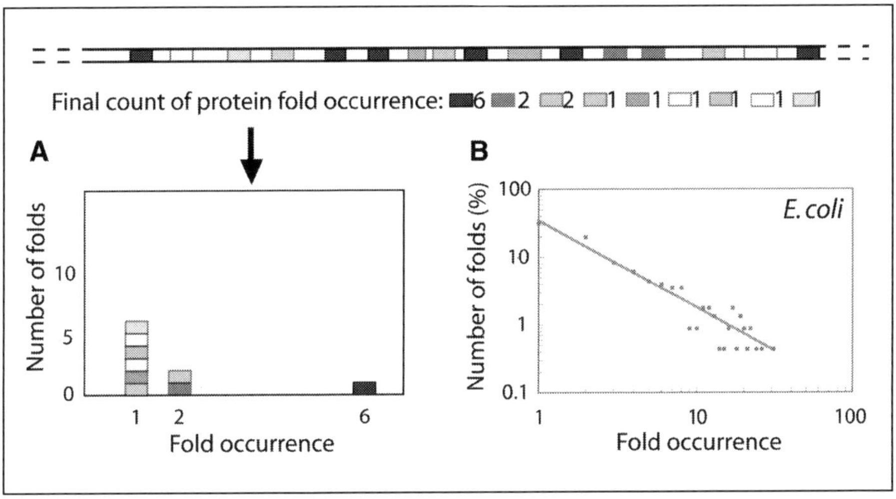

Figure 1. The occurrence of protein folds in genomes. A) The structural contents of a genome can be represented by counting the number of times different protein folds occur and grouping together those with similar occurrences. B) The relationship is described by a power-law function.

folds, either by ab initio creation or horizontal gene transfer.[18-22] Simulations replicated the genomic distributions very accurately, and provided insight into the rate at which different organisms acquired new folds and the origins of a common ancestral genome. Although our work focused on the distribution of protein fold populations, the model applied equally well for other gene classifications such as sequence families, and SCOP superfamilies.[23] The behavior also applies for alternative protein classification systems such as Interpro families and protein superfamilies.

The purpose of the current work is two-fold. First, we propose new models based on our previous model by fully incorporating two additional processes in evolution: gene deletion and selective pressure. These major biological processes were beyond the scope of our previous work, and it is important to test their effects on the outcome of the model. Second, we provide full analytical and numerical analyses; in doing so we explore the mathematical and biological significance of the model, and explore the relative effects that the different evolutionary processes (gene duplication, acquisition, deletion and selective pressure) have on the final appearance of different genomes. In the previous paper, our results are only based on simulations. In contrast, the analytical approach is also employed in this work.

Minimal Model: Gene Duplication and New Fold Acquisition

Suppose that the initial genome consists of N_0 distinct folds at time $t = 0$, i.e., the number of genes equal the number of folds. The growth of the genome in our model occurs by randomly duplicating existing genes, and by incorporating new folds into the genome at a constant rate. Both of these processes are assumed to operate independently and continually over time. We assume that at every instant, all genes are equally likely to be chosen for duplication and that on average, one duplication event happens per unit time. As a result, large folds, i.e., ones that are coded by many genes, are more likely to grow over time than smaller folds. We assume that R new folds of size 1 are always incorporated in to the genome per unit time, i.e., the acquisition of new folds is not stochastic.

Let $F(m, t)$ be the expected number of folds of a given size m at time t. The fold histogram determines both the expected total number of distinct folds $F(t)$ and the expected total number of genes $G(t)$:

$$F(t) = \sum_{m=1}^{\infty} F(m,t)$$
$$G(t) = \sum_{m=1}^{\infty} mF(m,t) \qquad (1)$$

Under these growth assumptions, the Markovian dynamics governing $F(m, t)$ are given by:

$$\frac{\partial F(m,t)}{\partial t} = \frac{(m-1)F(m-1,t)}{G(t)} - \frac{mF(m,t)}{G(t)} \quad (m>1)$$
$$\frac{\partial F(1,t)}{\partial t} = R - \frac{F(1,t)}{G(t)} \qquad (2)$$

Although duplication occurs at the gene level, it is more convenient mathematically to work directly with the fold histogram $F(m, t)$.

The intuition behind these equations is as follows. If the gene selected for duplication that originally is a member of a fold of size $m-1$, then after duplication that fold will now be a fold of size m, and the population of $F(m-1,t)$ and $F(m, t)$, will decrease and increase, respectively, by one. The probability for this particular gene selection is $(m-1)F(m-1,t)/G(t)$.

These equations ensure the appropriate expected growth rates for the total number of folds. A direct summation of (2) leads to:

$$\frac{\partial F(t)}{\partial t} = \frac{\partial}{\partial t}\sum_{m=1} F(m,t)$$
$$= R \qquad (3)$$

and hence: $F(t)= N_0 + Rt$. Similar manipulations show that the expect number of genes also grows as required: $G(t)= N_0 +(R+1)t$. It is important to note that evolution equations enforce the correct overall normalization for the histogram; there is no need to impose normalization conditions separately.

The complete analytical solution for the coupled equations (2) can be found by standard methods. Full details are included in Appendix A.

The biological interpretation of the analytical solution is best appreciated by examining two important limiting cases. If there is no acquisition of new genes ($R = 0$), the solution simplifies considerably:

$$F(m,t) = N_0\phi^{-1}\left(1-\phi^{-1}\right)^{m-1} \qquad (4)$$

where $\phi(t)$ relates the passage of time to the expected number of genes:

$$\phi(t) = \frac{G(t)}{N_0} = 1 + \frac{(R+1)t}{N_0}, \qquad (5)$$

Therefore, gene duplication alone leads to an exponential distribution of fold occurrence: $\log F(m, t) = m \log(1-\phi^{-1}) + \psi(t)$, with $\psi(t)$ independent of m.

The other revealing limit concerns the behavior for large times ($t \to \infty$)when new genes are acquired at a nonzero rate ($R \neq 0$). The asymptotic limit is given by:

$$F(m,t) \to A_m\phi(t) = A_m\left(1+\frac{(R+1)t}{N_0}\right) \quad \text{as } t \to \infty \qquad (6)$$

with coefficients A_m that depend only on R and N_0, and not on time:

$$A_m = \frac{RN_0}{R+2} \prod_{i=1}^{m-1} \frac{i}{R+2+i} \tag{7}$$

Consequently, the probability distribution of fold sizes, i.e., the normalized histogram, is determined by solely by the A_m:

$$p(m,t) = \frac{F(m,t)}{\sum_i F(i,t)} \to \frac{A_m}{\sum_i A_i} = \frac{R+1}{R+2} \prod_{i=1}^{m-1} \frac{i}{R+2+i} \tag{8}$$

and furthermore this asymptotic probability distribution depends only on R—the dependence on initial cluster size N_0 is removed by the normalization.

An examination of the leading large m behavior of A_m reveals that

$$\log A_m \sim -(R+2) \log m \tag{9}$$

Therefore, for large m, the terminal probability distribution (8) resembles a power-law with exponent R + 2. For small m, the coefficients decrease less rapidly with m and do not resemble power-law dependence. This observation is relevant for estimating R from empirical data or even numerical results.

It is also worth pointing that a power-law distribution that decays too slowly will lead to an infinite expected number of genes. A power-law distribution will that holds asymptotically for large m: $N(m) \sim 1/m^{\alpha}$ has to be described by an exponent $\alpha > 2$ for the sum $G(t) = \sum_{m=1}^{\infty} mN(m)$ to converge. The asymptotic limit of the exact solution, a power-law with exponent $R + 2$, satisfies this condition.

For nonzero R and times other than zero and infinity, the fold distribution will not be strictly exponential, nor will it conform to the limiting distribution (8). For small times, the analytic solution confirms what would be expected intuitively: the histogram behavior is dominated by duplication events involving the initial N_0 genes. To characterize the "crossover" behavior of the solution from the exponential to approximate power-law regime we have calculated the similarity of the exact probability distribution at different times to both the best fitting exponential distribution and to the limiting asymptotic distribution (8). The difference between any two probability distributions is measured by the sum of squared differences (the standard L^2 metric).

We have characterized the crossover time T_c for a range of values for both R and N_0 and find that the crossover time displays two distinct regimes. Within each regime it is approximately inversely proportional to R and directly proportional N_0: $T_c \sim N_0/R$, with a different proportionality constant for each regime. Details of this analysis can be found in Appendix B. The numerical results indicate that crossover occurs roughly when the number of new fold introductions: RT_c, becomes comparable to the initial genome size N_0, as might be expected intuitively.

So far, we have assumed that the starting genome contains just one copy of each fold. In fact, it is reasonable to expect the initial genome to have several copies of particular fold types (for example those involved in protein synthesis) when the evolutionary process described by the model was initiated. By definition, genomes in our model have a comparatively small starting state, and so the difference between the most and least common folds would be minimal, i.e., some occurring three or four times at most. However, it is nonetheless of interest to investigate the effect that the appearance of the initial genome would have on the final distribution.

The solution we have derived for a particular initial genomic configuration—N_0 distinct folds consisting of one gene—can be extended to describe the evolution of an arbitrary initial fold distribution $N_{init}(m)$ that is made up of N_0 genes: $\sum_m mN_{init}(m) = N_0$. The solution is similar to the special initial condition of N_0 distinct folds and is presented in detail in Appendix C.

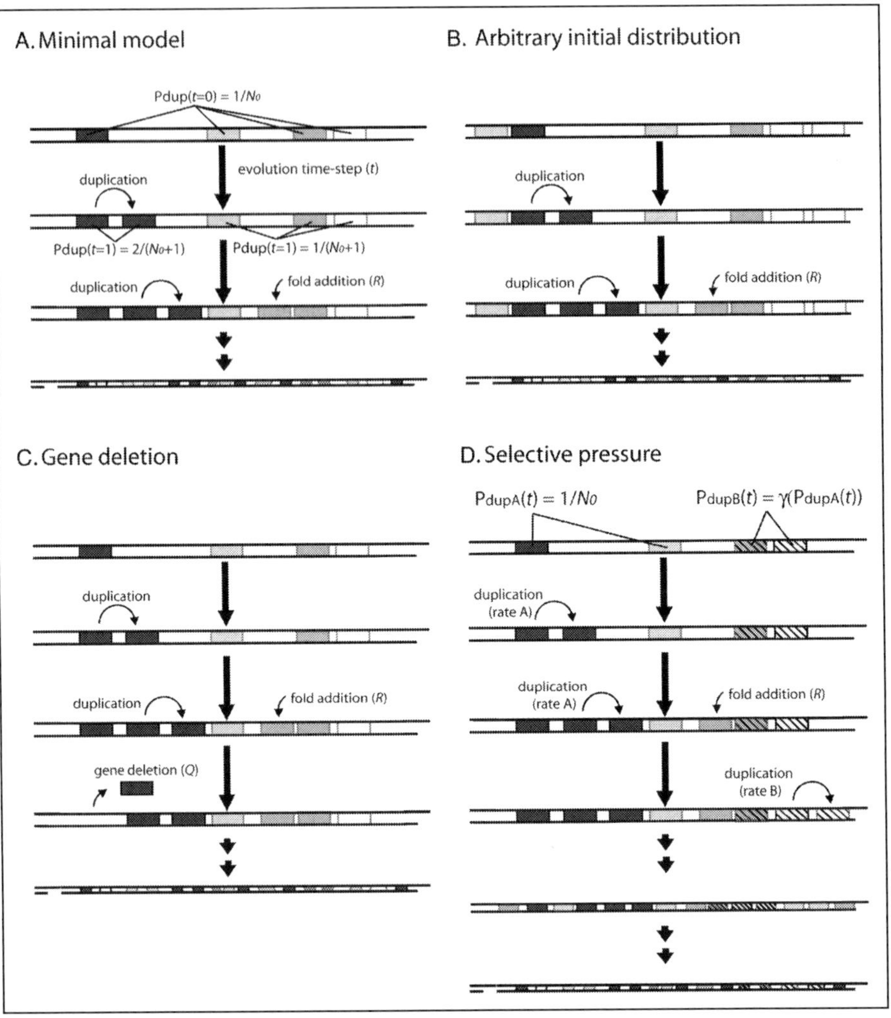

Figure 2. Three models: A) minimal model with uniform initial distribution, B) minimal model with an arbitrary initial distribution, C) gene deletion, and D) selective pressure.

One important conclusion may be drawn from the generalized model: all initial distributions ultimately lead to the same limiting distribution determined by the A_m. Just as before, the dependence on the initial fold distribution $N_{init}(m)$ decays with time, leading to the same asymptotic distribution as was found for an initial distribution of N_0 folds of size 1 in (9), reflecting the dominance of fold introduction over gene duplication for large times. Of course, the details of how and when the crossover happens will depend on the particular form of $N_{init}(m)$.

Extended Model: Including the Effects of Random Gene Deletion

Gene deletion is a major factor in evolution and is discussed briefly by Qian et al.[11] In this section we incorporate an additional parameter, Q, that represents gene deletion.

The most natural extension of (2) that accounts for random gene deletion at rate Q would be the following:

$$\begin{aligned}\frac{\partial F(m,t)}{\partial t} &= \frac{(m-1)F(m-1,t)}{G(t)} - \frac{mF(m,t)}{G(t)} + Q\frac{(m+1)F(m+1,t)}{G(t)} - Q\frac{mF(m,t)}{G(t)} \quad (m>1)\\ \frac{\partial F(1,t)}{\partial t} &= R-(1+Q)\frac{F(1,t)}{G(t)} + Q\frac{F(2,t)}{G(t)}\end{aligned} \tag{10}$$

The terms proportional to Q encode the dynamics for gene deletion, which are very similar to gene duplication: on average, Q deletions occur for every duplication event and the gene to be deleted is chosen randomly from all the genes in the genome. In this way the population of a given bin m can either decrease due to gene deletion if the gene to be deleted is from bin m itself, or it can increase as a result of a deletion in the neighboring bin $m+1$.

In this extended model, gene growth occurs at the uniform rate one would expect: $G(t) = N_0 + (1 + R - Q)t$. In contrast, the behavior of the expected number of folds is more complicated:

$$\begin{aligned}\frac{\partial F(t)}{\partial t} &= \frac{\partial}{\partial t}\sum_{m=1} F(m,t)\\ &= R - Q\frac{F(1,t)}{G(t)}\end{aligned} \tag{11}$$

Folds of size 1 that are deleted disappear from the genome so $F(t)$ depends explicitly on the population of $F(1,t)$; unlike the $Q = 0$ case, the dynamics of $F(t)$can not be determined without knowing the full solution to (10).

The extended model is much more complicated mathematically, primarily because the difference equations are now second order. In the minimal model, the behavior of larger folds depends only on the behavior of smaller folds, so the full solution can be constructed inductively starting from the solution for $m = 1$. With gene deletion operating as well, the dynamics of different fold sizes are coupled together. In many respects, these dynamics are like those describing diffusion phenomena; when $Q = 0$ the genome exhibits growth due to drift, or directed movement alone, while nonzero Q introduces diffusive, or non-directional movement as well.

Analytic Results

We were able to derive a full analytical solution only in the absence of any new fold introduction: $R = 0$. In this case, only stochastic gene deletion and duplication operate. We will restrict our discussion to when gene duplication occurs at a higher rate than gene deletion, which requires $0 < Q < 1$, so the genome will still grow in size, at least in terms of number of genes: $G(t) = N_0 + (1 - Q)t$. Note that since $R = 0$, equation (11) shows that the number of folds will actually decrease with time. Losing folds while gaining genes is possible if the larger folds make up for the loss of genes from smaller folds.

An analytic solution exists for an initial distribution of N_0 different folds of size 1 and is worked out in detail in Appendix D. Once again, the distribution is exponential. Figure 3A shows histograms $F(m,t)$ corresponding to three values of Q and a fixed time.

Remarkably, the normalized distribution of fold size (the probability distribution) is independent of the gene deletion rate Q:

$$\begin{aligned}p(m,t) &= \frac{F(m,t)}{\sum_{i=1}^{\infty} F(i,t)}\\ &= \frac{N_0}{N_0+t}\left[\frac{t}{N_0+t}\right]^{m-1}\end{aligned} \tag{12}$$

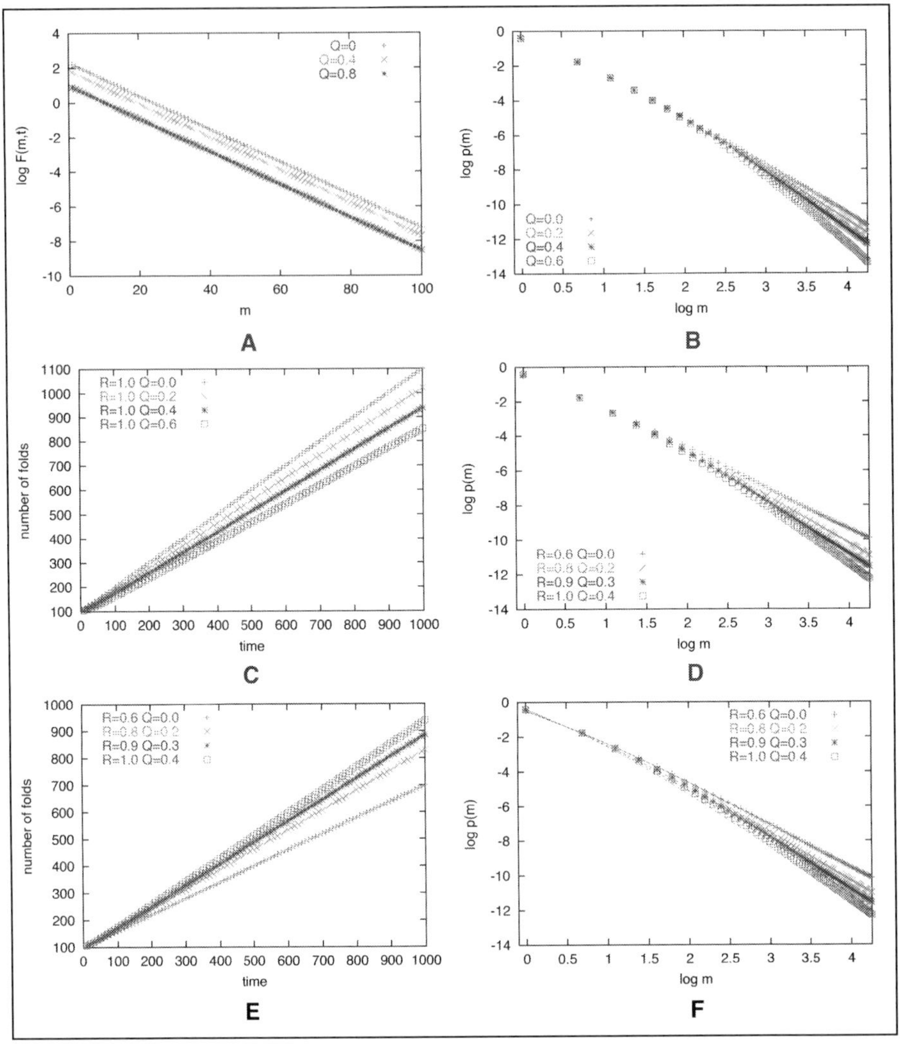

Figure 3. The effects of gene deletion: A) fold histogram $F(m, t)$ for $N_0 = 100$ and $t = 1000$ plotted for $Q = 0, 0.4, 0.8$ and $R = 0$; B) normalized large-time limiting fold distribution and C) the total number of folds as a function of time when $R = 1.0$ and $Q = 0, 0.2, 0.4, 0.6$; D) normalized large-time fold distribution and E) the total number of folds as a function of time for fixed overall gene growth: $1 + R - Q = 1.6$ and $Q = 0, 0.2, 0.3, 0.4$; F) analytic approximation, shown using solid lines, for parameters plotted in (A).

Hence gene deletion does not affect the shape of the distribution at all when $R = 0$, only the overall normalization is changed. This can also be seen directly from Figure 3B.

Although an exact analytical solution does not seem possible for arbitrary R and Q, it is nonetheless possible to derive analytic expressions for the higher moments of the fold distribution. Appendix E discusses how this is done and particular, includes an expression for the second moment that will prove useful when fitting the model to genomic data.

Numerical Results for Nonzero R and Q < 1

Numerical solution of (10) reveals for that large times, the normalized histograms of fold size approach a time-invariant limit that depends solely on R and Q. Figure 3B shows the probability distributions for a fixed rate of new fold acquisition, R =1.0, and increasing rates of gene deletion: Q = 0, 0.2, 0.4, 0.6. The power-law character of the distributions is retained even for large values of Q. Quite reasonably, higher rates of gene deletion encourages the dominance of smaller folds, leading to a more rapid decline of $p(m)$ with fold size m. Common folds require repeated gene duplication and an avoidance of gene deletion events to proliferate. As the probability of avoidance is proportional to $1 - Q$, the probability of multiple avoidance is suppressed as a power of $1 - Q$.

Figure 3D explores the effect of deletion when the overall gene growth rate is kept constant: $1 + R - Q = 1.6$. In this way, we can contrast the effects of deletion and fold acquisition in a controlled manner. Note that a commensurate increase in R does not overcome an increase in Q, as large folds are suppressed more than small folds. This means that the exponent that best describes the power-law decay is not merely a function of $R - Q$.

On the other hand, the effect of gene deletion is not dramatic; not only is similarity to a power-law retained the actual change in exponent is not large. Even for fold of large size, there isn't much difference between the curves even for a fairly large gene deletion rate. In practice, this makes it difficult to estimate Q statistically from the shape of fold histograms derived empirically from genomic data. While the effective gene introduction rate: $1 + R - Q$, should be easy to deduce from the data, an identification of Q itself from the rate of decay would require reliable occurrence data for very large folds.

When there is no gene deletion, the expected number of folds increases linearly with time at rate R. Equation (11) suggests gene deletion will lead to a less simple time dependence for $F(t)$. Perhaps surprisingly, $F(t)$ remains, to a good approximation, linear in time, with a slope that is no longer R, as can be seen in the numerical results of Figure 3C. Here $F(t)$ is plotted for fixed R = 1.0 and different values of the gene deletion rate: Q = 0.0, 0.2, 0.4, 0.6. In fact, the slope in each of these cases is less than R and decreases with Q, which is consistent with the analytic solution for $F(t)$ when $R = 0$, derived in Appendix D (see equation (42)).

If again we choose parameters that fix the growth rate for the expected number of genes (1 + $R - Q$), a commensurate increase in both R and Q leads to a greater increase in the expected number of folds, as can be seen in Figure 3E. This is entirely reasonable: in our model, the new folds that are continually acquired at rate R are all distinct, so a genome with large R and Q will end up with many small folds, each coded by only a few genes. In contrast, a genome with small R and Q will lead to fewer but larger folds.

Analytic Approximation Based on Perturbation Theory

The numerical results show that gene deletion, even for fairly large values of Q does not dramatically change the growth pattern of the genome, certainly qualitatively and to some extent, even quantitatively. Moreover, the analytic results when $R = 0$ showed that gene deletion is remarkably benign: in the absence of new gene acquisition, but with gene duplication operating, gene deletion does not change the probability distribution of fold occurrences, but does change expected total number of folds in the genome.

Here we consider an analytic approximation that attempts to capture the effects of gene deletion perturbatively by constructing an approximation around the $Q = 0$, $R > 0$ solution as an expansion in powers of Q. The perturbation expansion has to be handled carefully since a naive expansion, one that considers contributions only up to some finite power of Q, will not converge for all fold sizes m. The failure of conventional perturbation theory is explored in Appendix F.

To go beyond naive perturbation theory, we have adopted the following approach: (1) the dominant contribution at every order (or power) of Q is identified, (2) the dominant contribution is approximated, and (3) the resulting new infinite series in Q is summed exactly to arrive at an approximate solution that remains finite for all Q and m. The details are presented in Appendix F. Although not rigorous, this type of rescue or augmentation of perturbation theory is practiced routinely and often quite successfully on a variety physical models, such as models of phase transitions from statistical physics.[24]

This approach leads to the following approximation for the limiting fold distribution:

$$p_m = \frac{R+1+QR}{R+2+QR}\prod_{i=1}^{m-1}\frac{i}{R+2+QR+i} \tag{13}$$

Note that the approximation includes as a special case the exact distribution derived previously for $Q = 0$ (8). In fact, the approximate distribution for Q nonzero is obtained from the exact solution for $Q = 0$ by the substituting $R \rightarrow R + QR$. This correspondence also makes it clear that for large m, the $Q \neq 0$ probability distribution will resemble a power law with exponent $R + QR + 2$, just as $Q = 0$ distribution approached a power-law with exponent $R + 2$.

The true test of the effectiveness of the approximation rests with a comparison to the numerical results, which is done in Figure 3F. There seems to be good qualitative agreement, and fairly good quantitative agreement as well, even for $Q = 0.4$. As expected from the nature of the approximation, there is better agreement for large m in all cases. An approximation for the expected number of folds $F(t)$ within the same framework is given in Appendix F.

The Effects of Selection Pressure

Selective pressure plays an important role in evolution. It is well known that different genes have different duplication rates due to the selective pressure.[25] So far we have assumed that when genes are duplicated, or deleted, the target gene is chosen with equal probability from all the genes in the genome. A more realistic model would of course allow for favoritism in the selection process: presumably, genes that are useful or necessary are less likely to be deleted and perhaps more likely to be duplicated than genes that are less important. Note, however, that our model is not a differential survival model.

We explore the effects of selection pressure by extending the minimal model to allow for different duplication rates among genes. Suppose now that genes are not only identified with particular folds but also by their duplication types. For simplicity, assume that there are only two types: type "A" and type "B", and that "B" genes are γ times more likely to be chosen for duplication than "A" genes. There will still be one duplication event, on average, per unit time, so the total expected number of genes will remain the same, but the allocation of the total between types "B" and "A" will depend on γ. We will assume that $\gamma > 1$, so it is the "B" types that are more likely to be duplicated.

To keep track of the fold population we now need two histograms: $F_A(m, t)$ and $F_B(m, t)$ to distinguish between the duplication types. The full fold histogram is the sum of both sub-histograms: $F(m, t) = F_A(m, t) + F_B(m, t)$. Similarly, let $G_A(t)$ and $G_B(t)$ represent the total number of genes for each type and define a new variable $G_\gamma(t)$:

$$G_\gamma(t) = G_A(t) + \gamma G_B(t) \tag{14}$$

The evolution equations that extend (2) are:

$$\begin{aligned}
\frac{\partial F_A(m,t)}{\partial t} &= \frac{(m-1)F_A(m-1,t)}{G_\gamma(t)} - \frac{mF_A(m,t)}{G_\gamma(t)} \quad (m>1)\\
\frac{\partial F_A(1,t)}{\partial t} &= R_A - \frac{F_A(1,t)}{G_\gamma(t)}\\
\frac{\partial F_B(m,t)}{\partial t} &= \gamma\frac{(m-1)F_B(m-1,t)}{G_\gamma(t)} - \gamma\frac{mF_B(m,t)}{G_\gamma(t)} \quad (m>1)\\
\frac{\partial F_B(1,t)}{\partial t} &= R_B - \gamma\frac{F_B(1,t)}{G_\gamma(t)}
\end{aligned} \tag{15}$$

Note that we allow new folds to be acquired at different rates for each type: R_A can be different from R_B although we will restrict our numerical examples to the when they are equal.

The equations for the total number of genes of both types follow from the full dynamics (15) and are given in Appendix G. These confirm that the overall duplication rate is still one gene per unit time.

Once again, analytical solutions are possible for the two special parameter values addressed previously: (1) when there is no introduction of new folds, so $R_A = R_B = 0$; and (2) the limiting distribution when $t \to \infty$. When there is no introduction of new folds, a simple extension of the methodology employed in Appendix A establishes that the each of the sub-histograms $F_A(m, t)$ and $F_B(m, t)$ follows an exponential distribution for all times. The full histogram is consequently a sum of exponential distributions:

$$\begin{aligned}
p(m,t) &= \frac{F_A(m,t)+F_B(m,t)}{\sum_i F_A(i,t)+F_B(i,t)}\\
&= \frac{N_0^A}{N_0^A+N_0^B}e^{-u}\left[1-e^{-u}\right]^{m-1} + \frac{N_0^B}{N_0^A+N_0^B}e^{-\gamma u}\left[1-e^{-\gamma u}\right]^{m-1}
\end{aligned} \tag{16}$$

The number of distinct folds of each type, present at $t = 0$ is given by N_0^A and N_0^B. The variable $u(t)$ is a rescaled time variable related to $G_\gamma(t)$; the exact form of the dependence appears in Appendix G but is unimportant for the present discussion.

Of greater interest is the other special case: the ultimate evolutionary fate of the genome. The analytic behavior for large times is much easier to derive than an exact solution itself. For large t, $G_\gamma(t)$ will grow linearly with time: $G_\gamma \sim t$, according to a constant C_γ that depends on the rate of fold acquisition and the differential rate of duplication (see Appendix G for details).

In a similar fashion, we define coefficients C_m^A and C_m^B, akin to the coefficients A_m of the solution to the minimal model (7), that describe the ultimate linear growth of the histogram bins: $F_A(m, t) \sim C_m^A t$, and similarly for $F_B(m, t)$. The normalized probability distribution corresponding to this limit is given by:

$$\begin{aligned}
p(m,t) &= \frac{C_m^A + C_m^B}{\sum_i C_i^A + C_i^B}\\
&= \frac{C_\gamma}{C_\gamma+1}\frac{R_A}{R_A+R_B}\prod_{i=1}^{m-1}\frac{i}{C_\gamma+i+1} + \frac{C_\gamma}{C_\gamma+\gamma}\frac{R_B}{R_A+R_B}\prod_{i=1}^{m-1}\frac{i\gamma}{C_\gamma+\gamma(i+1)}
\end{aligned} \tag{17}$$

The important conclusion to be drawn from (17) is that powerlaw-like distributions describe the ultimate fate of the genome even when there are different rates of gene duplication. The probability distribution is the sum of two powerlaw-like distributions, each similar to the powerlaw-like distributions of the minimal model, but characterized by its own effective exponent. Figure 4 shows a comparison of the predicted distribution and numerical results when $R_A = R_B = 0.5$ for $\gamma = 1$, which is corresponds to the minimal model, and $\gamma = 10$, so type "B" genes

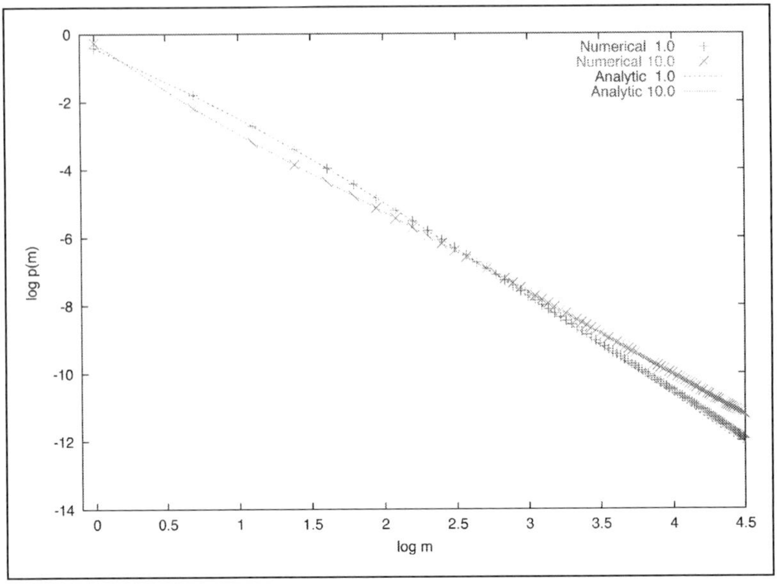

Figure 4. The effect of selective pressure on the model. Larger values of γ indicate larger differences in duplication rates between favored and unfavored protein folds. Large time limit for the fold probability distribution for $\gamma = 1$ and $\gamma = 10$. Numerical results are plotted as symbols; analytic results from Equation (16) as lines.

are ten times more likely to be selected for duplication. The two distributions are remarkably close to each other, even when there is an order of magnitude difference between the relative duplication rates of type "B" genes. We have found that the parameter has much less of an effect than differences between the gene introduction rates R_A and R_B.

We have also briefly considered the case of more than two duplication types. When there is no introduction of new folds into the genome equation (6) generalizes: the subhistogram for each duplication type is exponential. Furthermore, we have confirmed numerically that the terminal distribution is not dramatically affected by selection pressure, even when there are several families with significantly different rates of duplication. One particular example, involving four duplication types appears in Appendix G.

Fitting the Models to Genomic Data

Clearly, of greatest interest is to observe how our model compares with the genomic distributions. We start with the minimal model, for which we require estimates for the parameters t, N_0 and R for each organism.

Fitting the minimal model requires estimating three parameters: t, N_0 and R. We have determined these parameters separately for each organism by insisting that the minimal model match the number of folds: F, the number of genes: G, and the second moment of the actual fold histogram: H_2, to those predicted by the minimal model. The fitting procedure is greatly simplified by the linear relation that exists between the variables (N_0, t) and (F, G):

$$\begin{aligned} N_0 &= (R+1)F - RG \\ t &= G - F \end{aligned} \tag{18}$$

Table 1. Fit of the minimal model using genomic data from 20 organisms

Genome	Genes	Folds	G/F	t	N_0	R	Mismatch of of Third Moment (%)
M. genitalium	481	200	2.40	281	7	0.69	9.9
M. pneumonia	688	277	2.49	411	15	0.637	3.6
R. prowazeki	834	322	2.59	512	26	0.576	-13.1
C. trachomatis	894	344	2.60	550	29	0.574	-8.0
T. pallidum	1031	367	2.81	664	31	0.505	-14.1
C. pnemoniae	1052	390	2.70	662	34	0.538	-10.2
A. aeolicus	1522	357	4.26	1165	68	0.249	-3.0
H. pylori	1553	477	3.26	1076	52	0.395	0.6
B. burgdorferi	1638	559	2.93	1079	13	0.506	4.5
H. influenzae	1709	457	3.74	1252	70	0.31	0.9
M. jannaschii	1715	358	4.79	1357	34	0.239	3.8
M. thermoautotrophicum	1869	374	5.00	1495	35	0.227	-10.5
P. horikoshii	2064	450	4.59	1614	91	0.223	-5.7
A. fulgidus	2420	419	5.78	2001	72	0.173	-3.2
Synechocystis sp.	3169	558	5.68	2611	108	0.172	0.3
M. tuberculosis	3918	491	7.98	3427	118	0.109	-2.2
B. subtilis	4100	584	7.02	3516	153	0.123	-12.4
E. coli	4289	610	7.04	3679	141	0.127	-5.2
S. cerevisiae	6269	575	10.9	5694	128	0.078	2.8
C. elegans	19099	605	31.6	18494	120	0.026	-18.4

The estimation of R is aided by recasting the expression for H_2 (Eq. 47 in Appendix D) so that t no longer appears explicitly. Instead, the second moment can be expressed so that it depends directly on F, G and the unknown R:

$$\frac{H_2}{G} = \frac{R+1}{R-1} - \frac{2}{R-1}\left[\frac{\frac{G}{F}}{R+1-R\frac{G}{F}}\right]^{\frac{1-R}{1+R}} \tag{19}$$

This equation is well behaved and can be easily solved numerically. What threatened to be a coupled, nonlinear three dimensional estimation problem is actually nothing more than a single nonlinear equation and two linear equations. We have verified that this fitting procedure accurately recovers parameters values from distributions generated both numerically and from the exact solution.

The results appear in Table 1. As a measure of the quality of the fit, we also report the mismatch of between the third moment predicted by the minimal model and observed in the data, as a percentage of the observed value; a positive value indicates that the model moment is larger. Plots of the actual fits appear in Figure 5.

The parameter values are in fact very similar to those obtained in our previous work. The mismatch values range -13.1% to 9.9% and indicate that the distributions resulting from our model closely resembles the genomic distribution.

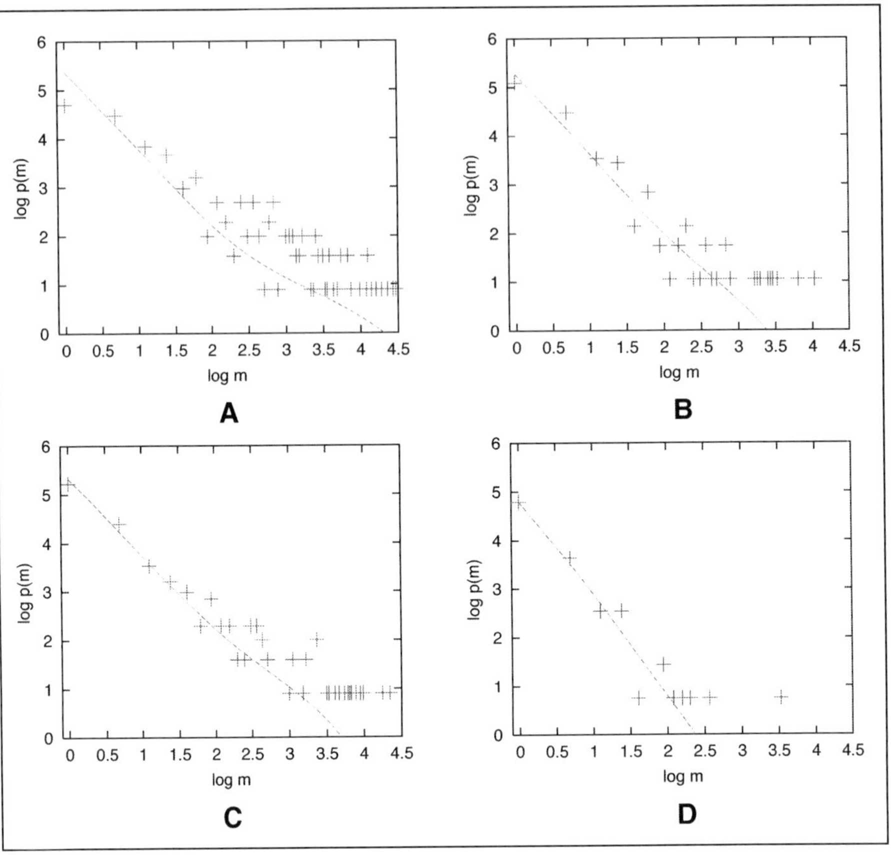

Figure 5. Minimal model fits for A) *C. elegans*, B) *A. fulgidus*, C) *M. tuberculosis*, and D) *M. genitalium* using parameters from Table 1.

Our attempts to fit the models that included gene deletion were not that informative. This is partly because, as we have seen already, the gene deletion parameter Q does not have a dramatic effect on the shape of the distribution. We had difficulties even trying to fit distributions generated numerically from the extended model. Unlike the equations describing the minimal model, these coupled equations are also nonlinear. Furthermore, since there is no exact analytic expression for $F(t)$, one of the variables itself has to be calculated numerically (We found that our analytic approximation for the number of folds given in Appendix F was not accurate enough to carry out the root-finding). We have found that naive multidimensional root-finding algorithms are either unable to distinguish between many approximate solutions, or find no solution at all at with increased sensitivity. The same difficulties were encountered in trying to discern evidence for selection pressure—there was too little dependence on the selection parameter γ to allow reliable estimation.

Conclusions

Here we propose two new models based on our previous model by fully incorporating two major processes in evolution: gene deletion and selective pressure. Both mathematically and biologically, including these effects are not slight. Mathematically, the derivations clearly show they are not trivial. Biologically, these effects provide a much more realistic model for genomic evolution than has been presented in any previous publications.[9,11,17] Furthermore, we provide analytical and numerical analyses of the original model and its extensions to explore the mathematical and biological significance of the models and to demonstrate the effects that the different evolutionary processes (gene duplication, acquisition, deletion and selective pressure) have on the final appearance of different genomes.

The field of the power-law distributions is controversial.[3-10,12,13] A number of fitting functions other than the power law were proposed to explain the observation. Our argument is that the question of which fitting function is the best should not be the central problem, because one can always find a function with more parameters fits the observation better than others.[2] Instead, we think biologically meaningful models are more helpful for understanding the origin of distribution and the analytical and numerical solutions shown in this work are vital for explaining the observation and further predicting the behaviour of the system.

The full analytical solution to this basic model revealed new facts that were unattainable from simulations only. As observed previously, gene duplication alone gives rise to an exponential distribution. However, the combined effect of duplication and acquisition changes the nature of genomic growth dramatically; beyond a sufficient length of evolutionary time, the fold distribution undergoes a transition from the exponential form, to a time-invariant limiting distribution that resembles a power law. The rate of fold acquisition (R) and the size of the initial genome (N_0) have distinct effects. Firstly, the cross-over time from the exponential to power-law phases is proportional to N_0 and approximately inversely proportional to R. This implies that the transition occurs when the number of new fold acquired becomes comparable to the initial size of the genome. Secondly, the decay rate of the power-law distribution i.e., the slope on a log-log plot is equal to R + 2 for large fold sizes. In fact, the final appearance of the distribution is independent of N_0, and is unaffected by the nature of the fold distribution in the starting genome. We find that the decay rate of the power-law distribution i.e., the slope on a log-log plot is equal to R + 2 for large fold sizes.

Note that we take R as a constant, and we regard this as the average rate of fold acquisition throughout the entire course of evolution. In reality, the value of R is likely to vary with time owing to a number of factors such as the decrease of available new protein folds. Further effects might be the increasing difficulty in horizontally transferring genes as the organism becomes more complex. These effects would generally lead to a decrease in rate of fold acquisition with time and this is perhaps reflected in the lower values of R for larger genomes.

We also studied extended models that fully incorporate the effects of random gene deletion and selective pressure. Gene deletion, represented by the parameter Q, does not significantly alter the qualitative behaviour found in the minimal model. The analytic solution showed that when there is no fold acquisition ($R = 0$), the distribution is again exponential and surprisingly, completely independent of Q. For cases where there is fold acquisition ($R > 0$), gene deletion had two main effects: firstly the final genome contained fewer fold types, and secondly all fold groups had smaller occurrences. Unsurprisingly, the extent of these effects was dependent on the size of Q. The final distribution nonetheless remains close to a power law, with a decay rate of $R + 2 + QR$.

The effects of selective pressure were incorporated into the minimal model by introducing favouritism into the gene selection process. This was done by having two groups of genes, one with a higher probability of selection than the other. In this case, the two sets of genes effectively evolve with two distributions, each undergoing a transition from the exponential to power-law phases. Therefore, the final fold distribution is the sum of two power-law distributions, which in fact still closely resembles the distribution when no selective pressure is present. This is true even for large differences in duplication probabilities between the two sets of genes. More generally, we could imagine an array of finer differences in duplication probabilities representing the full range of selection pressures for genes of distinct biological functions. For this, we conjecture that selective pressure, at least when modelled as a duplication bias, will lead to folds that co-exist and compete for prominence in the genome, each undergoing separate, but linked distributional transformation.

We compared our minimal model compares with the genomic data by fitting parameter values. Figure 5 and the mismatch values in Table 1 show, the fits between the model and genomic data are good. As discussed in our earlier work, the parameters can be interpreted in a biologically meaningful way.[11] We did not use the new models for simulating biological data for two reasons: (1) they do not greatly affect the final appearance of the distribution; (2) if we would be trying to fit a model with three additional free parameters, this would detract from the main results of the paper.

In conclusion, although our model considers a few of the many important processes underlying genomic evolution, it is significant that a simplistic model based on gene duplication and fold acquisition leads to distributions close to those observed in genomic data. The current genomes provide only a snapshot in evolutionary time, but through our model, we gain a glimpse into the biological processes that are most important. Furthermore, by estimating parameter values, we obtain quantitative estimates such as the rate of gene acquisition, which would be otherwise unattainable. Interesting expansions to our model in future may include allowing parameter values to vary during the course of evolution, and modelling the evolution of different genomes simultaneously and simulating their divergence into different organisms.

Appendix A: Analytic Solution of the Minimal Model

It helps to introduce a new parameterization of time:

$$u = \log \phi(t) = \log\left(1 + \frac{(R+1)t}{N_0}\right) \tag{20}$$

with associated derivative:

$$\frac{\partial}{\partial t} = \frac{R+1}{N_0 e^u}\frac{\partial}{\partial u} \tag{21}$$

With this definition, $u = 0$ corresponds to $t = 0$.

This new variable helps rid the differential equations (2) of explicit time dependence:

$$\begin{aligned} \frac{\partial F(m,u)}{\partial u} + \frac{mF(m,u)}{R+1} &= \frac{(m-1)F(m-1,u)}{R+1} \quad (m>1) \\ \frac{\partial F(1,u)}{\partial u} + \frac{F(1,u)}{R+1} &= \frac{N_0 R}{R+1}e^u \end{aligned} \tag{22}$$

Note that the equation the special bin $F(1,u)$ does not depend on any other $F(m, u)$, so it can be solved separately. Once it is known, the solution for any other m can be found by successive integration:

$$F(m+1,u) = \exp\left(-\frac{m+1}{R+1}u\right)\int_0^u \frac{dv}{R+1} mF(m,v)\exp\left(\frac{m+1}{R+1}v\right) \quad \text{for } m+1 = 2,3,\cdots \tag{23}$$

The solution for m serves as a "source" for $m + 1$. The relation (23) follows by multiplying both sides of (22) by $\exp\left(\frac{m+1}{R+1}u\right)$ and integrating. Note that this solution ensures that $F(m > 1, t = 0) = 0$, so the initial conditions are automatically satisfied. Our method of solving the differential equation is elementary and standard, see reference 26 for more details.

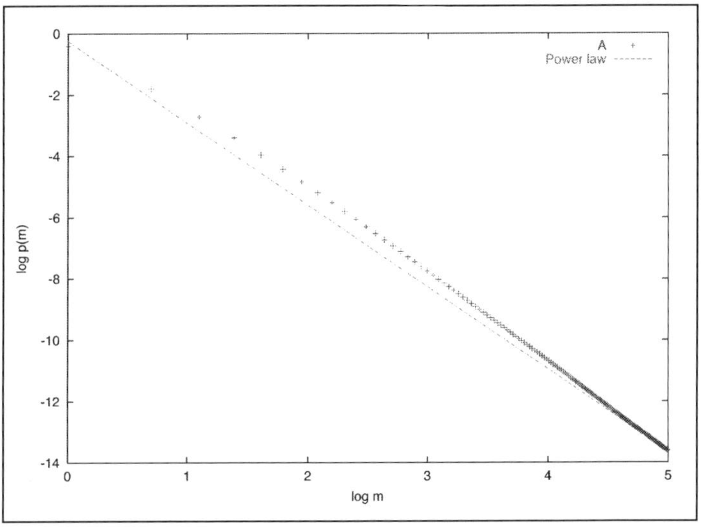

Figure 6. The normalized A_m coefficients (points) and a power-law fit (line), shown as a log-log plot as a function of size m, for $R = 1$.

The solution for $m = 1$ can be found in the same way:

$$\frac{\partial}{\partial u}\left[\exp\left(\frac{u}{R+1}\right)F(1,u)\right] = \exp\left(\frac{u}{R+1}+u\right)\frac{N_0 R}{R+1} \tag{24}$$

$$F(1,u) = N_0 \exp\left(-\frac{u}{1+r}\right) + \frac{N_0 R}{R+2}\left[\exp u - \exp\left(-\frac{u}{1+r}\right)\right] \tag{25}$$

The full solution follows by successive application of (22). There are two types of integrals that come up:

$$\exp\left(-\frac{m+1}{R+1}u\right)\int_0^u \frac{dv}{R+1}\left[m \exp v\right]\exp\left(\frac{m+1}{R+1}v\right) = \frac{m}{R+2+m}\left[\exp u - \exp\left(-\frac{m+1}{R+1}u\right)\right] \tag{26}$$

$$\exp\left(-\frac{m+n+1}{R+1}u\right)\int_0^u \frac{dv}{R+1}(m+n) \left[\exp\left(-\frac{mv}{R+1}\right)\left(1-\exp\left(-\frac{v}{R+1}\right)\right)^n\right]\exp\left(\frac{m+n+1}{R+1}v\right) = \frac{m+n}{n+1}\left[\exp\left(-\frac{mu}{R+1}\right)\left(1-\exp\left(-\frac{u}{R+1}\right)\right)^{n+1}\right] \tag{27}$$

The coefficients that emerge from these integrations define the recursion relations for A_m and β_n^m:

$$A_{m+1} = \frac{m}{R+2+m}A_m \tag{28}$$

$$\beta_{n+1}^m = \frac{m+n}{n+1}\beta_n^m \tag{29}$$

The full solution to (22), taking into account the initial conditions, is given by:

$$F(m,t) = N_0\phi^{-\frac{1}{1+R}}\left(1-\phi^{-\frac{1}{1+R}}\right)^{m-1} + A_m(\phi-\phi^{-\frac{m}{1+r}}) - \sum_{i=1}^{m-1} A_i\beta^i_{m-i}\phi^{-\frac{i}{1+R}}\left(1-\phi^{-\frac{1}{1+R}}\right)^{m-i}$$

$$A_m = \frac{RN_0}{R+2}\prod_{i=1}^{m-1}\frac{i}{R+2+i} = RN_0\frac{\Gamma(m)\Gamma(R+2)}{\Gamma(R+2+m)}$$
$$\beta^m_n = \prod_{k=1}^{n}\frac{m+k-1}{k} = \frac{(m+n-1)!}{(m-1)!n!} \tag{30}$$

with the understanding that an empty product is unity, i.e., $\prod_{i=1}^{0} f(i) = 1$.

Note that the coefficients A_m and β^m_n do not depend on time, and furthermore has no dependence on R or N_0. The product of coefficients $A_i\,\beta^i_{m-1}$ can be simplified:

$$A_i\beta^i_{m-i} = \frac{(m-1)!}{(m-i)!\prod_{j=1}^{i-1}(R+2+j)} \tag{31}$$

but it will be useful keep these coefficients separate when considering the solution for more general initial conditions. Note that we use the standard definition for the gamma function $\Gamma(x)$; see Appendix H.

Appendix B: Crossover Behavior

For nonzero R and times other than zero and infinity, the fold distribution will not be strictly exponential, nor will it conform to the limiting distribution (9). For small times, we would intuitively expect the histogram to be dominated by duplication events involving the initial N_0 genes. This is confirmed by the behavior of the analytic solution for small t:

$$\begin{aligned}F(m,t) \approx & N_0\left(1-\frac{t}{N_0}\right)\left(\frac{t}{N_0}\right)^{m-1} \\ & + A_m\left[1+\frac{R+1}{N_0}t-\left(1-\frac{t}{N_0}\right)^m\right] - \sum_{i=1}^{m-1}A_i\beta^i_{m-i}\left(1-\frac{t}{N_0}\right)^i\left(\frac{t}{N_0}\right)^{m-i}\end{aligned} \tag{32}$$

From this approximation, it is clear that the terms involving N_0 dominate for small times. Consequently, the fold distribution will resemble an exponential distribution more than the limiting distribution early on in the evolution of the genome. It is also clear that the histogram $F(m,\ t)$ will not approach the limiting distribution uniformly; the rate of convergence will depend on cluster size.

There are many possible ways of characterizing this transformation of the fold distribution, each suggesting a different notion of a "“crossover” time. We have looked at the convergence of the probability distribution as a whole. To quantify the extent to which the actual distribution $p(m)$ resembles a second distribution, say $p_A(m)$, we adopt the sum of the squared differences as our metric:

$$\eta_A = \sum_m (p(m) - p_A(m))^2 \tag{33}$$

$$\sum_m p(m) = \sum_m p_A(m) = 1 \tag{34}$$

Figure 7 tracks the evolution of $p(m)$ according to this metric when $R = 1.0$ and $N_0 = 100$. At each time, the closeness of $p(m)$ to the limiting distribution (9) is shown, as is the closeness to the best fitting exponential distribution for that time, obtained by a least-squares regression of $log\,p$ against m. For times greater than $t \approx 70$, the distribution of fold sizes resembles the final

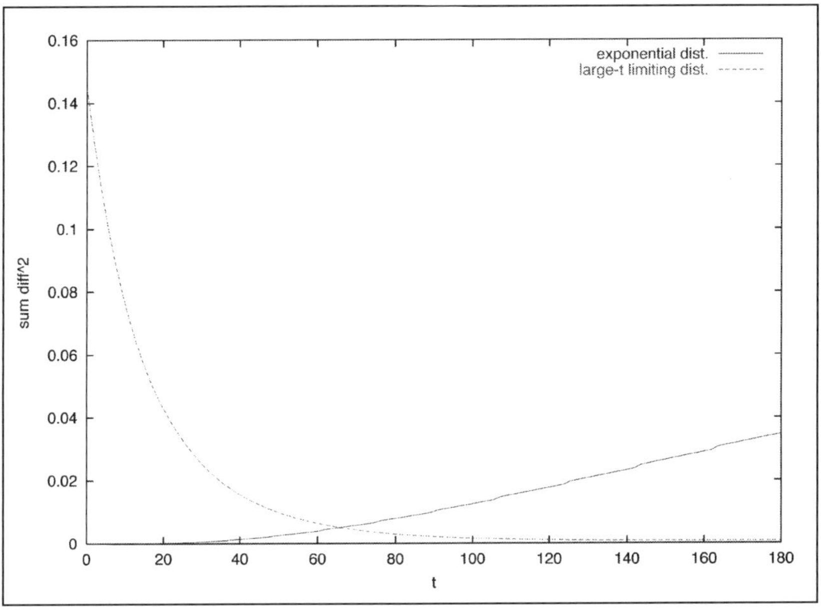

Figure 7. Crossover from exponential to large-time limiting distribution for $R = 1.0$ and $N_0 = 100$.

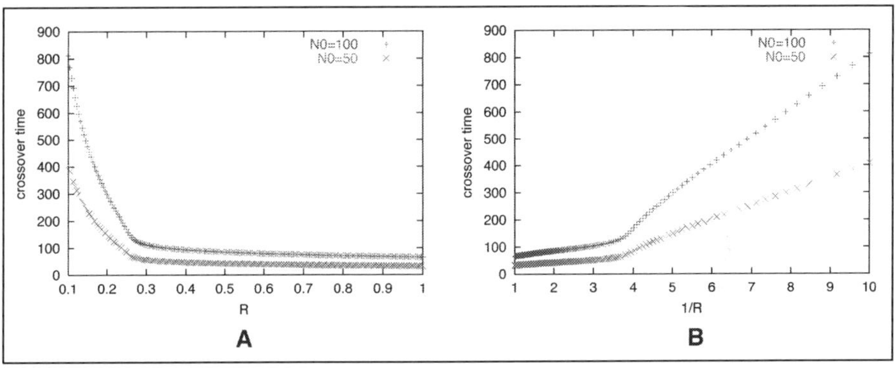

Figure 8. Crossover time for $N_0 = 100$ and $N_0 = 50$, plotted as a function of A) R, and B) $1/R$.

distribution more than any exponential distribution, this defines the crossover time for this set of parameters. The sum extends to cluster sizes large enough to ensure numerical convergence.

Figure 8 plots the crossover time as a function of R for two values of N_0. The range of R is chosen so that new fold acquisitions occur less frequently than (or as often as) gene duplication. The crossover time displays two distinct regimes. Within each regime it is approximately inversely proportional to R and directly proportional N_0. A different proportionality constant applies in each regime: $T_c \sim N_0/R$. These numerical results confirm that crossover occurs roughly when the number of new fold introductions: RT_c becomes comparable to the initial genome size N_0. The details of the dependence are not that important, as they are no doubt strongly affected by the choice of metric.

Appendix C: Arbitrary Initial Distribution

The solution for an arbitrary initial distribution: $N_{init}(m)$, requires solving (2) subject to different boundary conditions at $t = 0$; the terms proportional to A_m are the same, the term proportional to N_0 is replaced by the superposition of new terms describing the propagation of each bin of initial histogram:

$$F(m,t) = \sum_{i=1}^{\infty} N_{init}(i)\psi_i(m,t) + A_m(\phi - \phi^{-\frac{m}{1+r}}) - \sum_{i=1}^{m-1} A_i\psi_i(m,t)$$
$$\psi_i(m,t) = \begin{cases} 0 & \text{if } m < i \\ \beta^i_{m-i}\phi^{-\frac{i}{1+R}}\left(1-\phi^{-\frac{1}{1+R}}\right)^{m-i} & \text{for } m \geq i \end{cases} \tag{35}$$

with the same definitions for A_m and β_n^m as before. These are derived by following by successive integration in the same way as was done in Appendix A.

The fact that $\psi_i(m,t) = 0$ for $m < i$ reflects the fact that there is no gene deletion; genes that start in bin i may either stay put or advance to bins corresponding to larger fold sizes, but will never populate bins of fold size less than i.

One important conclusion may be drawn from the full solution: all initial distributions ultimately lead to the same limiting distribution determined by the A_m. Just as before, the dependence on the initial fold distribution $N_{init}(m)$ decays with time, leading to the same asymptotic distribution as was found for an initial distribution of N_0 folds of size 1 in Appendix A. Of course, the details of how the crossover happens will depend on the particular form of $N_{init}(m)$.

Appendix D: Solution to the Extended Model When $0 < Q < 1$ and $R = 0$

As one done in the solution for the minimal model, define $\phi(t)$:

$$\phi(t) = 1 + \frac{(1-Q)t}{N_0}, \tag{36}$$

and keep the association: $u = \log \phi(t)$. In terms of the time-like variable u, the fundamental evolution equations (10) now are:

$$(1-Q)\frac{\partial F(m,u)}{\partial u} = (m-1)F(m-1,t) - (1+Q)mF(m,t) + Q(m+1)F(m+1,u) \quad (m>1)$$
$$(1-Q)\frac{\partial F(1,u)}{\partial u} = -(1+Q)F(1,u) + QF(2,t) \tag{37}$$

Substituting the ansatz: $F(m, u) = f(u)g^{m-1}(u)$ into the equation for $m > 1$ leads to the following relation:

$$(1-Q)\left[\frac{\partial \log f}{\partial u}g + (m-1)\frac{\partial g}{\partial u}\right] = (m-1) + (1+Q)mg + Q(m+1)g^2 \tag{38}$$

Since neither $g(u)$nor $f(u)$ depend on m, this identity can only be satisfied if:

$$(1-Q)\frac{\partial g}{\partial u} = 1 - (1+Q)g + Qg^2$$
$$(1-Q)\frac{\partial \log f}{\partial u} = -(1+Q) + 2Qg \tag{39}$$

These equations can be solved by integration, together with the restriction that $f(t = 0) = 1$ and $g(t = 0) = 0$. It is easy to verify that the ansatz also works when $m = 1$.

$$\begin{aligned} F(m,t) &= N_0 f(t) g^{m-1}(t) \\ f(t) &= \phi^{-1}\left[\frac{1-Q}{1-Q\phi^{-1}}\right]^2 = \phi\left[\frac{N_0}{N_0+t}\right]^2 \\ g(t) &= \frac{1-\phi^{-1}}{1-Q\phi^{-1}} = \frac{t}{N_0+t} \\ \phi(t) &= 1 + \frac{(1-Q)t}{N_0} \end{aligned} \tag{40}$$

In fact, it is easy to solve for $F(t)$ in this special case:

$$\frac{\partial F(t)}{\partial t} = -Q\frac{F(1,t)}{G(t)} = -Q\frac{f(t)}{\phi(t)} \tag{41}$$

which can be integrated directly:

$$F(t) = N_0 \frac{N_0 + (1-Q)t}{N_0 + t} \tag{42}$$

The large-time asymptotic limit for $F(t)$ is $(1 - Q)N_0$ folds, which reflects the fact that some of the initial N_0 folds will ultimately be lost due to gene deletion. Equation (42) leads to a simple relation between the number of folds and the number of genes:

$$F(t) = \frac{G(t)}{1 + t/N_0} \tag{43}$$

Although $F(t)$ and $G(t)$ both depend on Q, their ratio does not.

The solution (40) we have derived for $0 < Q < 1$ is also the solution for $Q = 1$, which means that gene deletion and duplication occur at the same rate. Equations (42) and (43) for the total number of folds $F(t)$ are still valid for $Q = 1$, but now the expected number of genes is constant: $G(t) = N_0$. Although we will not do so here, analytic solutions can be derived when deletion dominates duplication so the genome shrinks in size.

Appendix E: Analytical Results for Higher Moments

Higher moments of the distribution, defined as $H_n(t) = \sum_m m^n F(m, t)$, for $n \geq 2$ in the extended model satisfy the following differential equation:

$$G(t)\frac{\partial H_n}{\partial t} = RG(t) + \sum_{m=1}^{\infty} F(m)\left[m(m+1)^n - (1+Q)m^{n+1} + Q(m-1)^n m\right] \tag{44}$$

In particular, the equations for the second and third moment are:

$$G(t)\frac{\partial H_2}{\partial t} = RG(t) + 2(1-Q)H_2(t) + (1+Q)G(t) \tag{45}$$

$$G(t)\frac{\partial H_3}{\partial t} = RG(t) + 3(1-Q)H_3(t) + 3(1+Q)3H_2(t) + (1-Q)G(t) \tag{46}$$

Higher moments depend on all lower moments except for the zeroth moment, the expected number of folds $F(t)$. This is fortuitous, since equation H for $F(t)$ could not be solved analytically due to its explicit dependence on the population of smallest folds: $F(1,t)$.

The solution for the second moment is given by:

$$H_2(t) = \begin{cases} N_0 \exp\left(\frac{2(1-Q)}{1+R-Q}u(t)\right) + N_0 \frac{1+R+Q}{R-1+Q}\left[\exp u(t) - \exp\left(\frac{2(1-Q)}{1+R-Q}u(t)\right)\right] & R \neq 1-Q \\ N_0 \exp\left(u(t)\right)\left[1 + \frac{u(t)}{2R}\right] & R = 1-Q \end{cases} \quad (47)$$

where the variable $u(t)$ is related to the expected number of genes:

$$u(t) = \log\left[1 + \frac{(R+1-Q)t}{N_0}\right] \quad (48)$$

This result will be important in fitting actual genomic data to the models.

Appendix F: Perturbation Theory Approximation for the Extended Model

As before, relate time and the number of genes through $\phi(t)$:

$$\phi(t) = 1 + \frac{(R+1-Q)t}{N_0}, \quad (49)$$

This extends the previous definition (36); the variable u is still defined as before: $u = \log \phi(t)$.

Recall that when $Q = 0$ and $R > 0$ the long-term behavior of $F(m, t)$ is determined by the coefficients A_m, as shown in equation (30). Assume that the large-time solution in the presence of gene deletion is determined by new coefficients B_m:

$$F(m,t) \to B_m \phi(t) = B_m \exp\left(u\right) \quad \text{as} \quad t \to \infty \quad (50)$$

Substituting this ansatz into the fundamental equations (10) leads to:

$$\begin{aligned} (1+R-Q)B_1 &= RN_0 - (1+Q)B_1 + 2QB_2 \\ (1+R-Q)B_m &= (m-1)B_{m-1} - (1+Q)mB_m + Q(m+1)B_{m+1} \end{aligned} \quad (51)$$

Motivated by the numerical results, we will develop the perturbation around a new variable γ_m:

$$B_m = \gamma_m A_m \quad (52)$$

that relates B_m to the $Q = 0$ solution (A_m) as closely as possible. Using the explicit form of A_m from (30) in (51) leads to:

$$\begin{aligned} \gamma_1 &= 1 + \frac{2}{(R+2)(R+3)}\gamma_2 \\ \gamma_m &= \gamma_{m-1} + Q\frac{(1-m)}{R+1+m}\gamma_m + Q\frac{m(m+1)}{(R+1+m)(R+2+m)}\gamma_{m+1} \end{aligned} \quad (53)$$

It is easy to see that when $Q = 0$, $\gamma_m = 1$, which means $B_m = A_m$ for all m. The perturbation theory approach expands γ_m for each m as a power series in Q:

$$\gamma_m = \sum_{i=0}^{\infty} Q^i \gamma_m^{(i)} \quad (54)$$

From the solution when $Q = 0$ we immediately know the first term in the expansion: $\gamma_m^{(0)} = 1$. The remaining terms are determined order-by-order by substituting into (53) and collecting terms with the same power of Q:

$$\begin{aligned}\gamma_1^{(i)} &= 1 + \frac{2}{(R+2)(R+3)}\gamma_2^{(i-1)} \\ \gamma_m^{(i)} &= \gamma_{m-1}^{(i)} + \frac{(1-m)}{R+1+m}\gamma_m^{(i-1)} + \frac{m(m+1)}{(R+1+m)(R+2+m)}\gamma_{m+1}^{(i-1)}\end{aligned} \tag{55}$$

The first-order ($i = 1$) equations are easy to solve since the zeroth-order solutions are just unity:

$$\begin{aligned}\gamma_1^{(1)} &= 1 + \frac{2}{(R+2)(R+3)} \\ \gamma_m^{(1)} &= \gamma_1^{(1)} + \sum_{i=2}^{m} g(i) \\ g(i) &= \frac{2+2R+R^2}{1+R+i} - \frac{2+3R+R^2}{2+R+i}\end{aligned} \tag{56}$$

An important limitation of the perturbation expansion is revealed by the first order solution. Consider the behavior of the sum:

$$\begin{aligned}\sum_{i=2}^{m} g(i) &\approx \int_2^m dx g(x) \\ &= (2+2R+R^2)\log\frac{m+R+1}{R+3} - (2+3R+R^2)\log\frac{m+R+2}{R+4}\end{aligned} \tag{57}$$

For large m, the sum increases in magnitude logarithmically with m:

$$\sum_{i=2}^{m} g(i) \approx -R\log m \tag{58}$$

This means that no matter how small Q is, for large enough m the first order expansion will fail. This reflects a limitation of the perturbation expansion itself for this problem—stopping the expansion at any finite order will lead to a series valid only up to some maximum size m.

The only way to obtain a consistent expansion is to sum all orders of the series. Unfortunately, the equations (55) are difficult to solve exactly, and even if they were possible to solve, it would be even more difficult to carry out the summation. However, it isn't difficult to figure out the dominant contribution at each order. It helps to first look at the equations for $i = 2$:

$$\begin{aligned}\gamma_m^{(2)} &= \gamma_{m-1}^{(2)} + g(m)\sum_{i=2}^{m} g(i) + \frac{m(m+1)}{(R+1+m)(R+2+m)}g(m+1) \\ \Longrightarrow \gamma_m^{(2)} &= \gamma_1^{(2)} + \sum_{i=2}^{m} g(i)\sum_{j=2}^{i} g(j) + \sum_{i=2}^{m}\frac{i(i+1)}{(R+1+i)(R+2+i)}g(i+1)\end{aligned} \tag{59}$$

The first summation dominates the second in the above equation; the first grows like $\log^2 m$, while the second grows like $m \log m$.

The same pattern emerges at all orders—the dominant contribution can be isolated as:

$$\begin{aligned}\gamma_m^{(i)} &\sim \gamma_1^{(i)} + \sum_{j_1=2}^{m} g(j_1) \sum_{j_2=2}^{j_1} g(j_2) \cdots \sum_{j_i=1}^{j_{i-1}} g(j_i) \\ &\approx \gamma_1^{(i)} + \frac{1}{i!}\left(\sum_{j=2}^{m} g(j)\right)^i\end{aligned} \tag{60}$$

The sum of the dominant contributions remains finite:

$$\gamma_m \sim \exp\left[Q\sum_{j=2}^{m} g(j)\right] \sim \exp\left(-QR\log m\right) \tag{61}$$

and suggests that for large m, γ_m will decay as a power-law with exponent QR.

Motivated by this observation, and recalling that for large m, $A_m \sim 1/m^{R+2}$ (from equation (8)), we suggest the following approximation for B_m, valid for all values of m, not just when m is large:

$$B_m = C\frac{RN_0}{R+2+QR}\prod_{i=1}^{m-1}\frac{i}{R+2+QR+i} \tag{62}$$

where C is a constant that is independent of m. The above expression for B_m is derived by replacing R by $R + QR$ in the denominator of the product that defines A_m (equation (8)) This is really nothing more than informed guesswork; this is the simplest expression for B_m that recovers a power-law with exponent $R + QR$ for large m and reduces to A_m when $Q = 0$.

In order to determine $F(t)$, the total number of folds at time t, equation (11) has to be solved using the approximate solution (62). First, the a choice has to be made for the constant C—since the equation is an approximation, there is freedom in the choice. One way is to enforce the consistency of equation (53) for $m = 1$:

$$\begin{aligned}\gamma_1^{(1)} = \frac{B_1}{A_1} = C\frac{R+2}{R+2+QR} = 1 + \frac{2}{(R+2)(R+3)} \\ \Longrightarrow C = \left(1+\frac{2}{(R+2)(R+3)}\right)\left(1+\frac{QR}{R+2+QR}\right)\end{aligned} \tag{63}$$

As $F(t)$ is directly affected by B_1, it is natural to focus on $m = 1$. Note that for small Q, $C \approx 1 + 2 / (R + 2)(R + 3)$.

Equation (11) can be integrated to give an approximation for $F(t)$:

$$F(t) \approx N_0 + R\left(1 - \frac{QC}{R+2}\right)t \tag{64}$$

Using the identity of Appendix H, the normalized coefficients are given by:

$$p_m = \frac{B_m}{\sum_{m=1}^{\infty} B_m} = \frac{R+1+QR}{R+2+QR}\prod_{i=1}^{m-1}\frac{i}{R+2+QR+i} \tag{65}$$

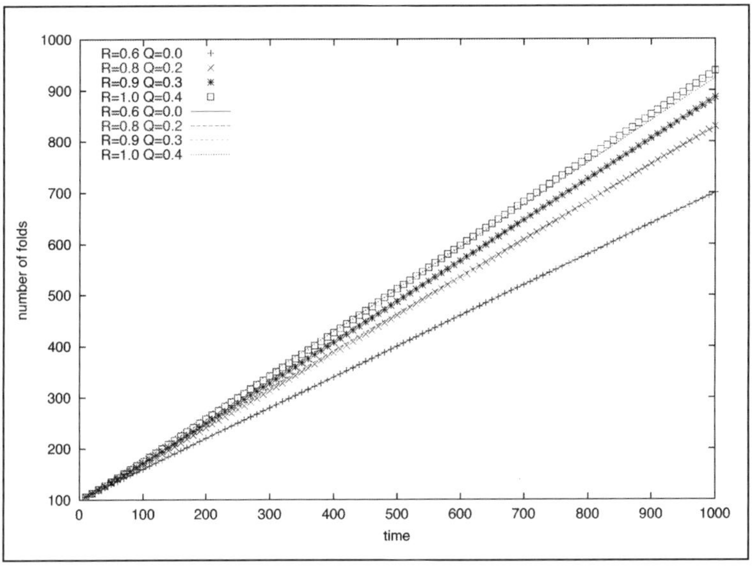

Figure 9. Analytic approximation for the total number of folds compared to numerical results of Figure 3.

$$F(t) = N_0 + R\left(1 - \frac{QC}{R+2}\right)t$$
$$C = \left(1 + \frac{2}{(R+2)(R+3)}\right)\left(1 + \frac{QR}{R+2+QR}\right) \tag{66}$$

In the presence of gene deletion, the approximation for $F(t)$ shows linear growth with time at a rate less than R. As expected, a greater rate of gene deletion reduces the growth of $F(t)$. However the approximation predicts that the number of folds will always increase with time, which can be verified by taking the uppermost limit, $Q = 1$. For small Q, the constant C itself can be approximated more simply: $C \approx 1 + 2 / (R + 2)(R + 3)$.

Figure 9 confirms these observations. The approximation for the expected number of folds seems to work quite well and could be useful in trying to infer both R and Q from genomic data. Certainly the impact of gene deletion is easier to identify through $F(t)$ and $G(t)$ than through the shape of the histogram $F(m, t)$.

Appendix G: The Effects of Selection Pressure

Recall that we have assumed that there are only two duplication types: type "A" and type "B", and that "B" genes are γ times more likely to be chosen for duplication than "A" genes. There will still be one duplication event, on average, per unit time, so the total expected number of genes will remain the same, but the allocation of the total between types "B" and "A" will depend on γ. We will assume that $\gamma > 1$, so it is the "B" types that are more likely to be duplicated.

To keep track of the fold population we now need two histograms: $F_A(m, t)$ and $F_B(m, t)$ to distinguish between the duplication types. The full fold histogram is the sum of both sub-histograms: $F(m, t) = F_A(m, t) + F_B(m, t)$. Similarly, let $G_A(t)$ and $G_B(t)$ represent the total number of genes for each type and define a new variable $G_\gamma(t)$:

$$G_\gamma(t) = G_A(t) + \gamma G_B(t) \tag{67}$$

The evolution equations that extend (2) are:

$$\begin{aligned}
\frac{\partial F_A(m,t)}{\partial t} &= \frac{(m-1)F_A(m-1,t)}{G_\gamma(t)} - \frac{mF_A(m,t)}{G_\gamma(t)} \quad (m>1)\\
\frac{\partial F_A(1,t)}{\partial t} &= R_A - \frac{F_A(1,t)}{G_\gamma(t)}\\
\frac{\partial F_B(m,t)}{\partial t} &= \gamma\frac{(m-1)F_B(m-1,t)}{G_\gamma(t)} - \gamma\frac{mF_B(m,t)}{G_\gamma(t)} \quad (m>1)\\
\frac{\partial F_B(1,t)}{\partial t} &= R_B - \gamma\frac{F_B(1,t)}{G_\gamma(t)}
\end{aligned} \tag{68}$$

Note that we allow new folds to be acquired at different rates for each type: R_A can be different from R_B although we will restrict our numerical examples to the when they are equal.

As before, we derive equations for the total number of genes from the full dynamics (68):

$$\begin{aligned}
\frac{\partial G_A(t)}{\partial t} &= \frac{\partial}{\partial t}\sum_{m=1} mF_A(m,t) = R_A + \frac{G_A(t)}{G_A(t)+\gamma G_B(t)}\\
\frac{\partial G_B(t)}{\partial t} &= \frac{\partial}{\partial t}\sum_{m=1} mF_B(m,t) = R_B + \gamma\frac{G_B(t)}{G_A(t)+\gamma G_B(t)}\\
\frac{\partial G(t)}{\partial t} &= \frac{\partial G_A(t)}{\partial t} + \frac{\partial G_B(t)}{\partial t} = R_A + R_B + 1
\end{aligned} \tag{69}$$

This confirms that the overall duplication rate is still one gene per unit time. The evolution of $G_\gamma(t)$ is more complicated:

$$\frac{\partial G_\gamma(t)}{\partial t} = R_A + \gamma R_B + 1 + \gamma\left[1 - \frac{G(t)}{G_\gamma(t)}\right] \tag{70}$$

It is possible to establish the distributional properties of the genome without having to solve (68) explicitly for the special parameter values encountered previously: (1) the case when there is no introduction of new folds, so $R_A = R_B = 0$; and (2) the limiting distribution when $t \to \infty$. When there is no introduction of new folds, a simple extension of the repeated integration employed in Appendix A establishes that the each of the sub-histograms $F_A(m, t)$ and $F_B(m, t)$ follows an exponential distribution for all times:

$$\begin{aligned}
F_A(m,t) &= N_0^A \exp(-u(t))\,[1-\exp(-u(t))]^{m-1}\\
F_B(m,t) &= N_0^B \exp(-\gamma u(t))\,[1-\exp(-\gamma u(t))]^{m-1}
\end{aligned} \tag{71}$$

The number of distinct folds of each type, present at $t = 0$ is given by N_0^A and N_0^B. The variable $u(t)$ is determined by $G_\gamma(t)$:

$$u(t) = \int_0^t \frac{ds}{G_\gamma(s)} \tag{72}$$

The full histogram is consequently a sum of exponential distributions:

$$\begin{aligned}
p(m,t) &= \frac{F_A(m,t)+F_B(m,t)}{\sum_i F_A(i,t)+F_B(i,t)}\\
&= \frac{N_0^A}{N_0^A+N_0^B}e^{-u}\left[1-e^{-u}\right]^{m-1} + \frac{N_0^B}{N_0^A+N_0^B}e^{-\gamma u}\left[1-e^{-\gamma u}\right]^{m-1}
\end{aligned} \tag{73}$$

The large time behavior of the solution is much easier to derive than an exact solution. For large t, $G_\gamma(t)$ will grow linearly with time: $G_\gamma \sim C_\gamma t$, according to a constant C_γ that depends on the rate of fold acquisition and the differential rate of duplication:

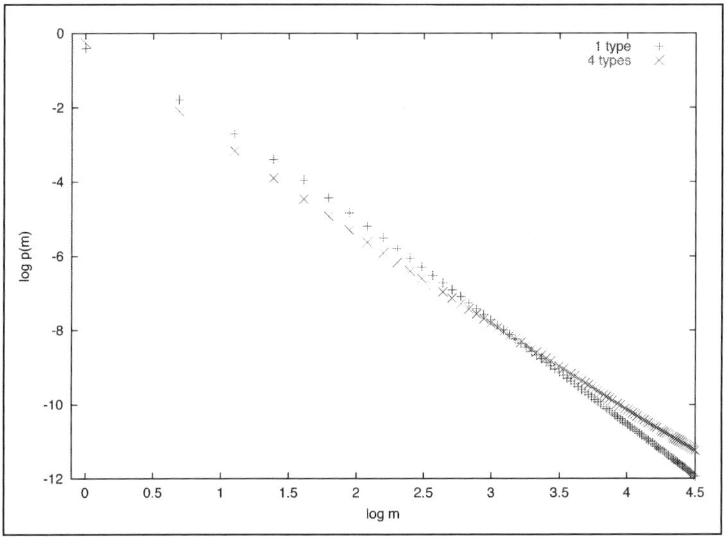

Figure 10. Large time limit for the fold probability distribution for the minimal model (one duplication type) and four duplication types: $B = 4$, $C = 8$, $D = 16$. The total rate of new fold acquisition is the same for both genomes.

$$C_\gamma = \frac{1}{2}(R_A + \gamma R_B + 1 + \gamma) + \frac{1}{2}\sqrt{(R_A + \gamma R_B + 1 + \gamma)^2 - 4\gamma(R_A + R_B + 1)} \qquad (74)$$

In a similar fashion, we define coefficients C_m^A and C_m^B, akin to the coefficients A_m of the solution to the minimal model (7), that describe the ultimate linear growth of the histogram bins: $F_A(m, t) \sim C_m^A t$, and similarly for $F_B(m, t)$. The form of the coefficients is very similar to the minimal model's A_m:

$$\begin{aligned} C_m^A &= \frac{R_A}{C_\gamma + 1} \prod_{i=1}^{m-1} \frac{i}{C_\gamma + i + 1} \\ C_m^B &= \frac{R_B}{C_\gamma + \gamma} \prod_{i=1}^{m-1} \frac{i\gamma}{C_\gamma + \gamma(i+1)} \end{aligned} \qquad (75)$$

The normalized probability distribution corresponding to this limit can be found using the same normalization identity that was helpful in deriving the probability distribution in the minimal model (Appendix H):

$$\begin{aligned} p(m,t) &= \frac{C_m^A + C_m^B}{\sum_i C_i^A + C_i^B} \\ &= \frac{C_\gamma}{C_\gamma + 1}\frac{R_A}{R_A + R_B} \prod_{i=1}^{m-1} \frac{i}{C_\gamma + i + 1} + \frac{C_\gamma}{C_\gamma + \gamma}\frac{R_B}{R_A + R_B} \prod_{i=1}^{m-1} \frac{i\gamma}{C_\gamma + \gamma(i+1)} \end{aligned} \qquad (76)$$

We have also briefly considered the case of more than two duplication types. When there is no introduction of new folds into the genome, the same argument behind equations (72) and (73) generalizes: the sub-histogram for each duplication type is exponential. Furthermore, we have confirmed numerically that the terminal distribution is not dramatically affected by selection pressure, even when there are several families with significantly different rates of duplication. One particular example, involving a four duplication types appears in Figure 10. In this rather extreme case, types "B", "C" and "D" are 4.0, 8.0 and 16.0 times more likely to be duplicated than type "A". The total rate of new fold acquisition is the same for both genomes.

Appendix H: A Useful Normalization Identity

A series whose terms z_m, $m = 1, 2, \cdots$ are defined by a recursion relation:

$$z_m = \prod_{i=1}^{m-1} \frac{i}{\alpha + i} \tag{77}$$

can be summed exactly as follows.

Rewrite z_m as:

$$z_m = \frac{\Gamma(m)\Gamma(\alpha+1)}{\Gamma(\alpha+m)} \tag{78}$$

with the usual definition for the gamma function:

$$\Gamma(z) = \int_0^\infty dt\, t^{z-1} e^{-t} \tag{79}$$

The integral representation of the beta function $B(x, y)$ provides the key identity to carry out the sum:

$$B(x,y) = \frac{\Gamma(x)\Gamma(y)}{\Gamma(x+y)} = \int_0^1 dt\, t^{x-1}(1-t)^{y-1} \tag{80}$$

Combining these relations leads to:

$$\sum_{m=1}^{\infty} z_m = \alpha \int_0^1 \sum_{m=1}^{\infty} t^{m-1}(1-t)^{\alpha-1} \tag{81}$$

$$= \alpha \int_0^1 (1-t)^{\alpha-2} \tag{82}$$

$$= \frac{\alpha}{\alpha - 1} \tag{83}$$

Acknowledgement

We thank the NIH for support through the PSI Initiative.

References

1. Zipf GK, ed. Human Behaviour and the Principle of Least Effort. Cambridge: Addison-Wesley, 1949.
2. Luscombe NM, Qian J, Johnson T et al. Power-law behaviour applies to a wide variety of genomic properties. Trends Genet, submitted.
3. Mantegna RN, Buldyrev SV, Goldberger AL et al. Linguistic features of noncoding DNA sequences. Phys Rev Lett 1994; 73(23):3169-72.
4. Konopka AK, Martindale C. Noncoding DNA, Zipf's law, and language. Science 1995; 268(5212):789.
5. Israeloff NE, Kagalenko M, Chan K. Can Zipf distinguish language from noise in noncoding DNA? Phys Rev Lett 1996; 76(11):1976.
6. Bonhoeffer S, Herz AV, Boerlijst MC et al. No signs of hidden language in noncoding DNA. Phys Rev Lett 1996; 76:1977.

7. Voss RF. Comment on "Linguistic features of noncoding DNA sequences". Phys Rev Lett 1996; 76(11):1978.
8. Gerstein M. A structural census of genomes: Comparing bacterial, eukaryotic and archaeal genomes in terms of protein structure. J Mol Biol 1997; 274(4):562-76.
9. Huynen MA, van Nimwegen E. The frequency distribution of gene family sizes in complete genomes. Mol Biol Evol 1998; 15(5):583-9.
10. Koonin EV, Wolf YI, Aravind L. Protein fold recognition using sequence profiles and its application in structural genomics. Adv Protein Chem 2000; 54:245-75.
11. Qian J, Luscombe NM, Gerstein M. Protein family and fold occurrence in genomes: power-law behaviour and evolutionary model. J Mol Biol 2001; 313:673-681.
12. Jeong H, Tombor B, Albert R et al. The large-scale organization of metabolic networks. Nature 2000; 407(6804):651-4.
13. Park J, Lappe M, Teichmann SA. Mapping protein family interactions: intramolecular and intermolecular protein family interaction repertoires in the PDB and yeast. J Mol Biol 2001; 307(3):929-38.
14. Rzhetsky A, Gomez SM. Birth of scale-free molecular networks and the number of distinct DNA and protein domains per genome. Bioinformatics 2001; 17(10):988-96.
15. Taverna DM, Goldstein RA. The distribution of structures in evolving protein populations. Biopolymers 2000; 53(1):1-8.
16. Shakhnovich EI. Protein design: a perspective from simple tractable models. Fold Des 1998; 3(3):R45-58.
17. Yanai I, Camacho CJ, DeLisi C. Predictions of gene family distributions in microbial genomes: evolution by gene duplication and modification. Phys Rev Lett 2000; 85(12):2641-4.
18. Lawrence JG, Ochman H. Molecular archaeology of the Escherichia coli genome. Proc Natl Acad Sci USA 1998; 95(16):9413-7.
19. Doolittle WF. Phylogenetic classification and the universal tree. Science 1999; 284(5423):2124-9.
20. Ochman H, Lawrence JG, Groisman EA. Lateral gene transfer and the nature of bacterial innovation. Nature 2000; 405(6784):299-304.
21. Kidwell MG. Lateral transfer in natural populations of eukaryotes. Annu Rev Genet 1993; 27:235-56.
22. de la Cruz I, Davies I. Horizontal gene transfer and the origin of species: lessons from bacteria. Trends Microbiol 2000; 8(3):128-133.
23. LoConte L, Ailey B, Hubbard TJ et al. SCOP: a structural classification of proteins database. Nucleic Acids Res 2000; 28(1):257-9.
24. Itzykson C, Drouffe JM, eds. Statistical Field Theory, Vols. 1 and 2. Cambridge: Cambridge University Press, 1989.
25. Patthy L, ed. Protein Evolution. London: Blackwell Science, 1999.
26. Simmons GF, ed. Differential Equations with Applications and Historical Notes. Englewood Cliffs: Prentice-Hall, 1994.

Chapter 11

The Protein Universes:

Some Informatic Issues in Protein Classification

S. Rackovsky*

Abstract

We discuss some informatic problems in protein classification. We first address a neglected problem in sequence classification—information loss resulting from alphabet contraction. Since the use of reduced alphabets is a standard bioinformatic tool, this is a significant issue. We review recent work in which it was shown that information theoretic methods can be used to quantitate the amount of structural information carried by a specified sequence representation. These tools are then used to construct reduced alphabets of specified size which retain the maximum possible amount of structural information. We then turn to structure classification. After briefly reviewing previous work in this field, we discuss the fact that sequence and structure classification give different pictures of the protein space. We outline ongoing research in which new parameters are sought which explicitly encode architecture choice by protein sequences.

Introduction

Within the last few years, the quantitative classification of protein structures and sequences has progressed from being an area of great interest to being an area of critical importance. When there were a couple of thousand sequences available, and the structures of 400-500 proteins had been experimentally determined, it became clear that there is considerable scientific merit in this exercise. With the advent of genome-scale sequencing and high-throughput structure determination, workers in bioinformatics and proteomics are faced with a very practical problem- the need to organize a vast and growing body of data, so that information of interest and important correlations are readily accessible.

Progress on several problems of fundamental biological importance depends on the development of appropriate classification methodology. The first is the delineation of functional relationships between sequences. The completion of any new genome yields a flood of sequences of proteins whose structure and function are unknown. The most reliable method of assigning a function to a new protein is to demonstrate appropriate relationships to known molecules. A second problem is the prediction of protein structure. Although significant progress has been made in ab initio methods for protein structure prediction, the most reliable methods remain those in which a model can be constructed based on a demonstrated homology to a

*S. Rackovsky—Department of Pharmacology and Biological Chemistry, and Center for Biomathematics, Mount Sinai School of Medicine of New York University, One Gustave L. Levy Place, New York, New York 10029, U.S.A. Email: shelly@camelot.mssm.edu

Power Laws, Scale-Free Networks and Genome Biology, edited by Eugene V. Koonin, Yuri I. Wolf and Georgy P. Karev.

protein of known structure. A third problem is the elucidation of the factors which determine fold choice in proteins.[1] While this problem is very fundamental, its solution has important practical implications for genome analysis.[2]

One can classify either structures[3-6] or sequences.[7-11] A central observation in this field is the fact that the two classifications do not give the same results. Sequences which have no demonstrable homology are observed to occur in the same fold. In fact, a general problem in the classification of proteins is that the structure one observes for the protein universe as a whole depends on the features which one uses to classify the molecules. It is therefore useful to think of proteins in a "many worlds" context, and to ask how the structure of the space of proteins depends on classification criteria. It is intended in this chapter to concisely survey some topics of current importance in the protein classification problem. We first give a broad outline of the methods which have been developed. Recent work on some underlying problems will then be highlighted.

General Methodology

The first requirement for the classification of a set of objects is a function which defines a degree of difference between the objects. Once such a function has been defined, it is possible to organize the differences in a distance matrix, each element of which gives the degree of dissimilarity between two members of the set. [An alternative approach, which has found some application in sequence studies, is the use of a measure of similarity rather than difference. These two approaches are essentially equivalent, and it is possible to convert from similarity to difference representations in a straightforward manner.] Once a distance matrix is available, it is both possible and conceptually useful to think of the set of proteins as occupying of a set of points in a space. This space will, in general, be of high dimensionality, and the specified set of distances may not obey the constraints which characterise a Euclidean space. The study of such systems is the province of Multivariate Analysis, which provides several approaches which can be used to delineate the structure of this type of space. Among those which have proven useful in various protein contexts are clustering (both hierarchical and partitioning methods), principal component analysis[5] and graph-theoretical methods.[3,12] Whichever tools are utilized, this analysis reveals the underlying organization of the set of proteins under consideration. This organization must be rationalized, by relating it to some set of known properties of the proteins. Properties which may be reflected in the organization of a given space include biochemical function, fold family, and evolutionary relationship.

Another way of thinking about the structure revealed by the foregoing approach is to regard the resulting space as a network of nearest-neighbor relationships between the proteins in the database. The physical significance of the network depends on the meaning of the distances between molecules. Furthermore, the structure of a network encodes important information about the process which gave rise to it. It is therefore important to develop precise descriptions of networks. The statistical mechanics of networks has been studied extensively in recent work,[13] and several general classes of network have been delineated, with strikingly different mathematical properties, carrying correspondingly diverse physical implications. A surprising result of this work is the demonstration of the widespread occurence of scale-free networks in diverse systems.

Scale-free networks are characterized by a power-law distribution of the number of links experienced by network nodes.[13] This indicates the presence in these systems of a small number of nodes with many links and a large number with only a few, a fact which has significant implications both for the mechanism of network generation (which is specific to a particular system) and for the resistance of the network to disruption (a general property of scale-free systems). These properties of scale-free networks have been reviewed in detail.[13]

Protein Sequences

Sequence alignment is one of the cornerstones of bioinformatics. It forms the basis for homology searching, in which the structure and/or function of a protein whose sequence alone is known is surmised by detecting similarity with other molecules of known structure and function. The ultimate objective of alignment is to blur the distinction between sequence and structure, by making it possible to relate three-dimensional structures using one-dimensional information. The alignment of sequence pairs can be used to produce quantitative descriptors of the similarity between them, and can therefore be used to produce distance matrices and to delineate the organization of sequence space. It is clear that the structure of the space (or equivalently of the network of protein sequences) depends entirely on the detailed characteristics of the individual modules which together comprise the alignment process.

Sequence alignment is a modular task.[14-16] The modules are:

An equivalence matrix—An objective function is necessary to determine the quality of an alignment. This function is calculated as the sum of similarities between residues which are declared to be in some sense equivalent. The investigator must therefore choose criteria for assigning a degree of similarity to amino acids in different sequences. These criteria are determined by those amino acid characteristics which are considered to be important in the context of the specific problem. Historically, the view was taken that one measured the probability that a given amino acid would be replaced by another in an evolutionary/mutation process. This led to the development of the PAM250 matrices.[17] The use of a large number of blocks of aligned sequence segments to count replacements gave the BLOSUM substitution matrices.[18] Other criteria are possible. Those based on physical rather than evolutionary properties are of particular interest, since they don't presuppose specific models or degrees of evolutionary distance.[19] These criteria include similarity with respect to a specified physical property, or similarity in some structural sense.[20-24] It should be noted that alignment can be performed using either similarity or distance matrices. The advantage of the former is that they make possible local alignment,[25] which is the preferred approach to database searching.

A set of gap parameters—It is often found that better correspondence between two sequences is obtained if account is taken of the presence of insertions and deletions in one or both. Such indels must be accounted for in calculating the alignment objective function, and this is usually done by means of a gap penalty function.[14,26] A common form of the penalty is the affine penalty function, in which a fixed penalty is counted for the initiation of a gap, and a length-dependent propagation penalty is added to account for the size of the gap.

An alignment algorithm—A method is required by which alignments between sequences are generated, a corresponding figure of merit calculated from the objective function, and the best alignment (and some of the runners-up[27-31]) selected. There are two general types of alignment- global, in which entire sequences are aligned, and local, in which a region of one sequence is matched with a region of another. The classical algorithm is some variant of dynamic programming.[32] More recent methods include FASTA[33] and BLAST,[34] hidden Markov models,[35,36] and variants of these.[37] Simultaneous consideration of multiple sequences introduces a set of additional problems into the alignment algorithm, which are rapidly exacerbated as the number of sequences grows. This problem has been addressed in several ways. Methods of calculating limited subsets of the dynamic programming matrix have been developed,[38] and genetic algorithms have been investigated.[39]

A criterion for success—A broadly accepted standard for successful sequence alignment is given by structure alignment.[40-42] If a sequence alignment method, applied to two proteins with known structure, produces an alignment similar to that resulting from an independent, structurebased identification of equivalent residues, the sequence alignment method is considered able to identify biophysically (or evolutionarily) relevant correspondences in the two sequences. An alternative, alignment-free approach to defining a distance between sequences is based on counting N-residue fragments.[43,44] In this approach a fingerprint for each sequence is provided by the distribution of frequencies of N-mers, and a distance function is constructed which measures the similarity of two frequency distributions. This approach has certain advantages over the alignment approach.

Since normalized distributions are compared, it is straightforward to define a distance between sequences of different molecular weight.

No gap parameters need be defined, since the comparison method automatically includes the effects of insertions and deletions.

On the other hand, this approach does not identify residues which may be functionally or structurally equivalent, since no alignment is produced. It has been shown[44] that the distances produced by this approach are equivalent to those arising from alignmentbased methods. There have been several classifications of large databases of protein sequences.[7-11,45] These have been directed almost entirely toward the goal of structure and function elucidation, and little if any attention has been paid to the network organization of the space. This is an important distinction because, as we remarked above, and will note again in connection with structure classification, the overall structure of a protein network is believed to carry important information about evolutionary processes.

In this section we wish to devote particular attention to some recent studies of an important but neglected aspect of sequence classification- the problem of information loss in sequence comparison. This point directly concerns the first module in the sequence alignment algorithm, and is equally relevant to alignment-free distance methods. We note that the construction of an equivalence matrix between amino acids is closely related to the use of reduced amino acid alphabets in protein studies. Adoption of an equivalence matrix is an implicit declaration that two amino acids, hitherto considered to be informatically distinct, are to some degree interchangeable. This step leads to the loss of information, and information loss must inevitably distort the structure of the sequence space/network. It is therefore of extreme interest to ask whether reduced alphabets (or amino acid equivalency matrices) can be constructed which are optimized to retain maximal information. The first point which must be decided is what kind of information one wishes to retain. Since we are considering information content in sequences, the natural choice is to maximize the retention of structural information. In recent publications, we have developed methods for constructing reduced alphabets which encode the maximum possible amount of local structural information.

In our initial studies[46] on the structural information content of sequence representations we used information theoretic and statistical methods, and protein sequence and structure data, to demonstrate the following points:

It is possible to quantitatively calculate the amount of structural information made available by knowledge of local sequence alone.This number depends on the representations used for both sequence and structure.

A contracted amino acid alphabet of any specified size can be constructed in a manner which retains maximal structural information. The loss of information resulting from optimized alphabet contraction was calculated, and it was shown that, in practical applications, this loss is offset by statistical improvements resulting from the greatly decreased number of distinct sequence fragments.

The optimal mapping of the 20 amino acids onto the reduced alphabet depends on the structure representation used. Structurally optimized alphabets as a function of size were produced for both the DSSP (secondary structure) and GBM (alpha-carbon backbone) representations. Examination of details of the clustering optimization process reveals that the former representation is only able to detect low-resolution properties of the amino acids, related to secondary structure preference and hydrophobicity. The GBM representation, on the other hand, gives reduced alphabets which reflect subtle conformational nuances of the amino acids.

In subsequent work,[47,48] we have extended this approach to consider the constraints imposed on the optimization of representations by the finite size of the databases from which we derive information. A serious problem in this regard is the presence of rare sequence fragments, for which it is not possible to construct statistically meaningful structure distributions. This problem was addressed by representing the structure distribution associated with a given sequence fragment as the superposition of two distributions- one specific to the sequence in question, and the other a background distribution with lower sequence specificity. The relative weights of these two components in the final structure distribution depend on the number of rare sequence fragments in the data set. When there are few rare fragments, the actual observed distribution is heavily weighted. When there are many, the background distribution is weighted more heavily, reflecting the lack of sequence-specific information. A Monte Carlo procedure was developed which makes it possible to simultaneously optimize distribution weights and amino acid clustering for a given alphabet size. The information-theoretic and statistical machinery developed in the course of these studies give optimized reduced alphabets, and associated structural distributions, which have the following characteristics:

1. They compensate for the scarcity of structural data;
2. They use multi-residue (context specific) information;
3. They contain the maximum amount of local structure information which the underlying data set allows.

It was demonstrated that the maximum structural information is encoded in sequence fragments six residues long. This length scale represents the optimal compromise between additional sequence information, intrinsic in longer fragments, and statistical deterioration due to the finite size of the protein database. The distribution of the amount of structural information encoded in local sequences was analyzed, and it was shown that there is at least a 35% variance in structural entropy among different sequence fragments. The result of these investigations is a set of contracted amino acid alphabets optimized to encode the maximum possible amount of structural information available in several commonly used structural representations. By construction, the structure of a sequence space based on these alphabets should represent structural relationships between sequences as accurately as possible. In work currently in progress,[49] these alphabets are being incorporated into both alignment-based and alignment-free sequence distance functions, and the effect of this optimization on sequence classification is being explored.

Protein Structures

Structure comparison, like sequence comparison, is one of the cornerstones of protein bioinformatics. The need for appropriate tools is clear in both the experimental and theoretical domains. It is of obvious interest to compare a newly determined structure to those which have already been solved, in order to correctly trace evolutionary and functional relationships between molecules. The efficacy of a structure prediction algorithm can only be evaluated by comparing the predicted and actual structures of test molecules. An elegant and perceptive review of conformational comparison methods has been given by Brown et al.[50] The earliest approaches to structure comparison were based on optimal superposition. One structure is translated and rotated relative toanother, fixed structure until a chosen figure of merit is optimized. In most cases, the figure of merit is the root-mean-square deviation (RMSD) between corresponding atoms of the two structures. This approach continues to be widely used in various incarnations. There are, however, a number of alternative metrics which can be used, and these have been compared recently by Wallin et al.[51]

There are several important concerns which should be noted here. The first is a general problem in the comparison of structures: The result which one obtains for a structure comparison depends critically on the method which is adopted, because structure comparison is a length-scale-dependent problem. The meaning of this point can be made clear by a simple example. Consider a protein consisting of two domains, and imagine generating an alternative conformation of the molecule by rotating one of the domains relative to the other around a single connecting bond. An attempt to compare the resulting conformation to the original conformation of the protein by optimized superposition will give a poor result, because it is no longer possible to bring the two domains into superposition simultaneously. Superposition algorithms are designed to operate on a length scale approximating the size of the molecules being compared.

Imagine, however, comparing the two conformations using an algorithm which compares **local** conformations along the two chains. The result will be a chain plot which indicates identity of the two conformations everywhere except at the single bond around which rotation occurred. A **qualitatively** different answer is generated by a method which considers the problem at a different length scale. This is a feature of the comparison problem which investigators must keep in mind.

The second point which must be made is specific to the superposition method. While the method gives meaningful results when the molecules being compared are reasonably similar, it is somewhat difficult to know how to interpret results for the comparison of proteins which differ significantly in molecular weight and/or structure.

What are the corresponding atoms in two unrelated structures? What is the meaning of a RMSD for such molecule pairs? It is in fact a question, as Godzik has pointed out,[52] whether there is a unique answer to the optimal superposition problem. In order to address these problems, other methods of comparison have been developed. An early alternative was based on the application of differential geometric (DG) methods to the description of protein conformation.[53-60] It was first demonstrated that a DG-based representation of chain structure can be defined, and that it is possible to base a distance function on that representation which describes local differences in chain folding. The output of the algorithm is a chain plot in which conformational differences at corresponding sites are revealed. This approach makes possible the detailed comparison of chain fragments of equal length, and can be used to compare chains of different lengths using a moving-window method. It was then demonstrated that the distribution of differential-geometric parameters can be used to define a length-independent fingerprint for any chain of known structure, and that these fingerprints can be used to compare the structures of proteins of different molecular weights. The method was used to carry out an all-against-all comparison of a small group of structures, giving a sparse description of the structure of structure space. This was the earliest quantitative comparison of protein structures, and the earliest attempt to quantitatively delineate the characteristics of structure space, known to this author. A limiting characteristic of the DG approach is the fact that it operates on a single, defined length scale within the molecule- the 4-alpha-carbon scale. All parameters are calculated from the coordinates of successive fragments of that size, and all information is therefore limited to structure at that scale. Attempts to create a representation in which the length scale is a definable parameter led to the development of the Generalized Bond Matix (GBM) representation.[3] This representation of backbone structure is far more flexible than the differential geometric representation, in that fragments of any length, defined using any chemical or virtual bonds of interest, can be used as a basis for structure description and comparison. It shares with the DG representation several advantages over superposition methods. Both allow the definition of normalized (molecular weight independent) structure fingerprints, making it possible to compare chains of arbitrarily different molecular weight. Both representations share with the alignment-free sequence comparison methods discussed above the characteristic that the presence of insertions and deletions is accounted for automatically, without the necessity for defining gap initiation and propagation penalties. At the same time, sequence ordering information present in superposition algorithms is lost, or at least obscured, in the distribution-based methods. (There is some evidence that, if the sequence fragments considered are sufficiently long, the correlations necessary to reconstruct the sequence from the fragment distribution are, in fact, preserved.[61]) The GBM representation was used[3,62] to carry out a detailed classification of a database of structures which well represented the known structure universe at the time. This involved the used of techniques from graph theory to study clustering in the space. One hundred and twenty three structures were classified, and it was shown that structure space can be represented as a nonuniform continuum of structures, grading from all-helical structures at one edge of the space to sheet/barrel structures at the other. Details of the distribution of structures were investigated, as were the effect of the length scale and resolution of the chosen representation on the structure of the space.

A method for comparing structures based on intramolecular distance matrices was developed by Yee and Dill,[4] and used to reanalyze structure space. Although this method is very different from the GBM approach, the anatomy of structure space which it revealed is substantially similar to that discussed in our own work.[3] Holm and Sander[63,64] have also used distance matrices to compare structures. An excellent summary and review of earlier studies of protein classification, together with a discussion of the coarse-grained statistical properties of protein space, has been given by Brenner et al.[65] Nussinov, Wolfson and collaborators have developed an approach to structure comparison based on tools of pattern recognition. The method is based on a hashing algorithm first applied to computer vision studies, and is able to carry out sequenceorder- independent comparisons. This method has been used[66] to construct a nonredundant dataset of structures and investigate characteristics of the resulting space. In later work, the same group has used hashing methods to carry out multiple alignments[67,68] and detect common structural motifs.[67] Recently, Hou et al[5] have revisited the structure of fold space using a method based on the DALI comparison algorithm,[63] which is distance-matrix-based. Using a factor analysis of the resulting protein-protein distances, they constructed a picture in which the high-dimensional fold space was contracted to its three most significant dimensions. The folds cluster into disjoint regions corresponding to the classical low resolution definition of fold types- α, β, α/β and $\alpha+\beta$. The authors report that domain size is an important determinant of the structure of the space.

The statistical significance of a given comparison is an important point to address. Levitt and Gerstein[69] have given a general framework for the statistical validation of both sequence and structure comparisons. A number of workers have investigated the network properties of structure space, which carries important implications for fold evolution. One approach to this question is to analyze the distribution of domain family sizes, and this has been studied by Qian et al,[70] Kuznetsov[71] and Karev et al.[72] The network of domain relationships has also been constructed directly by Dokholyan et al[12] using DALI-generated distances. All evidence suggests that the distribution of contact numbers follows the power law characteristic of scale-free networks. Several dynamic simulations of the evolutionary process have been developed[12,72,73] which give this type of scale-free behavior in an model proteome. Perhaps the central problem in protein science is the fact that sequence and structure classifications do not give the same picture of protein space. Various manifestations of this fact have been known for many years. It is widely recognized, for example, that some of the more common architectures are adopted by large groups of protein sequences, many of which exhibit no detectable mutual sequence similarity. An understanding of the mechanism which determines architecture choice will lead directly to a solution of the classical folding problem- the prediction of structure from sequence. This would therefore seem to be a problem worth study. In recent work we have addressed this question.[1] Our approach is based on the following observations:

The set of proteins folding to a specified architecture frequently includes molecules which are not only unrelated by homology, but also differ widely in molecular weight.

The choice of architecture must be determined by physical properties of the amino acids in the sequence.

The second of these points suggests that the architecture signal is expressed in some pattern of physical characteristics. The first suggests that the signal must scale with sequence length. In order to investigate this problem, we therefore need to express protein sequences in terms of amino acid properties. There are many property sets available, and an arbitrary choice of properties leads to a twofold problem: The set chosen can be simultaneously incomplete and correlated. This problem has been solved by Scheraga and collaborators,[74,75] who carried out a factor analysis of all available sets of amino acid attributes. They showed that the entire attribute dataset can be described by 10 property factors. Four major factors correspond essentially to individual amino acid properties, and the remaining six are superpositions of a limited number of properties. The 10 factors together carry 86% of the variance for the entire dataset. In mathematical terms this result means that the physics of the amino acids can be embodied in a set of 20 10-vectors, each of which gives the weights of the 10 factors for a particular amino acid. It follows that an N-residue protein sequence can be described by a set of 10 N-number strings, each of which traces the value of a particular property factor along the chain.

The next step is to construct a database of proteins suitable for the problem. Our approach is to assemble sets of proteins which fold to a common architecture but exhibit low sequence homology. We chose two architectures which are sufficiently populated that statistically meaningful samples of this type can be constructed- the TIM barrel and Immunoglobulin folds. For each of these folds, an ensemble of sequences was chosen with pairwise similarities well below the homology limit. Having written the amino acid sequence in a property-related numerical form, we wish to extract scalable signals which can be associated with protein architecture. We therefore carried out a Fourier analysis of the property strings for each protein in a specified architecture group. Note that the Fourier transform of a property string, for any wave number, is a function of the entire string. A consequence of this fact is that chain length is not a relevant variable in Fourier space, and the Fourier power spectra of chains of different sequence length can be directly compared. The Fourier analysis was followed by signal averaging over all proteins in the architecture group, which enabled us to distinguish Fourier components which are common to all members of the group from those which are characteristic of specific sequences. It is important to ask whether the common Fourier components detected are statistically significant. In order to address this concern, randomized protein sequence groups were generated, by independently permuting the sequences of each of the proteins in the original architecture group. The entire Fourier analysis/signal averaging process was repeated on the permuted sequence groups. This was iterated 10,000 times for each architecture group, and each Fourier component arising from the actual sequence was compared to the average Fourier coefficient and standard deviation arising from the ensemble of random sequence groups. Only those Fourier components of the actual sequences which exceeded the average by two standard deviations were regarded as significant. It was found[1,76] that Fourier components which satisfy this requirement do indeed exist. A particularly dramatic result is observed in the TIM barrel group, in which a composite power spectrum signal was found at $k = 21$ which is 18σ above the average. A set of signals in the range 5σ-6σ was also found in the Immunoglobulin group. A particularly fascinating insight into the mechanism of architecture selection emerges when we ask in which physical properties these signals are expressed. It is found that, while essentially all the proteins in a given architecture group exhibit statistically significant signals at the values of k identified by the signal averaging procedure, these signals are expressed in different properties in the various proteins of the group. This suggests that an architecture can be generated by a well-defined set of periodicities, but that these periodicities can be expressed in a wide variety of physical properties. This constitutes a degeneracy in the architecture code. The existence and characteristics of the architectural signals provide an understanding of certain fundamental observations about protein architecture and folding. The fact that proteins with no apparent sequence homology fold to common architectures[77] is an immediate consequence of the degeneracy of architecture signals, which guarantees that there are many, dissimilar ways to produce a given architecture. It has also been noted[78–82] that proteins with similar architecture but no mutual homology fold with similar rates. A connection between this observation and the properties of architecture signals is readily made. The wavelength associated with a sinusoidal signal of wave number k in a sequence of length N is N/k. The sequence is composed of a set of k segments of this length, each of which contains a region in which the associated physical property of the amino acids is strongly expressed, flanked by regions in which it is weakly expressed. Note that, while the relevant physical property and the length of the segments differ in the various proteins of an architectural group, the number of segments is the same in all the sequences. These observations are consistent with a scenario[83,84] in which folding is governed by a number of early nucleation events, distributed over the entire sequence, each of which takes place in a relatively short, localized region—the segments delimited by the architectural signal. In each protein of a particular architecture group, the nature of these events is determined by the properties in which the folding signal is expressed. This suggestion is supported by the experimental demonstration85 that proteins of similar architecture can fold by different mechanisms. It is possible1, but not mandatory, that the segments defined by the architectural signals might be correlated with structures visible in the native fold. This possibility, and other implications of the present results, will be explored in forthcoming work. These results suggest an alternate view of the classification problem. The reason that sequence and structure classification reveal different, parallel universes is that sequence classification, as currently practiced, is based on "incorrect" parameters. If one takes the reasonable (and widely held) view[40-42] that structure classification is the more fundamental process, it becomes clear that we should be searching for those sequence-related variables which give the closest correspondence possible between the two protein spaces. The approach we have just outlined, in looking for physical signals which encode architecture in sequence, represent a step in this direction, and away from a search for codes based on residue identity.

Acknowledgements

Our work in these areas was supported by Grant LM06789 from the National Library of Medicine of the National Institutes of Health. The author would like to acknowledge the outstanding contributions of Drs. Igor Kuznetsov and Armando Solis to the work summarized herein.

References

1. Rackovsky S. "Hidden" Sequence periodicities and protein Architecture. Proc Nat Acad Sci USA 1998; 95:8580-8584.
2. M Gerstein. A structural census of genomes: Comparing bacterial, eukaryotic, and archaeal genomes in terms of protein structure. J Mol Biol 1997; 274:562-576.
3. Rackovsky S. Quantitative organization of the known protein X-ray structures. I. Methods and short length-scale results. Proteins: Structure, Function and Genetics 1990; 7:378-402.
4. Yee DP, Dill KA. Families and the structural relatedness among globular proteins. Prot Sci 1993; 2:884-899.
5. HOU J, Sims GE, Shang C et al. A global representation of the protein fold space. Proc Nat Acad Sci USA 2003; 100:2386-2390.
6. Holm L, Sander C. Dali/FSSP classification of three-dimensional protein folds. Nucleic Acid Res 1997; 25:231-234.
7. Gonnet GH, Cohen MA, Benner SA. Exhaustive matching of the entire protein sequence database. Science 1992; 256:1443-1445.
8. Linial M, Linial N, Tishby N et al. Global self-organization of all known protein sequences reveals inherent biological signatures. J Mol Biol 1997; 268:539-556.

9. Gracy J, Argos P. Automated protein sequence database classification. I. Integration of compositional similarity search, Local similarity search, and multiple sequence alignment. Bioinformatics 1998; 14:164-173.
10. Wang H-C, Dopazo J, De La Fraga LG et al. Self-organizing tree-growing network for the classification of protein sequences. Prot Sci 1998; 7:2613-2622.
11. Yona G, Linial N, Linial M. Proto Map: Automatic classification of protein sequences, A heirarchy of protein families, and local maps of the protein space. Proteins: Structure Function and Genetics 1999; 37:360-378.
12. Dokholyan NV, Shakhnovich B, Shakhnovich EI. Expanding protein universe and its origin from the biological big bang. Proc Nat Acad Sci USA 2002; 99:14132-14136.
13. Albert R, Barabási A-L. Statistical mechanics of complex networks. Rev Mod Phys 2002; 74:47-97.
14. Myers EW. Seeing conserved signals: Using algorithms to detect similarities between biosequences. In: Lander ES, Waterman MS, eds. Calculating the Secrets of Life. Washington, DC: National Academy Press, 1995.
15. Barton GJ. Protein sequence alignment techniques. Acta Cryst 1998; D54:1139-1146.
16. Smith TF. The art of matchmaking: Sequence alignment methods and their structural implications. Structure 1999; 7:R7-R12.
17. Dayhoff MO, Eck RV. Atlas of Protein Sequence and Structure. Silver Spring, MD: NBRF Press, 1996:2.
18. Henikoff S, Henikoff J. Amino acid substitution matrices from protein blocks. Proc Nat Acad Sci USA 1992; 89:10915-10919.
19. Altschul SF. A protein alignment scoring system sensitive at all evolutionary distances. J Mol Evol 1993; 36:290-300.
20. Naor D, Fischer D, Jernigan RL et al. Amino acid pair interchanges at spatially conserved locations. J Mol Biol 1996; 256:924-938.
21. Russell RB, Saqi MAS, Sayle RA et al. Recognition of analogous and homologous protein folds: Analysis of sequence and structure conservation. J Mol Biol 1997; 269:423-439.
22. Johnson MS, Overington JP. A structural basis for sequence comparison: An evaluation of scoring methodologies. J Mol Biol 1993; 233:716-738.
23. Prlic A, Domingues FS, Sippl MJ. Structurederived substitution matrices for alignment of distantly related sequences. Protein Engineering 2000; 13:545-550.
24. Blake JD, Cohen FE. Pairwise sequence alignment below the twilight zone. J Mol Biol 2001; 307:721-735.
25. Smith TF, Waterman MS. Identification of common molecular subsequences. J Mol Evol 1981; 147:195-197.
26. Altschul SF. Generalized affine gap costs for protein sequence alignment. Proteins: Structure Function and Genetics 1998; 32:88-96.
27. Argos P, Vingron M, Vogt G. Protein sequence comparisons: Methods and significance. Protein Eng 1991; 4:375-383.
28. Saqi M, Sternberg M. A simple method to generate Nontrivial alternate alignments of protein sequences. J Mol Biol 1991; 219:727-732.
29. Zuker M. Suboptimal sequence alignment in molecular biology: Alignment with error analysis. J Mol Biol 1991; 221:403-420.
30. Agarwal P, States D. A bayesian evolutionary distance for parametrically aligned sequences. J Comput Biol 1996; 3:1-17.
31. Vingron M. Near-optimal sequence alignment. Curr Opin in Struct Biol 1996; 6:346-352.
32. Horowitz E, Sahni S. Fundamentals of Computer Algorithms. New York, NY: Computer Science Press, 1978:198-247.
33. Pearson W, Lipman D. Improved tools for biological sequence comparison. Proc Nat Acad Sci USA 1988; 85:2444-2448.
34. Altschul S, Gish W, Miller W et al. Basic local alignment search tool. J Mol Biol 1990; 215:403-410.
35. Krogh A, Brown M, Mian J et al. Hidden markov models in computational biology: Applications to protein modeling. J Mol Biol 1994; 235:1501-1531.
36. Eddy S. Hidden markov models. Curr Opin Struct Biol 1996; 6:361-365.

37. Bucher P, Hoffman K. A sequence similarity algorithm based on a probabilistic interpretation of an alignment scoring system. In: States D, Gaasterland T, Hunter L, Smith R, eds. ISMB-4. Menlo Park: AAAI Press, 1996.
38. Lipman DJ, Altschul SF, Kececioglu J. A tool for multiple sequence alignment. Proc Nat Acad Sci USA 1989; 86:4412-4415.
39. Notredame C, Higgins DG. SAGA: Sequence Alignment by Genetic Algorithm. Nucl Acids Res 1996; 24:1515-1524.
40. Brenner SE, Chothia C, Hubbard TJP. Assessing sequence comparison methods with reliable structurally identified distant evolutionary relationships. Proc Nat Acad Sci USA 1998; 95:6073-6078.
41. Sauder JM, Arthur JW, Dunbrack Jr RL. Large-scale comparison of protein sequence alignment algorithms with structure alignments. Proteins: Structure, Function and Genetics 2000; 40:6-22.
42. Panchenko AR, Bryant SH. A comparison of position-specific score matrices based on sequence and structure alignments. Prot Sci 2002; 11:361-370.
43. Blaisdell BE. A measure of the similarity of sets of sequences not requiring sequence alignment. Proc Nat Acad Sci USA 1986; 83:5155-5159.
44. Blaisdell BE. Average values of a dissimilarity measure not requiring sequence alignment are twice the averages of conventional mismatch counts requiring sequence alignment for a variety of computer-generated model systems. J Mol Evol 1991; 32:521-528.
45. Yona G, Levitt M. A unified sequence-structure classification of protein sequences: Combining sequence and structure in a map of the protein space. Tokyo: Proceedings of the Fourth Annual Conference on Computational Molecular Biology, 2000:308-317.
46. Solis AD, Rackovsky S. Optimized representations and maximal information in proteins. Proteins: Structure Function and Genetics 2000; 38:149-164.
47. Solis AD, Rackovsky S. Optimally informative backbone structural propensities in proteins. Proteins: Structure Function and Genetics 2002; 48:463-486.
48. Solis AD. Structural information from local sequence of proteins and DNA. Thesis, Mt. Sinai School of Medicine of New York University 2002; 148-191.
49. Kuznetsov IB, Solis AD, Rackovsky S. (work in progress).
50. Brown NP, Orengo CP, Taylor WR. A protein structure comparison methodology. Computers Chem 1996; 20:359-380.
51. Wallin S, Farwer J, Bastolla U. Testing similarity measures with continuous and discrete protein models. Proteins: Structure Function and Genetics 2003; 50:144-157.
52. Godzik A. The structural alignment between two proteins: Is there a unique answer? Prot Sci 1996; 5:1325-1338.
53. Rackovsky S, Scheraga HA. Differential geometry and polymer conformations. I. On the comparison of polymer conformations. Macromolecules 1978; 11:1168-1174.
54. Rackovsky S, Scheraga HA. Differential geometry and polymer conformations. II. Mathematical considerations and a conformational distance function. Macromolecules 1980; 13:1440-1453.
55. Rackovsky S, Scheraga HA. Intermolecular anti-parallel beta sheet: Comparison of predicted and observed conformations of gramicidin S. Proc Nat Acad Science USA 1980; 77:6965-6967.
56. Rackovsky S, Scheraga HA. Differential geometry and polymer conformations. III. Nearest-neighbor correlations and medium-range structure. Macromolecules 1981; 14:1259-1269.
57. Rackovsky S, Scheraga HA. Differential geometry and polymer conformations. IV. Conformational and nucleation properties of individual amino acids. Macromolecules 1982; 15:1340-1346.
58. Rackovsky S, Scheraga HA. Differential geometry and protein folding. Accounts of Chemical Research 1984; 17:209-214.
59. Rackovsky S, Goldstein DA. Differential geometry and protein conformation. V. Medium-range conformational influence of the individual amino acids. Biopolymers 1987; 26:1163-1187.
60. Rackovsky S, Goldstein DA. Protein comparison and classification: A differential geometric approach. Proc Natl Acad Sci USA 1988; 85:777-781.
61. Pevzner P. Personal communication.
62. Rackovsky S. Quantitative classification of the known protein X-ray structures. Polymer Preprints 1990; 31:205.

63. Holm L, Sander C. Protein structure comparison by alignment of distance matrices. J Mol Biol 1993; 223:123-138.
64. Holm L, Sander C. Mapping the protein universe. Science 1996; 273:595-602.
65. Brenner SE, Chothia C, Hubbard TJP. Population statistics of protein structures: Lessons from structural classifications. Curr Opin Struct Biol 1997; 7:369-376.
66. Fischer D, Tsai C-J, Nussinov R et al. A 3D sequence-independent representation of the protein data bank. Protein Engineering 1995; 8:981-997.
67. Leibowitz N, Fligelman Z, Nussinov R et al. Automated multiple structure alignment and detection of a common motif. Proteins: Structure Function and Genetics 2001; 43:235-245.
68. Dror O, Benyamini H, Nussinov R et al. MASS: Multiple structure alignment by secondary structures. Bioinformatics 2003; 19(Suppl.1):i95-i104.
69. Levitt M, Gerstein M. A unified statistical framework for sequence comparison and structure comparison. Proc Nat Acad Sci USA 1998; 95:5913-5920.
70. Qian J, Luscombe NM, Gerstein M. Protein family and fold occurrence in genomes: Power-law behaviour and evolutionary model. J Mol Biol 2001; 313:673-681.
71. Kuznetsov VA. In: Zhang W, Shmulevich I, eds. Computational and Statistical Approaches to Genomics. Boston: Kluwer, 2002:125-171.
72. Karev GP, Wolf YI, Rzhetsky AY et al. In: Galperin MY, Koonin EV, eds. Amsterdam, Horizon: Computational Genomics From Sequence to Function 2003:261-314.
73. Yanai I, Camacho C, DeLisi C. Predictions of gene family distributions in microbial genomes: Evolution by gene duplication and modification. Phys Rev Lett 2000; 85:2641-2644.
74. Kidera A, Konishi Y, Oka M et al. Statistical analysis of the physical properties of the 20 naturally occurring amino acids. J Prot Chem 1985; 4:23-55.
75. Kidera A, Konishi Y, Ooi T et al. Relation between sequence similarity and structural similarity in proteins. Role of important properties of amino acids. J Prot Chem 1985; 4:265-297.
76. Rackovsky S. work in progress.
77. Yang A-S, Honig B. An integrated approach to the analysis and modeling of protein sequences and structures. II. On the relationship between sequence and structural similarity for proteins that are not obviously related in sequence. J Mol Biol 2000; 301:679-689.
78. Alm E, Baker D. Matching theory and experiment in protein folding. Curr Opin Struct Biol 1999; 9:189-196.
79. Shea JE, Onuchic JN, Brooks IIIrd CL. Exploring the origins of topological frustration: Design of a minimally frustrated model of fragment B of protein A. Proc Nat Acad Sci USA 1999; 96:12512-12517.
80. Onuchic JN, Nymeyer H, Garcia AE et al. The energy landscape theory of protein folding: Insights in folding mechanism and scenarios. Adv Prot Chem 2000; 53:87-152.
81. Micheletti C, Banavar JR, Maritan A et al. Protein structures and optimal folding from a geometrical variational principle. Phys Rev Lett 1999; 82:3372-3375.
82. Abkevich V, Gutin A, Shakhnovich E. Specific nucleus as the transition state for protein folding: Evidence from the lattice model. Biochemistry 1994; 33:10026-10036.
83. Baldwin RL. Folding concensus? Nature Struct Biol 2001; 8:92-94.
84. Fersht AR. Transition-state structure as a unifying basis in protein-folding mechanisms: Contact order, chain topology, stability, and the extended nucleus mechanism. Proc Natl Acad Sci USA 2000; 97:1525-1529.
85. Burns LL, Dalessio PIM, Ropson IJ. Folding Mechanism of three structurally similar β-Sheet Proteins. PROTEINS: Structure, Function and Genetics 1998; 33:107-188.

CHAPTER 12

The Role of Computation in Complex Regulatory Networks

Pau Fernández and Ricard V. Solé*

Abstract

Biological phenomena differ significantly from physical phenomena. At the heart of this distinction is the fact that biological entities have computational abilities and thus they are inherently difficult to predict. This is the reason why simplified models that provide the minimal requirements for computation turn out to be very useful to study networks of many components. In this chapter, we briefly review the dynamical aspects of models of regulatory networks, discussing their most salient features, and we also show how these models can give clues about the way in which networks may organize their capacity to evolve, by providing simple examples of the implementation of robustness and modularity.

Introduction

As has been highlighted by John Hopfield, several key features of biological systems are not shared by physical systems. The origin of such difference stems from the relevance that information plays in the first, which is not shared by the second.[1] Although living entities follow the laws of physics and chemistry, the fact that organisms adapt and reproduce introduces an essential ingredient that is missing in the physical sciences.[2] Due to this fact, biological structures result from evolutionary pathways and as such they are contingent.[3]

Perhaps the clearest consequence of the role of information is the observation that biological entities perform **computations**: there is an evolutionary payoff placed on being able to predict the future. Typically, more complex organisms are better able to cope with environmental uncertainty because they can compute, i.e., they have memory or some form of internal plasticity, and they can also make calculations that determine the appropriate behavior using what they sense from the outside world.

Computation thus becomes a crucial ingredient when dealing with the description of biocomplexity and its evolution, because it turns out to be much more relevant than the underlying physics. Its dynamics is governed mainly by the transmission, storage and manipulation of information, a process which is highly nonlinear. This nonlinearity is well illustrated by the nature of signaling in cells: local events involving a few molecules can produce a propagating cascade of signals through the whole system to yield a global response. If we try to make predictions about the outcomes of these signaling events in general, we are faced with the

*Corresponding author: Ricard V. Solé—ICREA-Complex Systems Lab, Universitat Pompeu Fabra (GRIB), Dr Aiguader 80, 08003 Barcelona, Spain and Santa Fe Institute, 1399 Hyde Park Road, Santa Fe, New Mexico 87501, U.S.A. Email: ricard.sole@upf.edu

Power Laws, Scale-Free Networks and Genome Biology, edited by Eugene V. Koonin, Yuri I. Wolf and Georgy P. Karev.

inherent unpredictability of computational systems.[4] It is at this level where computation becomes central and where idealized models of regulatory networks seem appropriate enough to capture the essential features at the global scale.

Cells are probably the most complete example of this traffic of signals at all levels. They comprise millions of molecules that act coherently persisting far from equilibrium by the exchange of matter, energy and information with the environment. All this molecular processes, ultimately controlled by genes, take place at different points in space and time and involve the leading participation of proteins, which act as the nanomachines that drive cellular dynamics. The cellular network can be divided into three major self-regulated sub-webs:

- the **genome**, in which genes can affect each other's level of expression;
- the **proteome**, defined by the set of proteins and their interactions by physical contact; and
- the metabolic network (or the **metabolome**), integrated by all metabolites and the pathways that link each other.

All this subnetworks are very much intertwined since, for instance, genes can only affect other genes through special proteins, and some metabolic pathways, regulated by proteins themselves, may be the very ones to catalyze the formation of nucleotides, in turn affecting the process of translation.

It is not difficult to appreciate the enormous complexity that these networks can achieve in multicellular organisms, where large genomes have structural genes associated with at least one regulatory element and each regulatory element integrates the activity of at least two other genes. The nature of such networks started to be understood from the analysis of small prokaryotic regulation subsystems and the current picture indicates that even the smallest known webs that shape cellular behavior are indeed very complex.[5,6]

Luckily, all this extraordinary complexity can be abstracted, at least at some levels, to simplified models which can help in the study of the inner-workings of cellular networks. Overall, irrespective of the particular details, biological systems show a common pattern: some low-level units produce complex, high-level dynamics coordinating their activity through local interactions. Thus, despite the many forms of interaction found at the cellular level, all come down to a single fact: the state of the elements in the system is a function of the state of the other elements it interacts with. What models of network functioning try, therefore, is to understand the basic properties of general systems composed of units whose interactions are governed by nonlinear functions. These models, being simplifications, do not allow to make predictions at the level of the precise state of particular units. Their average overall behavior, however, can shed light into the way real cells behave as a system.

On the other hand, whereas the question of how networks of many components can achieve global order is very important, it is no less important to gain an understanding of how such networks could have been assembled step by step throughout the evolutionary process. It seems sensible to expect some properties of these networks to directly influence their capacity to smoothly integrate the changes that can make them fitter in the next generation. In this context, technology should immensely benefit from a deep knowledge of the processes behind biological evolution, since by design, engineered systems are not at all susceptible of blind tinkering. It is interesting, therefore, to explore how the same simplified models used to understand global dynamics can give hints as to how "evolvability" could be put into practice.

In summary, in this chapter we will explore the computational dimension of cellular networks. We will see that biological networks may be computationally **irreducible**, and hence why Boolean units are appropriate to understand their global properties. We will also briefly review the most important features of the Kauffman model, and their implications for computation. Finally, taking advantage of the Boolean approximation, we will show how important aspects of the capacity to evolve such as robustness and innovation could be implemented, through the use of simple, clear examples.

The Evidence for Computing Networks

Molecules, proteins and genes interact with each other in many ways, and the result of their interactions is the coordinated behavior we observe. The first step is, therefore, to identify the different kinds of elements which make up regulatory networks and to describe their forms of interaction.

Perhaps the most important units in regulatory networks are genes, which interact through gene regulation. Genes are translated into proteins by means of a transcription machinery that is controlled by multiple mechanisms. Interference with these mechanisms allows certain molecules to alter the level of expression of specific genes, as the diagram of Figure 1.2 shows. Transcription is basically initiated at the promoter region, which has usually a "TATA" sequence, marking the binding site of TBP ("TATA"-binding protein). This protein is the first of a series of proteins, known as general transcription factors, that help to position the RNA polymerase correctly at the promoter. The most basic regulation, therefore, involves DNA binding proteins, or regular transcription factors, that either block the promoter, obstructing transcription or increase the probability of attachment of the RNA polymerase, enhancing it. These proteins operate in the vicinity of the promoter and in a majority of cases form complexes made of many units that combinatorially bind to DNA.

In addition to the close binding of transcription factors, other mechanisms are known that play a significant role in transcription regulation, including modifications of this basic scheme like downstream and distal enhancers or totally different mechanisms such as insulation,[7] alternative splicing or post-transcriptional modification.[8] All this mechanisms affect translation and therefore determine the level of expression of a certain gene at a given instant, given the concentration of its multiple regulators. This level of expression produces a certain concentration of the protein molecules that are the products of translation. Actually, different proteins can be produced from the activation of a certain gene due to post-translational modifications, giving rise to different regulatory elements in the network. Figure 1.3 shows an example in which the direct product of a gene can turn into two different proteins, depending on the presence of another "scissor"-like protein that cleaves the initial molecule.

The concentration of each protein molecule is, however, not only regulated at the level of transcription. Very many of them have structures that can be greatly modified in the presence of other molecules such as metabolites or other proteins. This requires their separate treatment as regulatory elements, since the different shapes usually carry out different tasks. Figure 1.1 shows a transmembrane protein which, in the presence of some metabolite, changes its conformation and becomes active at another site. These processes are the basis of the functioning of the signaling network, which comprises membrane receptors, intracellular signaling proteins and the receivers of the messages, for instance enzymes and regulatory or cytoskeletal proteins. Many of the components of this network are proteins that can only be in one of two states, active or inactive. Other proteins are inactive alone but active while bound to others in complexes, as shown in Figure 1.4, up to very high levels of complication. As regulatory elements in their own right, these complexes also qualify as units in the regulatory network.

To summarize, many entities in cellular networks can be identified as the basic units of regulation, mainly distinguished by their unique roles with respect to interaction with other units. These basic units are genes, each of the proteins that the genes can produce, each of the forms of a protein, protein complexes, and all related metabolites. These units have associated values that either represent concentrations or levels of activation. These values depend on the values of the units that affect them due to the mechanisms discussed, plus some parameters that govern each special form of interaction.

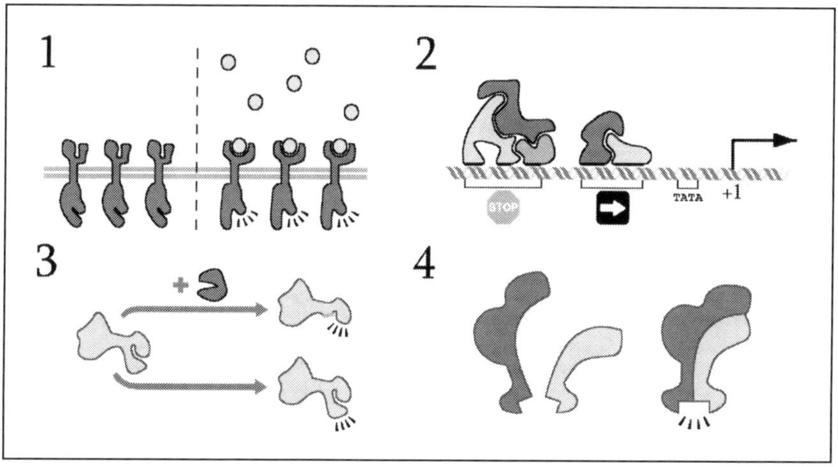

Figure 1. Several ways in which the units of cellular networks interact. 1) Signal transduction: a membrane protein becomes active if a certain metabolite is present outside the cell. 2) Gene regulation: genes are transcribed starting at the +1 site and are affected positively by an upstream activator sequence and negatively by a silencer sequence. Both sequences can be bound by protein complexes. 3) Postranslational modification: a given protein can be modified after transcription to yield two different forms depending on the presence of other proteins. 4) Complex formation: the union of two proteins exposes a new active site that makes the complex active only then.

Modeling

To make the description more concrete, is interesting to look at a complete, real example. In Figure 2 the circuit of the segment polarity network of *Drosophila melanogaster* is shown. The genes in this network are expressed throughout the life of the fly, and its pattern defines and, more importantly, maintains the borders of the segments since the first stages of development. This is a network in which all the elements discussed are present, displaying many forms of interaction, and in particular, the same 4 different mechanisms depicted in Figure 1 are highlighted by 4 boxes numbered accordingly. For example, *wg* (*wingless*) interacts with *en* (*engrailed*) in the neighboring cells by secreting a protein, *WG*, that binds to a membrane receptor *FZ* which, when activated, enhances the transcription of *en*. It is perhaps easier, looking at this diagram, to imagine how complex the dancing concentrations of genes, proteins or complexes are, all regulated through their input links and in turn regulating other elements.

Computer modeling of this network, however, has provided insight into various questions. A very important result is the fact that this network seems to be a conserved module. Evidence for this has been obtained by simulations demonstrating its robustness against the change of parameters. If the regulatory elements are modeled using a continuous-valued approach, a set of equations can be defined governing the rates of change in their populations, levels of expression, etc. Altogether, the unknown kinetic constants that have to be specified amounts to 48: half-lives of messenger RNAs and proteins, binding rates, cooperativity coefficients, etc. Surprisingly, from a huge number of possible combinations of parameters, many of them have actually a stable pattern that corresponds to the known pattern of activity of the genes, thus suggesting that the module is very robust.[10] In fact, other work has come to similar conclusions with respect to other mechanisms, such as adaptive responses in bacterial chemotaxis.[11] It seems that the topology of the network plays, in some cases, a more important role than the exact mechanisms at each node.

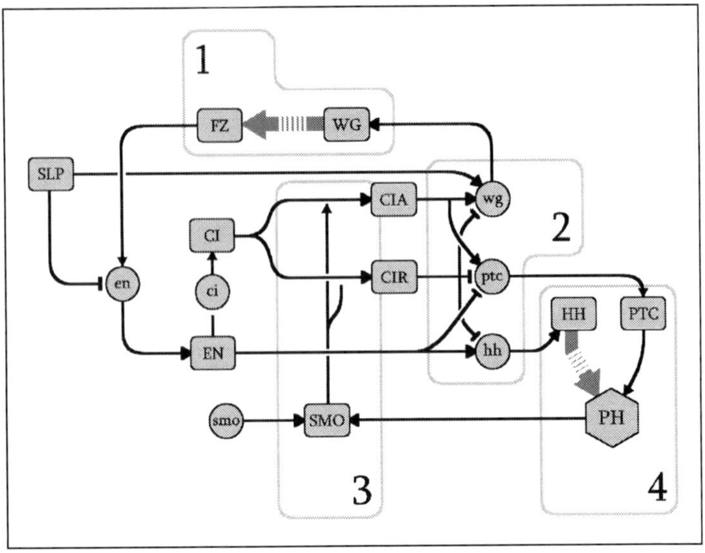

Figure 2. The network of interactions between the segment polarity genes (modified from ref. 9). Rectangles represent proteins, circles genes, and hexagons protein complexes, respectively. Special, thick arrows are for transmembrane links. Examples of different types of interaction are numbered from 1 to 4. 1) Signaling network: *WG* protein binds receptor *FZ*. 2) Transcription control: *wg*, *ptc*, and *hh* transcription is controlled by the activation of other genes. 3) Postranslational control: after translation, *CI* is transformed into *CIR* or *CIA* depending on the presence of *SMO*. 4) Complex formation: the complex *PH* is formed only when HH and PTC are present.

This is precisely the thesis of another work:[9] "our purpose here is to demonstrate that in one well-characterized system, knowledge of the interactions together with their signatures, by which we mean whether an interaction is activating or inhibiting, is enough to reproduce the main characteristics of the network dynamics". The Boolean network presented, albeit a simplification, seems to capture the essential features because it matches the patterns of activity not only of the wild-type embryo, but of some known mutants, and it also points to other possible effects of mutations that have not been observed. This is done through an exhaustive analytical treatment of the resulting equations, as well as simulations, that in addition reveal the important roles of some of the genes involved.[9] In addition to this, other work approaches gene networks within the context of the evolution of development.[12] It is true that the network may not be so crucial in other cases, but the results nevertheless suggest the importance of aggregated behavior.

In brief, the modeling of regulatory networks involves different methods that give answers to different questions,[13] but ultimately, these methods also illustrate that there is a deep, common pattern: simulation by computer seems to be the key to the solutions. As Venter puts it: "If we hope to understand biology, instead of looking at one little protein at a time, which is not how biology works, we will need to understand the integration of thousands of proteins in a dynamically changing environment. A computer will be the biologist's number one tool".* A crucial question, then, arises: Why do we need a computer to be able to study biology at all? Some insights into this question are in fact given by the theory of computation.

* As quoted in reference 14.

Irreducibility

One of the most important problems in the theory of computation is the halting problem.[15] It concerns the automatic verification of software in the following terms: one is given a computer program *A* and what the program is supposed to do, and the task is to design an automatic process that verifies the correctness of the program. In other words, another program *B* has to be written that, given a description of *A* and the correct outputs, predicts what are the outputs for each set of inputs, and just checks that the answers are correct. As simple as the problem seems, it is unsolvable: there is no such program *B*. The trap lies in the fact that, to solve it, a computer has to be "smarter" than another computer. Instead of testing the execution of program *A* by explicitly following it through, *B* has to be able to make some kind of shortcut that enables it to predict the outcome without having to follow each step, to avoid, for instance, the fact that *A* may enter a very long or complicated loop. Since *B* cannot exist, there are no such shortcuts to the long-term dynamics of computers, and their step-by-step evolution must be followed perforce. This impossibility to predict is called **irreducibility**, and has been hypothesized to be much more common that usually acknowledged.[4]

The fact that regulatory networks may be irreducible seems to be a plausible hypothesis, since computer modeling of regulatory networks seems to be the only way to deal with their complexity. Apart from that, there seems to be some awareness of this fact, since some authors have treated cellular networks with the tools of electronic design,[16] and compared molecules with computational elements:[17] "Putting aside for the moment the question of whether it is useful or even sensible to view them in this manner, it is nevertheless true that protein molecules are in principle able to perform a variety of logical or computational tasks". An additional reason may be seen in the fact that multistability (or bistability) is very often the mechanism behind some genetic circuits,[13] and that this switching behavior is the base for computational capabilities. As a consequence, the assumption that computational irreducibility characterizes regulatory networks makes simplified Boolean models sufficient to understand their relevant properties, since they have the minimal, essential ingredients. This is the view that we favor in this work. It is important, nevertheless, to emphasize two important points.

On the one hand, this kind of modeling consciously neglects the details of the precise functioning of particular units, not because they are irrelevant, but because they inherently cannot contribute to the understanding of the whole. The exact strengths of certain interactions are indeed very important to some physiology processes,[18] because these processes determine important aspects of cell functioning that need fine tuning. But in general, the details of the switching behavior of networks of many elements do not seem to be crucial to the overall patterns of activity, which otherwise would make the network too sensitive to particular parameters. Furthermore, irreducibility makes impossible to gain any understanding whatsoever of a process which involves big numbers of components: "Even if an ideal parameter set was provided (say, by software for automatic parameter optimization), the numerical solutions churned out by the computer would be just as inscrutable as the cell itself".[19]

On the other hand, in our opinion, no thorough understanding of all the processes in the cell can give hints as to why higher level behavior does occur. After all, understanding means the ability to explain a phenomenon, which is equivalent to be able to predict its behavior in all situations. In the case of the whole network, this seems virtually impossible. Moreover, even if all cell processes were known in detail, the resulting cell map would be useful for many purposes, such as designing very complex and specific drugs, but would otherwise leave open the question of how such a wonderful organization arose through the accumulation of small variations. An evolutionary explanation of the assembly of such a complex structure will surely be aided more by an understanding at the global level of the general dynamics of idealized Boolean networks than by a detailed study of all the real, discovered subnetworks. Essentially, there seems to be two levels of approach in regulatory networks, either at the level of small modules, or at the level of the whole system, with exclusive goals and providing answers to qualitatively different behavior.[20] We will focus on Boolean dynamics as the main tool for whole-system study.

The Boolean Idealization

In order to properly address the role of computation in regulatory networks, a exactly defined model is required. One possible approach is to see the networks as devices performing a definite task in an automatic, orderly manner. Given a set of inputs, such a device would react by performing a number of predefined operations, and yielding some output. If the device, either biological or artificial, has a minimal amount of memory, an appropriate description is provided by the so called discrete finite automata (DFA), a kind of abstract machines commonly used in the theory of computation.[15] These automata are also used in the design of logic circuits because they allow the designer to explicitly state the requirements of a circuit and they serve as the basis for optimization processes that minimize various parameters of it, including wiring and the number of memory units.[21]

Figure 3A, depicts the state diagram of a DFA, in this simple case an example of a machine that recognizes the pattern "011". This means that given a string of binary digits as input it will return as output a 1 whenever it detects this pattern and 0 otherwise. For this purpose, the automaton has three different states *A*, *B* and *C* and at each time step, it is given an input that makes it jump to another state, and yield some output value. For each state, two possible transitions are possible, denoted with an arrow in the diagram. On each arrow two values *a*/*b* are drawn, representing the input value of the transition, *a*, and the resulting value delivered by the machine, *b*. In our case, the *A* state represents the initial phase of the detection, in which the first 0 is detected. In fact, all states go to *A* if a 0 is given. Accordingly, *B* represents the middle phase of the detection and *C* the final one, being the initial state as well.

In Figure 3B, The simplest network that performs the task defined by the above automaton is shown. It has one input unit, *i*, one output unit, *o*, and two internal units, *a* and *b*. Given an initial state with all units set to 0, at each time step, all units compute their next output as a function of their present inputs, and switch to the new values at once. For units that have no inputs, the next value is assumed to be specified. It is not difficult to trace the values of the units through the detection sequence. First, upon the reception of a 0, *a* switches to 1, and the other units remain at 0. The unit *a* is then a testimony of a 0 in the input at the last time step. At the next time step, provided then that *a* is active, *b* turns to 1 only if the input was 1, thus implicitly detecting a 01 by means of the temporary memory of *a*. Finally, if *b* is 1 and the input is again 1, the output turns to 1, ending the detection process.

This network is a simple example of a general class of networks called Boolean networks, in which inputs perform Boolean functions.[22] The basic ingredients have been used already in the example: Boolean (i.e., on-off) states for the units, discrete time steps (synchrony), and general Boolean functions (a different specified output for each combination of inputs) at each unit. Its introduction was motivated by the questions raised in the modeling of the gene regulatory network by Kauffman,[23] although with a somewhat different perspective. In the last section, we have seen examples of the modeling of real networks with the aim of understanding particular parts of the cellular network. Kauffman adopted the complementary perspective of studying Boolean networks wired at random, with the hope of finding properties that would apply to the system in its entirety.[24] As we will see, he mostly succeeded.

Currently known as the Kauffman model, a system composed of N genes g_i interacting through Boolean functions f_i, with discrete time steps, has a dynamics defined by the following equation:

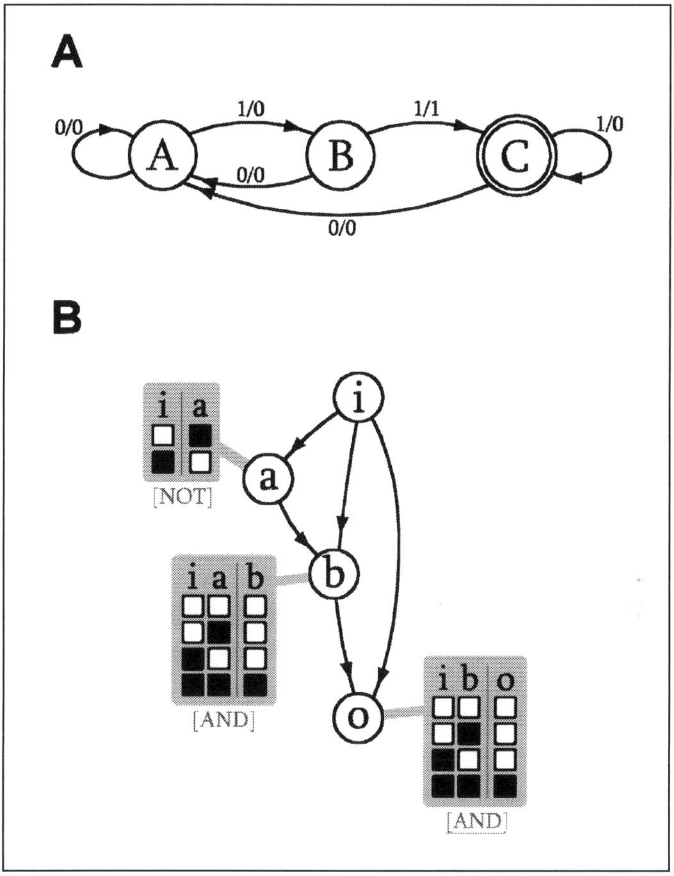

Figure 3. A) State diagram of a DFA (Discrete Finite Automaton) recognizing the sequence "011". The *C* state represents the initial and the final state of the recognition sequence and the labels at the links connecting states represent inputs/outputs of the automaton. B) Minimum Boolean network implementing the task defined by the automaton. Each unit in the Boolean network computes its next value using a table that tells, for each input combination, what should be the output.

$$g_i^{t+1} = f_i\left(g_{j_1}^t, \ldots, g_{j_K}^t\right). \tag{1}$$

To fully specify the network, the *K* inputs of each node are chosen at random among the *N* units of the system, and the functions are chosen so that the outputs have a 1 with probability *p* and a 0 with probability 1 - *p*, with no special units as inputs or outputs. Since it is specified at random, the network only has two parameters of interest: *K*, which defines the average connectivity between nodes; and *p*, which actually tunes the susceptibility of the function to changes in the input values: the closer *p* to 0.5, the easier it is that f_i changes if input *k* is reversed.

The global dynamics of Kauffman networks can be made clearer making the following observation. As already mentioned, at each time step, all nodes are updated synchronously using equation 1 from the values of their inputs. Therefore, we can treat the whole system as having a global state *S*, given by the composite state of all the units (or genes), that is,

$$S \equiv \left(g_1, g_2, \ldots, g_N\right).$$

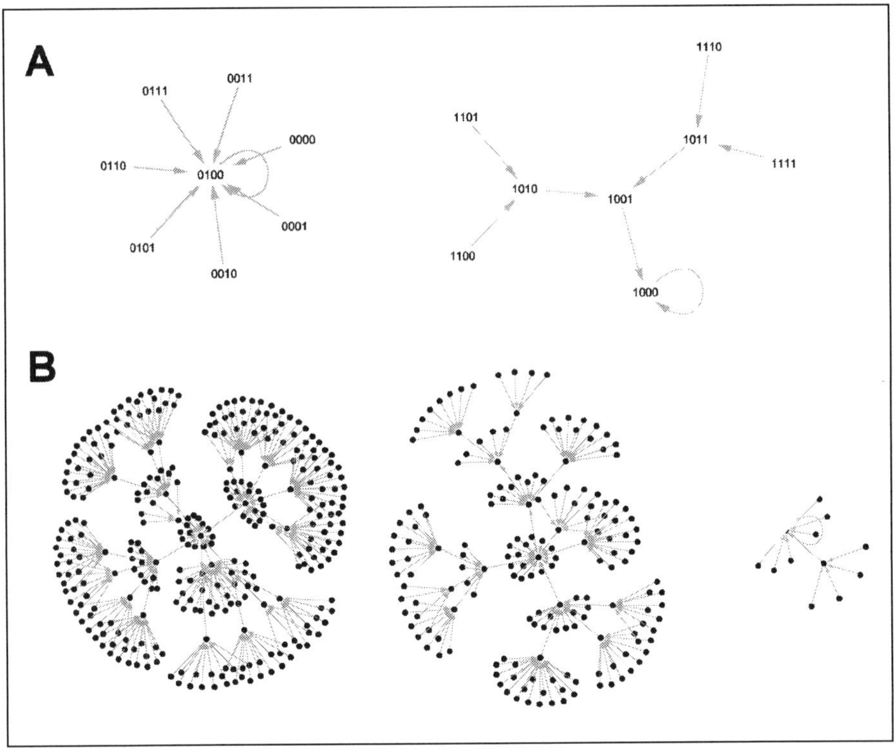

Figure 4. A) Basins of attraction of the circuit shown in Figure 3. The nodes in this graph are represented as the corresponding states of the Boolean network. B) Examples of the basins of attraction of a randomly generated network with 9 units and average connectivity $\langle K \rangle = 2$.

This global state S represents a point in the space S of all possible states, and at each time step it jumps to a different point following a trajectory given by the network configuration, starting at the chosen initial state.

Eventually, as time goes on, a state will be reached that has already been visited before, closing the trajectory into either a loop or a single state, if this state maps onto itself. To see the dynamics of a Boolean network at a glance, it is useful to examine a graph in which all the possible states of the network are linked to their successors in the dynamics. In such a chart, disjoint subgraphs represent different subsets of states that end in the same loop, called **basins of attraction**. Figure 4 depicts precisely the basin of attraction field of two networks: the example of Figure 3 in case A, and a random Boolean network with 9 units and connectivity $K = 2$ in case B. All points of the dynamics of the network are present, followed by their successors, and all possible trajectories are implicit in them, making the graph a very useful map. Indeed, software tools exist to draw basins of attraction fields for any specified network.[25]

Kauffman associated these basins of attraction with the different cell types specified by the underlying gene network, and made some arguments regarding the number of cell types (basins of attraction) present as a function of the network size N and the average connectivity K.[24] But the most important finding in relation with our discussion involves the dynamical properties of the system, and in particular, the propagation of errors. An important property of Boolean networks is, in fact, that depending on the connectivity K, errors have only three possible fates: either they die out, propagate to the whole system, or maintain themselves in the exact border between fading and exploding. This kind of behavior is a good example of a so called critical phase transition, a phenomenon well known in statistical physics.[26]

A simple explanation to understand this behavior can be given by means of a percolation argument,[27] and is general enough to include networks with a nonuniform distribution of links of average $\langle K\rangle$. Consider a given gene g_i in a Kauffman network of connectivity $\langle K\rangle$, and let us assume that the gene is externally flipped to the opposite state. The question asked is: how is this change going to be propagated through the network? Since the connectivity is $\langle K\rangle$, the change in g_i will arrive, on average, at the inputs of its $\langle K\rangle$ neighbors. It remains to be seen with which probability these nodes will propagate the change, which is the same as asking with what probability a random Boolean function changes its output when a single input is changed. Two possible propagation situations can take place, either the original output was 0 and shifts to 1 or the opposite occurs. Each of this situations has a probability $P = p(1 - p)$ (given the independence between values in the function f_i) and two of them are possible, thus the propagation probability is $P^* = 2p(1 - p)$. The average number of changes will be, then,

$$N_{ch} \equiv P^* \langle K\rangle = 2p(1-p)\langle K\rangle. \tag{2}$$

The three phases of behavior can be understood making the observation that N_{ch} represents the factor with which errors will multiply. If $N_{ch} < 1$ then changes will tend to disappear, at each time step the average number of changes diminishes. This is the so-called **ordered phase**, in which robustness is enough to cancel errors in the long term. If $N_{ch} > 1$ then errors will multiply and eventually the whole system will be affected by the avalanche. This is the **chaotic phase**, in which the state of the system in the future is governed by the uniform amplification of small events.

At the critical point, that is $N_{ch} = 1$, the number of errors does not have a tendency, so it will be impossible to predict what shall happen in the long run. In practice, this means that there will be a mixture of effects: some errors will die out, and some will propagate to the whole system. Using the equation 2, the critical point dictates the critical connectivity,

$$K_c = \frac{1}{2p(1-p)},$$

which simply leads to $K_c = 2$ for the case $p = 0.5$, as considered by Kauffman in its initial formulation. One simple implication of this formula is the fact that connectivity is rather low, i.e., that the network is **sparse**, an observed property in real networks.[28] It is also important to note that the connectivity $\langle K\rangle$ and the probability p alone determine the global behavior of the system. Although it does not make much sense to think that evolution can tune K or p directly, the accumulation of mutations will surely affect them, in turn affecting its mode of behavior with respect to the phase transition.

The importance of this transition lies in its intimate relationship with computation, and in particular, with the characteristics that computation requires to systems that implement it. These requirements have to do with the ability to process information, or in the words of Langton:[29] "First, the physics must support the *storage* of information, which means that the dynamics must preserve local state information for arbitrarily long times. Second, the physics must support the *transmission* of information, which means that the dynamics must provide for the propagation of information in the form of *signals* over arbitrarily long distances. Third, stored and transmitted information must be able to interact with one another, resulting in a possible modification of one or the other".* In addition, the issue of irreducibility plays an important role, because systems whose behavior can be predicted in the long run may not be able to implement complex tasks.

* Italics from the original.

In the light of these ideas, it does not seem probable that Boolean networks with computational utility could be in the ordered phase. Signals do not seem to be able to travel as far as needed, that is **arbitrarily** long distances. Although the analysis proposed is seen from the viewpoint of errors, a single unit that serves as input to the system and flips its state can be also seen as an external signal rather than an error, and then, the propagation of this error can be regarded as a signaling cascade. If the signal is unable to reach some parts of the system due to the network's inherent dynamics, many computations cannot be performed. On the other hand, computing Boolean networks do not seem to live in the chaotic phase either. Since regulatory networks are very noisy,[30] any computation that did work in the absence of noise would be surely disrupted by a single error. The critical phase, therefore, seems to have the suitable balance: it has the possibility of communicating any pair of units in the system, and it is not too sensitive to the values of all of them.[31]

Many authors have drawn attention to the fact that criticality in the dynamics of Boolean networks or cellular automata have desirable properties, and in two cases, properties directly related with computation. The major arguments in favor of criticality are the following:

- the capacity of systems at the critical point to exhibit arbitrarily large correlation lengths in space and time, supporting the basic mechanisms of storage, transmission and modification of information;[29]
- the undecidability (or the incapacity to predict without explicitly simulating the system) of the properties of systems in the critical phase as a basic characteristic of systems capable of computation;[32] and,
- that emergence of order "for free" in networks which are critical.[24]

There are also some arguments against this hypothesis. In reference 33, it is demonstrated that many cellular automata (a type of regular Boolean network embedded in space) with computational capabilities exist in the ordered and chaotic regions defined in reference 32. Their existence is indeed a significant result, but it does not say anything about the density of automata with computational capabilities in each phase, which may influence drastically the probability of reaching them by an evolutionary process. In reference 34, it is argued that Boolean networks with a scale-free degree distribution may provide, through their uneven distribution of connectivity, ways of making changes that have a significant impact on function, but allowing the network to remain in the ordered phase at the global scale.

Finally, in reference 35, an example of simulation of the evolution of cellular automata is shown that does not select automata with critical properties, suggesting that the critical phase does not have a higher density of systems with computational capabilities. In all cases, it is apparent that evolutionary properties are a very important ingredient in addition to dynamics. Overall, however, we are still ignorant about the applicability of these ideas in real regulatory networks, because current information includes more data with regard to the presence or absence of interactions than with their function.

The Evolutionary Point of View

To complement current understanding of the dynamics of Boolean networks, we also want to focus on the functional aspects of network evolution, again using Boolean networks. Very little is known about this subject, and yet simple examples can demonstrate the subtle differences in evolvability between variants of the same circuit.

Figure 5A shows a Boolean network implementing the discrete machine shown in Figure 3A. To make drawings simpler, we have chosen to follow the notation used by von Neumann,[36] which eliminates the use of Boolean tables. Although this notation also implies loosing some richness in the repertoire of Boolean functions, von Neumann proved its completeness in the specification of any computational device, and it is somewhat closer to actual regulation in

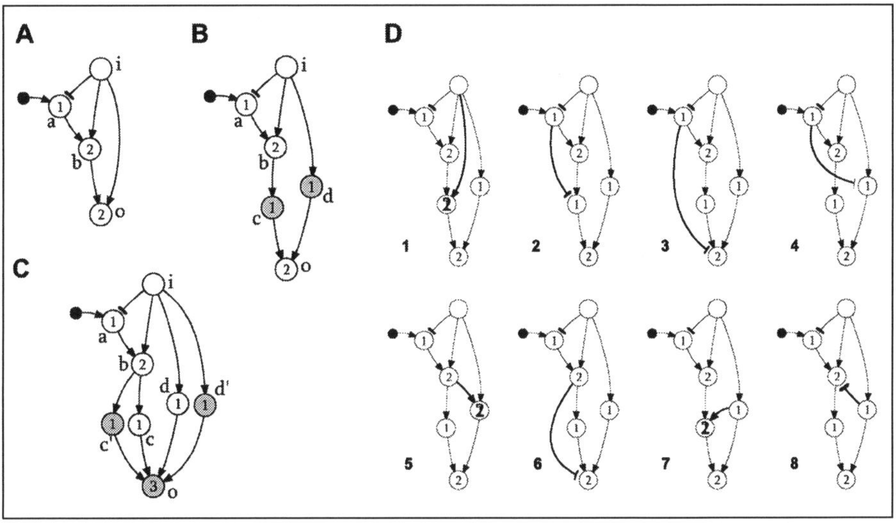

Figure 5. Several Boolean networks implemented with threshold units. Excitatory inputs end in a black arrow, and inhibitory ones in a terminating segment. The threshold is specified inside each unit, and the units are in gray when they represent changes made to another network. A) The equivalent of the network of Figure 3B. B) The same functional circuit as in A with two delay units *c* and *d* added. C) An example of the use of redundancy by the multiplication of lines. D) Several modifications to B that maintain functionality.

cells. The new units are also Boolean and synchronous, but determine their output by comparison to a threshold h. Inputs can either excite or inhibit a given unit, and the output of the unit at the time $t + 1$ will be in the excited state only if the sum of the number of excitatory inputs minus the number of inhibitory inputs at t is greater than or equal to h. This threshold h is represented as an integer number inside the circle that represents the unit. As usual in diagrams of genetic regulatory circuits, excitation is represented by a black arrowhead and inhibition by a terminating segment.

To complete this rather simple scheme we must augment it with the introduction of a simple unit assumed to be in a permanent excitatory state, denoted by a smaller black circle, to allow the possibility of negation. This is in fact what happens with unit *a* in the diagram of Figure 5A, to be compared with the circuit of Figure 3B. With respect to the new notation, it is not difficult to see how the "AND" gate behavior of units *b* and *o* is now implemented with units that have two excitatory states and a threshold $h = 2$, since "AND" gates are active only if both inputs are active. Conversely, an "OR" unit would be the same as an "AND" but with $h = 1$. Finally, before any discussion of the circuit properties we want to introduce a slight modification to the simplest implementation. The modification is shown in Figure 5B, and it just adds a delay of one time unit to the prior circuit, needing the introduction of units *c* and *d* whose role is simply to retransmit the values at their inputs. The behavior of the circuit is thus unaltered except for the one time step delay.*

*To make the discussion less involved, we have made some appropriate choices. First, to omit the initial state of the network, which might give positive output values in the intermediate steps of the detection. Starting from random values, it is sufficient to neglect the output in the first steps of the process, and after that all of it is correct. Second, we have chosen excitation links with preference, because they make the exposition clearer, although in all circuits the units can be made to function in reverse with a few changes.

Let us, then, examine this new circuit. One of the first evident properties is its complete lack of robustness: if the link connecting *c* and *o* happens to fail when transmitting the activation signal at the final steps of the detection process, *o* will never activate, and the overall result will be the failure at recognizing the input sequence "011". So it happens with the link from *b* to *c*, and many others in the circuit. Boolean logic is unable to cope with the failure of single components provided that the circuit represents a minimal implementation, as is the case the circuits we have seen. This fragility is also displayed by many man-made systems, in which the failure of individual components is assumed to be very infrequent. When a failure finally happens, the system is often not able to function at all. However, natural systems, and in our case cells, do have a great deal of robustness, motivated, basically, by two important sources of distress.

The first is thermal noise: the same process that makes molecules move and wander inside the cytoplasm introduces an inevitable stochasticity in the effects produced by them, for example at the level of gene expression.[37,38] The second, a byproduct of the first, is mutation: cells inherently accumulate changes in the genome through time, altering at random the networks they code for, a source of "permanent" noise. Despite the existence of these two sources of noise, cells behave in a very deterministic manner, compensating for its presence in some way. Deterministic responses also may include the explicit exploitation of noise to generate phenotypic variation, the only exception to its repression. At the level of molecules, cells have mechanisms to ensure that signals are received at the appropriate places.[30] At the level of genes, for instance, cells of *S. cerevisiae* do not display signs of a decrease in fitness in a 40 percent of null mutations to all genes in chromosome V.[39] The question is: how can cells achieve this powerful buffering?

Redundancy

A similar question was probably asked by von Neumann, albeit in a more abstract manner. He was searching for a logic system composed of unreliable components which worked in a reliable manner.[36] Many engineered systems require, in fact, high standards of reliability, such as, for instance, computerized bank accounting. The solution proposed by von Neumann, and still used today is to put **redundancy** into the system, or, stated plainly: to put many copies of the same thing. The idea is simple: if anyone of the copies fails, the copies that still work can compensate. In addition, a mechanism is needed to determine which are the copies that behave correctly. In the simplest case, the majority rule can be applied, which von Neumann implemented with his "majority organ".* As the name of the rule implies, in the face of mismatched behavior, the expected correct copies are assumed to be the most numerous, disregarding the others as wrong. In this way the failure of the whole system will happen only when, by chance, a number of copies bigger than half the number of available copies has failed, an event with a probability that can be made arbitrarily small as the number of copies grows, compared to the probability of failure of a single copy.

Figure 5C shows our circuit with redundancy implemented. Nodes *c* and *d* have been duplicated to create two redundant paths, *c'* and *d'*, one for each signal.** Upon arrival, *o* will activate with only 3 of them, allowing the failure of exactly one. As an example, the probability of failure of the unit *o* in this circuit can be calculated if we call *p* the probability of failure of its input links. Assuming that failure of links means not carrying a positive signal, three possible

* He called his units "organs".

** This duplication can be interpreted, in fact, as the duplication of genes *c* and *d*, a very usual mechanism for the creation of genes in eukaryotes.

events can occur that make o fail, which are the failure of 2, 3 or 4 links. Weighting by the number of combinations in which they can occur, we have the following equation:

$$p' = \binom{4}{2} p^2 (1-p)^2 + \binom{4}{3} p^3 (1-p) + \binom{4}{4} p^4. \tag{3}$$

With $p = 0.1$, the formula yields $p' = 0.052$. To obtain a higher gain, more parallel units could be introduced. It is worth mentioning, nevertheless, that the redundancy introduced is also useful to absorb the changes produced by the removal of units, a situation analogous to the knockout of genes. In the way this circuit is constructed, anyone of c, c', d or d' could be removed with no functional result whatsoever. In fact, it is clear that many implementations of networks detecting the pattern "011" are possible, each with some degree of redundancy placed in different points of the network. Therefore, it seems that the degree to which redundancy is found in biological systems must be the product of selection, at each generation adding or removing links that contribute positively or negatively to the robustness of the organism.

Degeneracy

But even if simple redundancy seems to suffice for the buffering of noise or mutation, its utility is of much less relative value than expected when put in an evolutionary context. Certainly, the introduction of duplicates protects organisms from mutation and noise, and studies exist that prove the stability of redundant genes under some conditions.[40] From the perspective of evolution, however, such a simple form of robustness would make organisms much less able to innovate. The reason for this difficulty is that all the copies of subparts that protect redundant systems probably have to be changed if a change in function is needed, making the adaptation process very awkward and frustrating. In fact, similar mechanisms can provide a source of robustness without the drawbacks of redundancy.

Figure 5D shows all possible circuits that are the same as 5B, but in which a single link has been added preserving the global function. In all cases, the new connections added basically crosscheck the detection sequence of the units in the simplest circuit. For instance, in the fifth one, unit d is modified to not only make sure that the last value of the input is 1, but also its coincidence in time with the activation of b, a detector of "01" in the past two values. The path leading from b to o is, in a way, duplicated, because the meanings of c and d overlap to a certain extent. The other cases involve other parts of the circuit but result in very similar modifications. These changes, in fact, can be seen as "neutral" mutations. Given the simplest, nude circuit, different combinations of this single modifications can provide a great deal of robustness, yet they do so in a different way, taking advantage of the multiple connections available that do not modify the behavior of the system. They also seem a more probable source of robustness, provided that mutation is random in nature.

This mode of robustness has already been defined and has been called **degeneracy**: "the ability of elements that are structurally different to perform the same function or yield the same output".[41] This applies to our system in the sense that different signaling paths can compute different subparts of the final pattern without being exact copies. Although first defined in the context of the nervous system,[42] degeneracy seems a good candidate for the implementation of robustness in biological systems in general. Redundancy, favored initially due to the existence of duplication in the genome, was rendered implausible by studies of duplicated genes showing an immediate and steady divergence of their sequences, implying that the major source of robustness is to be found in unrelated genes.[43] Again, for the same reasons mentioned above, the amount of degeneracy can be tuned by evolution to a suitable degree by making the appropriate changes to the network.

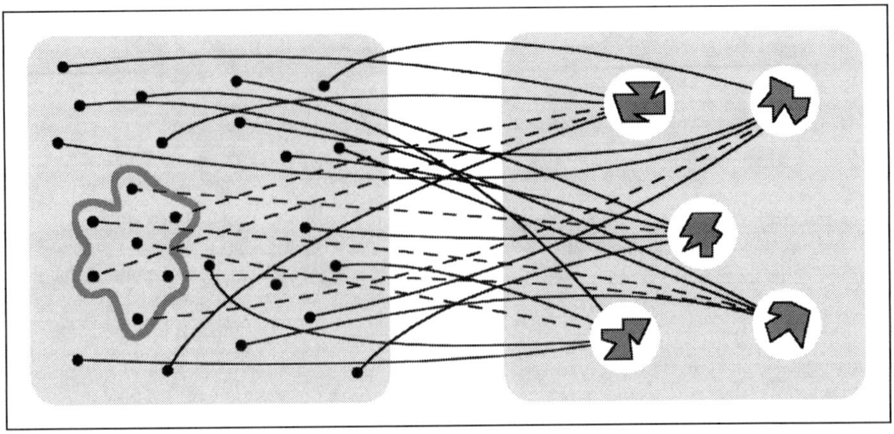

Figure 6. A diagram showing sequences and their associated shapes connected by a line. At the left, sequence space, at the right, shape space. The neighborhood marked in sequence space has many different sequences mapping into all shapes, as the dashed lines reveal.

Evolvability

This brings us to the issue of the capacity to evolve, or **evolvability**. Although mentioned only in the context of redundancy, the inability to innovate is not only related to the duplication of subparts but also with an excessive display of robustness, even if implemented using degeneracy. Evolvability has been discussed by many authors,[44,45] and it is defined as "the capacity to generate heritable, selectable, phenotypic variation". "This capacity may have two components: (i) to reduce the potential lethality of mutations and (ii) to reduce the number of mutations needed to produce phenotypically novel traits".[44] In relation to our discussion, it is clear that robustness contributes to the reduction of lethal mutations, but it is still unclear how to reduce the number of mutations needed to produce novelty.

Although in a somewhat different context, the study of the evolution of populations of RNA molecules can provide important insights into this question.[46] In particular, RNA molecules have the analogs of a genotype and a phenotype in their sequence and folding shape, respectively. Therefore, a genotype space (or sequence space) and a phenotype space (or shape space) can be defined. The studies of the landscapes that appear when linking genotype space with phenotype space tell us that sequence space is completely traversed by the so called **neutral networks**.[47] This networks comprise all sequences sharing a common shape that can be accessed by one point mutations, hence the name. The implications of this fact are more easily understood looking at Figure 6, in which sequence space and shape space are next to each other. The links that appear between the two spaces connect sequences with their corresponding shapes. Due to the existence of neutral networks, all shapes have connections from all of sequence space. As is immediately apparent, a very small neighborhood of a given sequence has connections with approximately all shapes, implying that many shapes are a few mutations away. This corresponds precisely to the idea of reaching novel traits through a small number of mutations, our second requirement for evolvability.

Even if the analogy with RNA has some risks, nothing prevents us in principle from applying these ideas to Boolean networks. In our context, sequence space is the analog of our circuit diagram (or genotype), and shape space is our function space (or phenotype). Neutral mutations have already been discussed in the context of degeneracy, where we have seen that many changes to a network do not alter the network's function. It is therefore a plausible idea that indeed whole

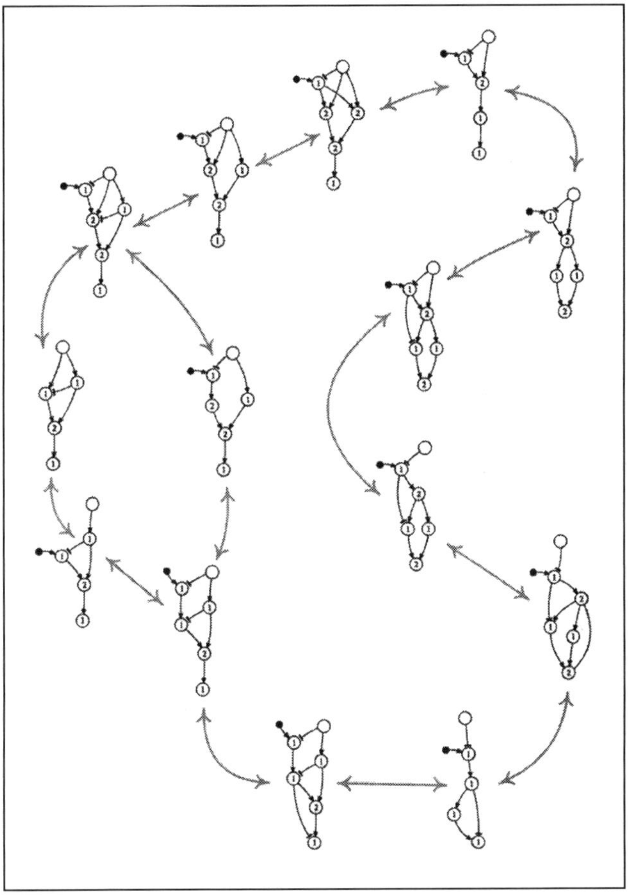

Figure 7. An example of a neutral network using simple Boolean circuits. All the circuits in the network perform the same task, which is the recognition of the "01" pattern. On each circuit, the input unit is at the top and output is at the bottom. Arrows represent mutations to the circuits, such as duplication or removal of units, and addition or removal of links, as well as changes in the activation thresholds.

networks of circuits with the same function can be accessed by single changes in their wiring, enabling a circuit to traverse circuit space and at the same time undergoing a complete rewiring. This is precisely what Figure 7 shows. The example circuit has been reduced to one signaling path that recognizes the subpattern "01". Following the arrows, at each step a single modification is made to the network that preserves function, including duplications, deletions, and addition or removal of links. Networks separated by many mutations have very few common links, and sometimes "homologous" links are part of functionally different signaling pathways.

Modularity

From another perspective, modularity also seems to contribute to the successful innovation in organisms. Many examples from evo-devo show that what makes sense is to study groups of genes in subnetworks responsible for traits[48-50] instead of isolated genes, and from an evolutionary viewpoint, modularity allows the adaptation of different traits with little or

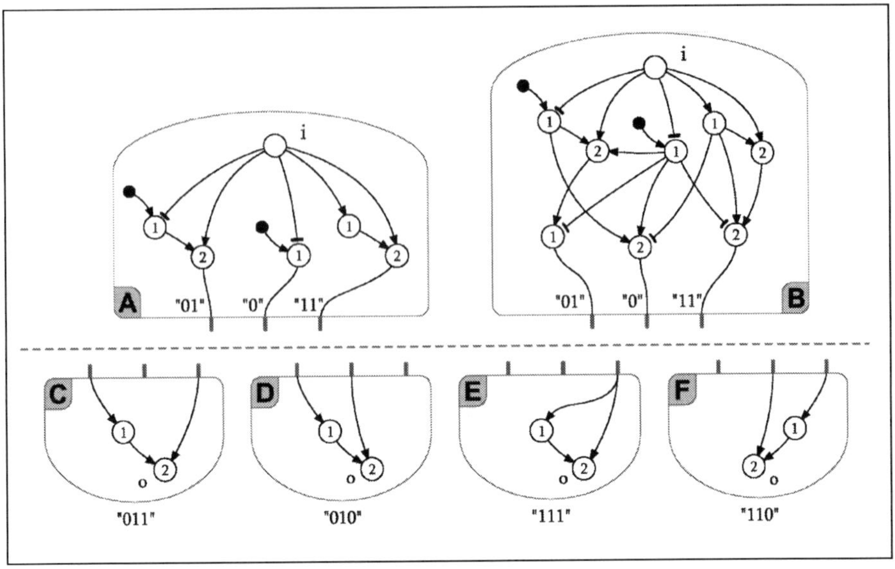

Figure 8. An example of modularity in Boolean networks. A) Module that makes useful "predetections". B) The same module with added robustness. C-F) Various ways to use the basic module to perform different functions.

no interference with each other.[51] Apart from the separation of functionally distinct traits, modularity also pervades molecular biology, with examples such as the recombination of domains in proteins,[52] or the combination of DNA sequences allowing the cis-regulation of genes.[53] In these cases, what matters most is the recombination of basic modules to form new structures in a much more rapid fashion, provided that these modules combinatorially allow any possible higher-level structure to be built. This gain in speed is widely used in engineering, in which often new systems are built using the components developed for their older brothers. Electronics, in particular, is a very good example of this, with the use of integrated circuits as building blocks that facilitate the construction of new, more complex circuits.

In Figure 8, an example of modularity is shown, again using the "011" detector circuit. As is implied by the shape of their boxes, any combination of upper and lower modules can be plugged to form a different circuit, with the upper modules performing basic functions, in this case the detection of particular subsequences, and the lower modules recombining the outputs of the upper modules to detect different input patterns. Upper modules perform the same subfunctions, so as to be compatible with the interface with lower modules, but differ in the degree of robustness. This fact illustrates an important point, in relation with the ideas discussed above, which is the following. As we have already seen, the degree of robustness can be tuned by an evolutionary process, giving more or less robustness to selected units in the network by the addition or removal of "degenerate" links. As the construction of the circuit progresses, a subset of units could be found to be useful as building blocks for higher-level processing and thus be made more robust, since the modifications necessary to generate a complete spectrum of behaviors would not involve this building blocks but the use of their precalculations in other parts of the network. Further evolution would be, therefore, speeded up by the finding of these modules. In this sense, the module *B* in Figure 8 could be one example of that process, resulting in an increased connectivity within the module.

Modularity in the topological sense is, in fact, measured in terms of these uneven patterns of connectivity, which produce clusters of nodes more densely connected. This feature, among others, is what some methods exploit to detect the modular structure of a network.[54-56] A simple enough picture of this kind of modularity, though, can be obtained by a coefficient C_i which measures the fraction of neighbors of this node that are neighbors themselves, that is,

$$C_i = \frac{2E_i}{k_i(k_i - 1)}.$$

In this formula, E_i is the number of edges present between neighbors of i, and k_i the actual number of neighbours, $k_i(k_i - 1)$ being the total number of possible links between neighbors of i. The average of C_i, that is, $\langle C \rangle$, describes in general the **clustering coefficient** of a network. This measure has been observed to be much higher in real networks than for random graphs in a variety of fields,[28] and in particular, it has also been shown to display a scale-free distribution.[57] This last fact demonstrates that modularity is indeed hierarchical, with small, strongly connected modules assembling into less cohesive, bigger modules in the upper level, and so on up to the whole network.

Discussion

In summary, we are still very much puzzled by the question of how complex regulatory networks are organized. But we think that the study of these networks with Boolean models can help understand the properties of general systems which, on the global scale, behave like real cells. The reasons for the success of this approximation might be found in the unsurmountable irreducibility of cellular processes, which behave in a manner similar to that of a computer. In the case of particular subnetworks, the Boolean approximation is successful in studying those mechanisms that are more "digital", and do not yield fine, graded responses. In the case of the whole system, these models can give important answers to questions regarding global, average dynamics.

In fact, two important aspects can be readily highlighted about Boolean networks. On the one hand, their dynamics undergoes a phase transition that enables us to classify its modes of behavior in three different zones, depending on a just two global parameters, such as the connectivity and the unit susceptibility. Looking at the properties of such modes of functioning, we find more probable that Boolean networks are in the critical phase, if they are to be capable of computation. As a consequence, networks must be sparse in connectivity, a feature which is present in real networks.

On the other hand, simple models of Boolean functions tell us that the degree of resistance to noise can be varied in a given network, mainly with the use of degeneracy, which adds neutral connections that perform parallel processing of the same information. Through a succession of single changes of this kind, a network can be rewired completely preserving its function at all times. This resistance to noise can also be considered as a resistance to mutation, which simply adds a form of coherent noise to the network. Although good for robustness, the resistance to change must not be too strong, because variation is also needed in evolution. Since degeneracy adds connections and their removal is related to sensitivity, an equilibrium between the two tendencies seems also to point to the idea that connectivity in regulatory networks has to be finely tuned to achieve evolvability.

In relation to it, modularity might emerge when parts of the network are found that enable further evolution in a quicker way by reusing their existing computations. If a suitable combination of useful modules is found, degeneracy can add protection to them, increasing connectivity within their subnetworks. This would implicitly direct the effects of mutations to the connections governing the combination of modules, which would avoid trying many worthless mutants. Although these ideas can be presented using simple examples, much work has to be done to thoroughly quantify them in models of networks with many units.

Acknowledgements

The authors would like to thank the members of the complex systems research group for useful discussions. This work was supported by a grant BFM2001-2154 (RVS) and the Generalitat de Catalunya (PFD, 2001FI/00732) and The Santa Fe Institute.

References

1. Hopfield JJ. Physics, computation, and why biology looks so different. J Theor Biol 1994; 171:53-60.
2. Hartwell LH, Hopfield JJ, Leibler S et al. From molecular to modular cell biology. Nature 1999; 402:C47-C52.
3. Gould SJ. The Stucture of Evolutionary Theory. Belknap: Harvard Press, 2003.
4. Wolfram S. Undecidability and intractability in theoretical physics. Phys Rev Lett 1985; 54:735-738.
5. Davidson EH et al. A genomic regulatory network for development. Science 2002; 295:1669-1678.
6. Lee TI et al. Transcriptional regulatory networks in Saccharomyces cerevisiae. Science 2002; 298(5594):799-804.
7. Bell AC, West AG, Felsenfeld G. Insulators and boundaries: Versatile regulatory elements in the eukaryotic genome. Science 2001; 291:447-450.
8. Albert B, Johnson A, Lewis J et al. Molecular Biology of the Cell. 4th ed. New York: Garland Science, 2002.
9. Albert R, Othmer HG. The topology of the regulatory interactions predicts the expression pattern of the drosophila segment polarity genes. J Theor Biol 2003; 223:1-18.
10. von Dassow G, Meir E, Munro E. The segment polarity network is a robust developmental module. Nature 2000; 406:188-192.
11. Barkai N, Leibler S. Robustness in simple biochemical networks. Nature 1997; 387:913-917.
12. Solé RV, Fernández P, Kauffman SA. Adaptive walks in a gene network model of morphogenesis: Insights into the cambrian explosion. IJDB 2003.
13. Hasty J, McMillen D, Isaacs F et al. Computational studies on gene regulatory networks: In numero molecular biology. Nature Rev Gen 2001; 2:268-279.
14. Butler D. Computing 2010: From black holes to biology. Nature 1999; 402:C67-C70.
15. Sipser M. Introduction to the Theory of Computation. PWS Publishing Company, 1997.
16. McAdams HH, Shapiro L. Circuit simulation of genetic networks. Science 1995; 269:650-656.
17. Bray D. Protein molecules as computational elements in living cells. Nature 1995; 376:307-312.
18. Sveiczer A, Csikasz-Nagy A, Gyorffy B et al. Modeling the fission yeast cell cycle: Quantized cycle times in wee1- cdc25delta mutant cells. PNAS 2000; 97:7865-7870.
19. Tyson JJ, Chen K, Novak B. Network dynamics and cell physiology. Nat Rev Mol Cell Biol 2001; 2:908-916.
20. Anderson PW. More is different. Science 1972; 177:393-396.
21. Hayes JP. Principles of Digital Logic Design. Reading: Addison-Wesley, 1993.
22. Aldana-González M, Coppersmith S, Kadanoff LP. Boolean dynamics with random couplings. In: Kaplan E, Marsden JE, Sreenivasan KR, eds. Perspectives and Problems in Nonlinear Science. A Celebratory Volume in Honor of Lawrence Sirovich. Springer: Applied Mathematical Sciences, 2003.
23. Kauffman SA. Metabolic stability and epigenesis in randomly constructed genetic nets. J Theor Biol 1969; 22:437-467.
24. Kauffman SA. The Origins of Order. New York: Oxford University Press, 1993.
25. Wuensche A. Genomic regulation modeled as a network with basins of attraction. In: Altman RB, Dunker AK, Hunter L, Klien TE, eds. Pacific Symposium on Biocomputi '98. Singapore: World Scientific, 1998.
26. Stanley HE. Introduction to Phase Transitions and Critical Phenomena. New York: Oxford University Press, 1971.
27. Luque B, Solé RV. Phase transitions in random networks: Simple analytic determination of critical points. Phys Rev E 1997; 55:257-260.

28. Dorogovtsev SN, Mendes JFF. Evolution of Networks. New York: Oxford University Press, 2003.
29. Langton C. Computation at the edge of chaos: Phase transitions and emergent computation. Physica D 1990; 42:12-37.
30. McAdams HH, Arkin A. It's a noisy business! genetic regulation at the nanomolar scale. Trends Genet 1999; 15(2):65-69.
31. Solé RV, Delgado J. Universal computation in fluid neural networks. Complexity 1996; 2(2):49-56.
32. Wolfram S. Universality and complexity in cellular automata. Physica D 1984; 10:1-35.
33. Dhar A, Lakdawala P, Mandal G et al. Role of initial conditions in the classification of the rule space of cellular automata dynamics. Phys Rev E 1995; 51:3032-3037.
34. Aldana M, Cluzel P. A natural class of robust networks. PNAS 2003; 100:8710-8714.
35. Mitchell M, Hraber PT, Crutchfield JP. Revisiting the edge of chaos: Evolving cellular automata to perform computations. Complex Systems 1993; 7:89-130.
36. von Neumann J. Probabilistic logics and the synthesis of reliable organisms from unreliable components. Shannon C, McCarthy J, eds. Automata Studies. Princeton: Princeton University Press, 1956.
37. Elowitz MB, Levine AJ, Siggia ED et al. Stochastic gene expression in a single cell. Science 2002; 297:1183-1186.
38. Blake WJ, Kaern M, Cantor CR et al. Noise in eukaryotic gene expression. Nature 2003; 422:633-637.
39. Smith V, Chou KN, Lashkari D et al. Functional analysis of the genes of yeast chromosome V by genetic footprinting. Science 1996; 274:2069-2074.
40. Nowak MA, Boerlijst MC, Cooke J et al. Evolution of genetic redundancy. Nature 1997; 388:167-171.
41. Gerald M, Edelman GM, Gally JA. Degeneracy and complexity in biological systems. PNAS 2001; 98:13763-13768.
42. Tononi G, Sporns O, Edelman GM. Measures of degeneracy and redundancy in biological networks. PNAS 1999; 96:3257-3262.
43. Wagner A. Robustness against mutations in genetic networks of yeast. Nat Genet 2000; 24:355-361.
44. Kirschner M, Gerhart J. Evolvability. PNAS 1998; 95:8420-8427.
45. Wagner GP, Altenberg L. Complex adaptations and the evolution of evolvability. Evolution 1996; 50:967-976.
46. Schuster P. How does complexity arise in evolution. Complexity 1996; 2:22-30.
47. Schuster P, Fontana W, Stadler P et al. From sequences to shapes and back: A case study in RNA secondary structures. Proc Roy Soc London B 1994; 255:279-284.
48. von Dassow G, Munro E. Modularity in animal development and evolution: Elements of a conceptual framework for evodevo. J Exp Zool 1999; 406(6792):188-192.
49. Solé RV, Salazar I, Garcia-Fernández J. Common pattern formation, modularity and phase transitions in a gene network model of morphogenesis. Physica A 2002; 305:640-647.
50. Salazar-Ciudad I, Newman SA, Solé RV. Phenotypic and dynamical transitions in model genetic networks i: Emergence of patterns and genotype-phenotype relationships. Evol Dev 2001; 3(2):84-94.
51. Wagner GP. Homologues, natural kinds, and the evolution of modularity. Am Zool 1996; 36:36-43.
52. Pawson T, Nash P. Assembly of cell regulatory systems through protein interaction domains. Science 2003; 300:445-452.
53. Buchler NE, Gerland U, Hwa T. On schemes of combinatorial transcription logic. Proc Natl Acad Sci USA 2003; 100:5136-5141.
54. Girvan M, Newman MEJ. Community structure in social and biological networks. PNAS 2002; 99:8271-8276.
55. Ihmels J, Friedlander G, Bergmann S et al. Revealing modular organization in the yeast transcriptional network. Nat Genet 2002; 31:370-377.
56. Zhou H. Distance, dissimilarity index, and network community structure. Phys Rev E 2003; 67:061901.
57. Ravasz E, Somera AL, Mongru DA et al. Hierarchical organization of modularity in metabolic networks. Science 2002; 297:1551-1555.

CHAPTER 13

Neutrality and Selection in the Evolution of Gene Families

Itai Yanai*

Abstract

Evolutionary relationships among genes, as revealed by sequence similarity, are used to characterize gene families. Surprisingly, a power-law can reasonably describe the distribution of sizes of a genome's gene families. Evolutionary models are able to reproduce the size distribution with simulations of a set of genes growing through duplications and modifications. Most conspicuously, positive selection is not included in the models, suggesting perhaps, that neutral forces determine gene family sizes. Here I advocate this notion with comparative genomic analyses and a review of recent research on the evolution of gene duplicates. I show that a power-law also relates the sizes of orthologous gene families across 66 known microbial genomes. Furthermore, singletons (gene families of size = 1) in one genome have orthologs that are themselves power-law distributed in other genomes. The signature of positive selection, however, is revealed in the fact that gene families of size six and more have a more skewed family sizes distribution across other genomes. The general pleiotropy of genes and the notion that gene duplicates may rapidly subfunctionalize support the conception of gene family growth without positive selection. Such a model runs contrary to Susumu Ohno's famous dictum that only "redundancy created" and suggests a novel view of the evolution of functional novelty.

Gene Family Sizes (GFS) Distributions

Unveiling the evolutionary processes that have shaped genomes is an efficient path towards their understanding. From such a standpoint, it is clear that the set of genes that comprise a genome is not arbitrary but contains a specific body of relationships. For example, the bacterium *E. coli* K12 has 3,762 genes which it shares with other sequenced microbes, and these can be clustered based upon sequence similarity into 2,131 families.[1] These sizes however, are not evenly distributed: most of the family clusters are of size one (one gene) while a few have over 30 members (Fig. 1B). As the linear relationship in the log-log plot of Figure 1B shows, the gene family sizes (henceforth, GFS) distribution can be summarized by a power-law relationship of the form, $y = Ax^{-b}$. Such a relationship signifies that genomes do not have a 'typical' family size. Interestingly, the same relationship has been recognized across all genomes examined thus far,[2-5] although larger genomes tend to have larger gene families, as may be expected, and thus a smaller power, *b*.

*Itai Yanai—Department of Molecular and Cellular Biology, Harvard University, Cambridge, Massachusetts 02138, U.S.A. Email: yanai@mcb.harvard.edu

Power Laws, Scale-Free Networks and Genome Biology, edited by Eugene V. Koonin, Yuri I. Wolf and Georgy P. Karev.

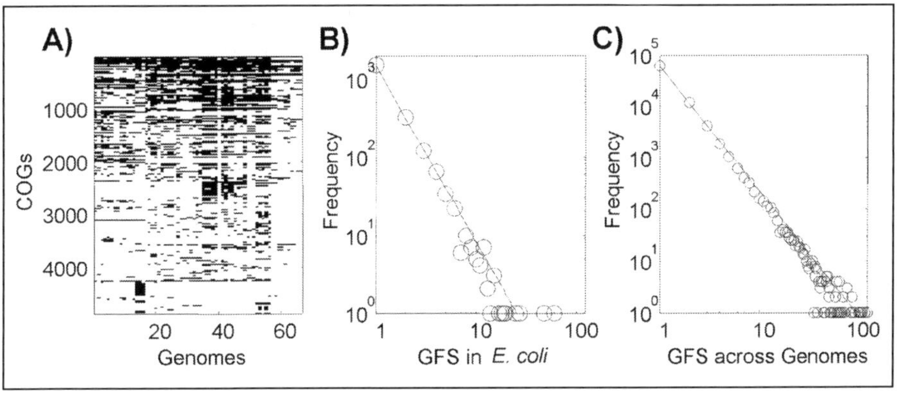

Figure 1. Gene family sizes (GFS) distributions. A) Based upon the COG database,[1] a matrix is constructed where the rows correspond to COGs (Clusters of Orthologous Groups) spread across the represented genomes (columns). Black and white here shows presence or absence, respectively, of a gene family in a particular genome. B) GFS distribution for the *E. coli* genome (a column of the matrix shown in A). The superimposed fit is of the power -2.27. C) Sum of the GFS histograms constructed for each COG (a row of the matrix shown in A). The superimposed fit is of the power -2.44.

GFS distributions showing power-laws are not limited to families within a genome but are found also 'across' genomes. Figure 1A shows a matrix of orthologous genes, where each row corresponds to a discrete gene family and relates its size in the various genomes (columns). The GFS distribution for a given genome corresponds to a histogram of the elements along a column of the matrix (Fig. 1B). Analogously, one can examine the distribution of sizes for a given gene family across the genomes (rows in the matrix). Since this currently amounts to a set of 66 genomes, a meaningful distribution is not possible on the scale of a single gene family. Cumulatively however, such distributions for **all** 4,873 gene families add up to another power-law distribution (Fig. 1C). Surprisingly, typical sizes for a certain gene family across genomes are also largely nonexistent. The significance of this distribution is discussed further below.

The interest in the GFS distributions stems from the ubiquity of power-laws across biological properties.[6] In protein-protein interaction networks, the number of interactions per protein is power-law distributed.[7,8] In metabolic pathways, the number of enzymes interacting with a given metabolite and the number of metabolites related to each enzyme are power-law distributed.[9,10] In protein domain networks, where domains are linked if they coappear in at least one other protein, the number of links per domain is power-law distributed.[11-13] The occurrence of structural folds in genomes also has a power-law relationship[14] as well as the number of genes in a genome of a given function—such as transcription factors—across genomes.[15] Is there a general mechanism common to these observations? Here, we focus on modeling and interpreting the GFS distribution, which may turn out to be applicable to other power-laws.

Modeling Genome Evolution

The most important insight into the composition of genes into gene families is its dynamic nature. It is clear that to model the phenomena one must include a time component by which gene family sizes can change. Gene duplication's role in this affair is uncontested and generally attributed to Susumu Ohno.[16] In the context of a growing network, the Barabasi and Albert model[17] has shown that two properties are sufficient to induce scale-free behavior in the degree distribution of a network: (1) evolution of the network instead of immediate conception and (2) preferential attachment of new nodes to popular nodes. The elegance of this model lies in its simplicity. Consider a network beginning with one node and that one by one, new nodes are added, each connecting to an existing node. If each node connects randomly, a scale-free network does not emerge. However, if the odds are shifted in favor of nodes that are already popular—such that the "rich get richer"—a power-law indeed will relate the distribution of connections per nodes.

The principle of preferential attachment can analogously be employed to model the evolution of gene family sizes by choosing for duplication any gene with equal probability. As an example, consider that a simulation beginning with two genes of different families, say α and β. A random choice of α for duplication results in three genes: α, β and α'. In the next round, when selecting a gene for duplication one has a two-thirds chance of selecting a gene from gene family α and one-third of selecting family β. The odds have turned in favor of the larger gene family. Thus, when randomly selecting a gene from a growing set, one is essentially invoking the principle of preferential attachment.

The second major issue in modeling GFS distribution is the origin of new gene families. Where do new gene families come from? One opportunity for generating new families is by importing them. Horizontal transfer is a dominant force in microbial genome evolution[18,19] and two models have modeled this kind of 'innovation'[3] or 'flux'[14] to capture this. However, a new family, although horizontally transferred, must have evolved somewhere and a model must allow for this if it is to be general.

An intuitive mechanism for generating new gene families is to evolve them from old families. Such instances are ubiquitously identified in the protein world (see the Interpro database).[20] For example, **Type II EGF-like signature domains** and **EGF-like calcium-binding** are two distinct types of protein domains, yet their sequences testify to their strong bond through an **EGF-like domain** ancestor. Such an origin for gene families can be modeled by specifying a threshold of similarity beyond which a pair of genes are considered members of distinct families. In our own model,[5] the time steps correspond to a mutation added to a randomly selected gene. As a simulation proceeds, mutations add up and consequently form novel families. Table 1 summarizes some features of the published models regarding the GFS distribution. In contrast to the universally accepted mechanism of duplication and (almost universally accepted) preferential attachment, the models differ in their mode of innovation.

What have we learned from the models? Although, the modeling of gene family sizes as a stochastic process of duplications and modification seems at first sight intuitive it actually corresponds to genomic heresy. A gene may spawn off a large family simply because it is lucky in a series of duplications and then by its sheer amount has the odds in its favor to duplicate further. In other words, the model seeks to explain family sizes without invoking **positive selection.** One may argue that selection is implied[6,13] but this does not seem compatible with the models. If the duplications are nonrandom but skewed in the form of some probability function of gene duplication other than a uniform distribution, the model's predictions are drastically altered.

Is the lack of positive selection in the models indicative of the difficulty involved in modeling it? In other words, are the models not unlike Nasruddin looking for the keys he lost in his house under the lamppost where there is more light. Or perhaps, are neutral forces actually shaping the GFS distribution?

Comparative Deconstruction of the Gene Family Sizes Distribution

While the GFS power-law distribution within a genome has received substantial attention, less research has focused on the lineage specific differences of gene families.[21] As shown here in Figure 1C, a power-law also relates the gene family size of orthologs across genomes. This means that when examining the sizes of a gene family across the genomes, the most

Table 1. Modeling genome evolution

Model	Expansion by Duplication	Innovation		Preferential Duplication	Comment
		Import	In-House		
Huynen and van Nimwegen, 1998	Yes	No	No	No	Begins with all gene families, each one copy. Simulation proceeds by a stringent mechanism where gene duplications and deletions behave coherently within a gene family.
Yanai et al, 2000	Yes	No	Yes	Yes	Each genome simulation begins with a number of precursor genes of distinct gene families. uplication rate is roportional to the number of point mutations. Duplicated gene inherits some number of mutations with respecto its ancestor, increasing with time. A new gene family is born when a gene's similarity with its ancestor is beyond the threshold.
Qian et al, 2001	Yes	Yes	No	Yes	Begins with a number of genes each one of a different 'fold' analogous to family). Each time step, a gene is randomly selected for duplication. Every number of time steps a new fold is inserted.
Karev et al, 2002	Yes	Yes	No	Yes	Birth, death and innovation model. A formalized version of Qian (Qian et al. 2001).
Dokholyan et al, 2002	Yes	No	Yes	Yes	Network approach. Simulation begins with one protein. At each time step a gene is duplicated that is different by a random number from its parent. A link between the child and parent is drawn if their similarity is below the threshold.
Unger et al, 2003	Yes	No	Yes	Yes	Start with a single protein (random sequence of amino acids). Each time point, a random sequence is randomly selected and duplicated with a certain level of mutations. With two thresholds the protein is either deleted, is added to the family or, forms a new family.

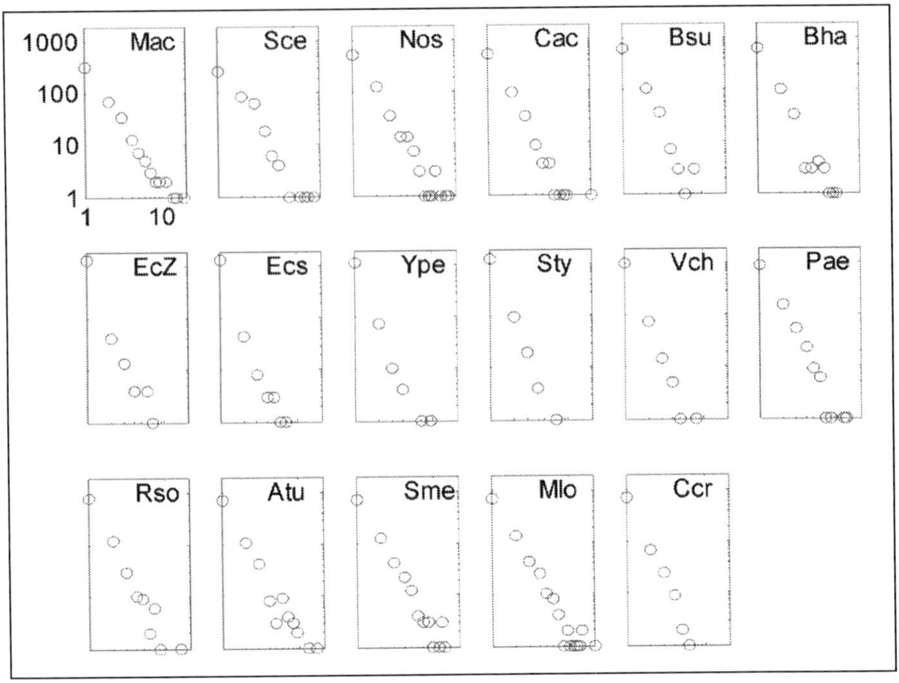

Figure 2. Gene family sizes distribution of orthologs of *E. coli* K12 singletons in 17 microbial genomes with over 3,000 genes in COGs. Organisms are abbreviated as in COGs: Mac= *Methanosarcina acetivorans* str.C2A; Sce= *Saccharomyces cerevisiae*; Nos= *Nostoc* sp. PCC 7120; Cac= *Clostridium acetobutylicum*; Bsu= *Bacillus subtilis*; Bha= *Bacillus halodurans*; Ecz= *Escherichia coli* O157:H7 EDL933; Ecs= *Escherichia coli* O157:H7; Ype= *Yersinia pestis*; Sty= *Salmonella typhimurium* LT2; Vch= *Vibrio cholerae*; Pae= *Pseudomonas aeruginosa*; Rso= *Ralstonia solanacearum*; Atu= *Agrobacterium tumefaciens* strain C58 (Cereon); Sme= *Sinorhizobium meliloti*; Mlo= *Mesorhizobium loti*; Ccr= *Caulobacter vibrioides*.

probable gene family size is one, although the gene family can grow particularly large in a few genomes. This is a surprising result since gene families are expected to have a characteristic size. However, since the distribution shown in Figure 1C is a cumulative histogram for all families, its resolution is too low to offer conclusive insight into the evolution of gene families.

A better view of the fate of orthologous gene families across other genomes comes from an 'orthology' deconstruction of the GFS power-law distribution for a given genome. There are 1,513 singletons (genes of family size = 1) in *E. coli* K12 that are also found in other known genomes. What is their family size in those other genomes? If one conceives that genes without paralogs are genes with inherently less potential in spawning a gene family, it is expected that the orthologs should also appear as singletons in other genomes. Strikingly the family sizes of these genes are power-law distributed across all genomes (though more noticeably in larger genomes) (Fig. 2). In other words, singletons in one genome may be found as large families in other genomes. Furthermore, the families of largest size is not fixed but varies greatly among the phyla.

The behavior of *E. coli*'s genes across phyla correlates with the size of the gene families examined. Figure 3A shows the deconstruction of the *E. coli* K12 GFS distribution as reflected by its orthologs in the genome of *Nostoc spp* PCC 7120. The same distribution is not observed for all of *E. coli*'s family size although its power-law nature appears coherent until family size four. With increasing gene family size, the slope of the distribution flattens until it becomes

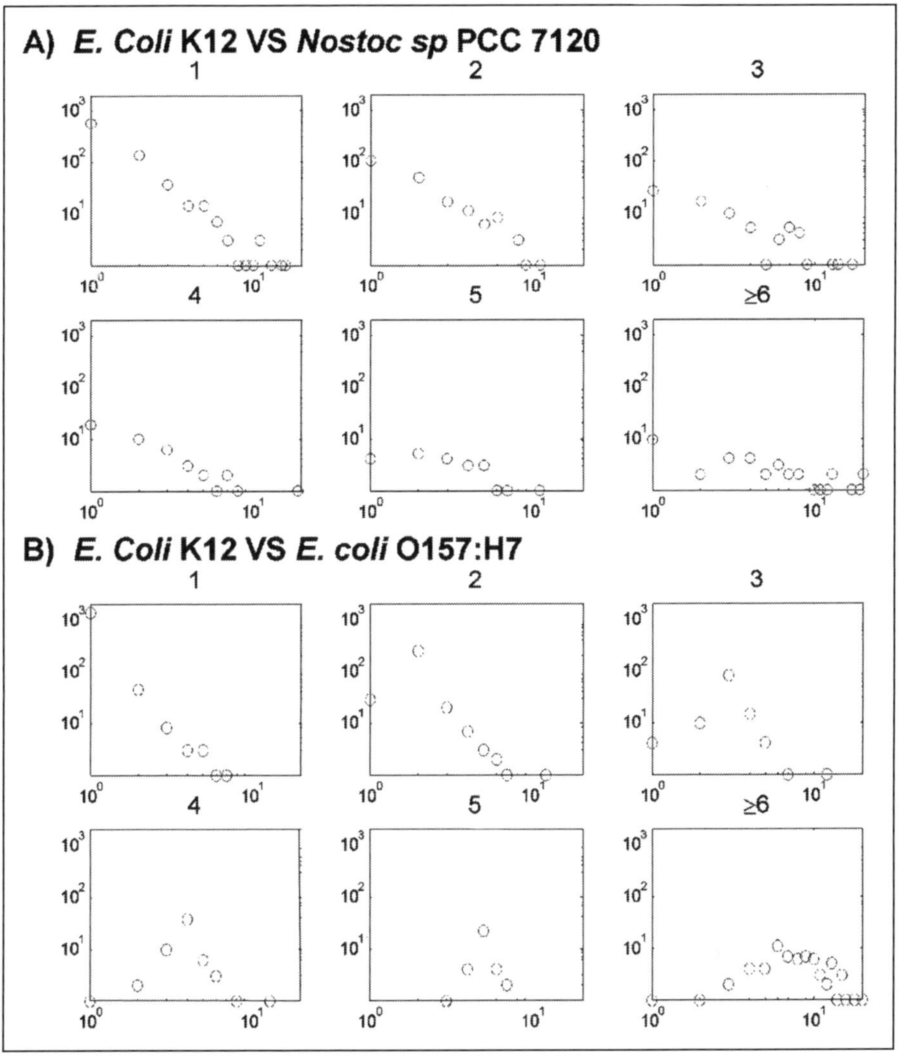

Figure 3. Orthology deconstruction of the *E. coli* K12 gene family size distribution. A) Each caption shows the GFS distribution in the bacteria *Nostoc spp* PCC 7120 genome of the orthologs of *E. coli* singletons, doublets, triplets, quadruplets, quintuplets, and sextuplets and greater, respectively. B) The same as in A for *E. coli* strain K12 deconstructed onto strain O157:H7.

nearly uniform at gene family size six and larger. The change in distributions with increasing family size can be unmistakably attributed to selection. That is, larger gene families in *E. coli* also tend to be of larger family sizes in other genomes because of the inherent properties of the gene to be advantageous as part of a larger family.

The situation appears different for genomes of closely related organisms. Figure 3B shows the deconstruction of *E. coli* K12 GNF distribution against another strain, *E. coli* O157:H7. These distributions differ markedly and can be characterized as normal distributions. Future work could be directed at estimating the strength of natural selection upon duplications from such distributions.

To summarize, knowing a gene family size in one genome actually confers very little predictive power about its size in a distant organism. Collectively, the results suggest a comprehensive model for the GFS distribution. As the simulations summarized in Table 1 suggest, neutral forces may be responsible for the persistence of the shape of the GFS distribution as exemplified by the distribution of a genome's singletons abroad. A smaller fraction of genes actually do depend on the size of the family and these receive special care from positive selection. As the sizes of these genes (family size ≥ 6) tend to be uniformly distributed they might only shift the amplitude of a power-law distribution set by selectively neutral duplications. Thus, overall, neutral duplications may lend the GFS distribution its distinctive shape.

Pleiotropy → Duplication → Subfunctionalization

The main sticking point of the models is the abstraction from the genes, modeled as anonymous entities. It is difficult to imagine that **any** gene in the genome can be duplicated and that the duplicate survives with equal probability. After all, survival is conditioned upon the probability of the duplicate finding a new function which presumably should vary greatly among genes. Such a mode of thought descends from Susumu Ohno who postulated that novel functions begins with the genomic redundancy that follows a gene duplication event.[16] The duplicated gene, freed from the 'policing' of selection, can then evolve freely and either adopt a new function that earns its keep in the genome or degenerate into a pseudogene. However, recently a flurry of research has suggested that Ohno's model of new-functionalization requires revision with important implications for the GFS model.

The main incongruity with Ohno's predictions and reality as reflected by the genome sequences is the fraction of duplications that survive and are maintained in the genome.[22,23] Although Ohno's prediction calls for a vast number of pseudogenized duplications, unaccounted for are a substantial fraction of duplicate survivors. What could explain their survival? A particularly persuasive model has been put forth which may shed light on this. Its starting point is that genes tend to have numerous functions each. This is supported by metabolic networks,[24] protein-protein interaction networks,[25] gene expression data,[26] and transcriptional regulation networks.[27] Consequently, it is conceivable that, following a duplication event, degenerate mutations which normally would be purified by selection, would become accessible in light of the duplicate's backup.[28] Certain series of such mutations can be envisaged that yield a situation where the duplicates are no longer completely redundant and that both are necessary to carry out the original function.[29] In other words, the original gene's function has been **subfunctionalized** by the duplicates.

A simple way to picture subfunctionalization is to imagine a gene's functional profile encompassing, for example, its context of expression, its protein-protein interactions, the chemical reactions that it aids in catalyzing and so forth (Fig. 4B). A mutation, for instance, in the promoter of a gene may disable its expression under a certain condition for which its transcription has been selected. Such a mutation can still rise to fixation by genetic drift because its duplicate acts as a backup. If an additional mutation disables another condition in the duplicate and rises in frequency to fixation, under the same logic, the two have now become subfunctionalized (Fig. 4A). Thus, a duplicate may avert pseudogenization if the pair manage to subfunctionalize. What is particularly convincing about this model is its complete independence of positive selection—only purifying selection is required to keep the two in survival.[28,30]

The notion of widespread pleiotropy along with the subfunctionalization model enables a simple explanation for the neutral forces that may drive neutral proliferation of gene families. Interestingly, the notion of subfunctionalization leads to a new model for neo-functionalization. Susumu Ohno had envisaged that the main problem involved in evolving a novel function is simultaneously maintaining the old. Genetic redundancy appeared the perfect solution to this conundrum. However, as evidence spills forth that genes are

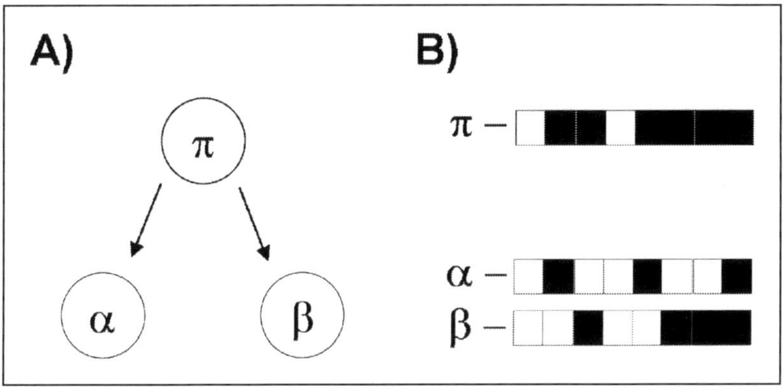

Figure 4. Subfunctionalization following gene duplication. A) Gene π undergoes duplication becoming genes α and β. B) Functional profiles of genes π, α, and β where each box represents a possible function (such protein-protein function with protein γ), the square is black if the gene carries out the specific function. A series of degenerating mutations can 'knock-out' a function until neither gene α or β is redundant and are both required to carry out the function originally carried by π. Based upon the model of Force et al.[28]

fundamentally pleiotropic, a picture emerges that neo-functionalization may generally occur before duplication[31-33]—only later to specialize by subfunctionalization. A schematic of this process is shown in Figure 5.

In the past few years, genomics has witnessed the rise of a view which encompasses the force of neutrality, alongside selection and in particular the fine interplay between them (see for example, refs. 34-37). There is no question that Motoo Kimura and King and Juke's neutral

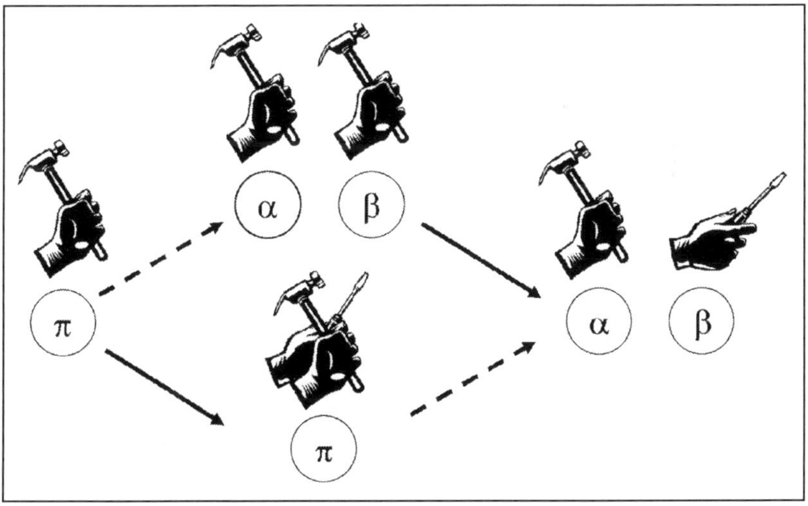

Figure 5. Alternative paths to neo-functionalization. As in Figure 4, circles represent genes. Solid arrows represent neo-functionalization and dashed lines represent gene duplication. The traditional Ohno model postulated that gene duplication precedes neo-functionalization—the screwdriver function evolves from the hammer function following its duplication. An alternative model attractive in light of the pleiotropy of genes suggests that neo-functionalization occurs in addition to existing functions. These can then subfunctionalize (see Fig. 4) following gene duplication.

mutation theory[38,39] now forms an integral way in which genomes are interpreted. The evolution of gene families has been well studied in the field of population genetics.[40-43] Here, the GNF distribution was studied and it has been argued that a neutral process plays a dominant role in its shape. Similarly, Wagner has suggested that the global structure of the protein-protein interaction network may be explained by a simple algorithm of duplication and removal of interactions without invoking selection.[8] On the local scale, the effects of natural selection can be detected by the occurrence of network motifs.[44-46] However, one may conjecture that a large-scale phenomena like the power-law so pervasive in biology, is the hallmark of neutral forces.

References

1. Tatusov RL, Fedorova ND, Jackson JD et al. The COG database: An updated version includes eukaryotes. BMC Bioinformatics 2003; 4:41.
2. Huynen MA, van Nimwegen E. The frequency distribution of gene family sizes in complete genomes. Mol Biol Evol 1998; 15:583-589.
3. Karev GP, Wolf YI, Rzhetsky AY et al. Birth and death of protein domains: A simple model of evolution explains power law behavior. BMC Evol Biol 2002; 2:18.
4. Luscombe NM, Qian J, Zhang Z et al. The dominance of the population by a selected few: Power-law behaviour applies to a wide variety of genomic properties. Genome Biol 2002; 3:RESEARCH0040.
5. Yanai I, Camacho CJ, DeLisi C. Predictions of gene family distributions in microbial genomes: Evolution by gene duplication and modification. Phys Rev Lett 2000; 85:2641-2644.
6. Koonin EV, Wolf YI, Karev GP. The structure of the protein universe and genome evolution. Nature 2002; 420:218-223.
7. Jeong H, Mason SP, Barabasi AL et al. Lethality and centrality in protein networks. Nature 2001; 411:41-42.
8. Wagner A. How the global structure of protein interaction networks evolves. Proc R Soc Lond B Biol Sci 2003b; 270:457-466
9. Jeong H, Tombor B, Albert R et al. The large-scale organization of metabolic networks. Nature 2000; 407:651-654.
10. Wagner A, Fell DA. The small world inside large metabolic networks. Proc R Soc Lond B Biol Sci 2001; 268:1803-1810.
11. Apic G, Gough J, Teichmann SA. Domain combinations in archaeal, eubacterial and eukaryotic proteomes. J Mol Biol 2001; 310:311-325.
12. Wuchty S. Scale-free behavior in protein domain networks. Mol Biol Evol 2001; 18:1694-1702.
13. Yanai I, DeLisi C. The society of genes: Networks of functional links between genes from comparative genomics. Genome Biol 2002; 3:research0064.
14. Qian J, Luscombe NM, Gerstein M. Protein family and fold occurrence in genomes: Power-law behaviour and evolutionary model. J Mol Biol 2001; 313:673-681.
15. van Nimwegen E. Scaling laws in the functional content of genomes. Trends Genet 2003; 19:479-484.
16. Ohno S. Evolution by gene duplication. New York: Springer-Verlag, 1970.
17. Barabasi AL, Albert R. Emergence of scaling in random networks. Science 1999; 286:509-512.
18. Doolittle WF. Phylogenetic classification and the universal tree. Science 1999; 284:2124-2129.
19. Koonin EV, Makarova KS, Aravind L. Horizontal gene transfer in prokaryotes: Quantification and classification. Annu Rev Microbiol 2001; 55:709-742.
20. Mulder NJ, Apweiler R, Attwood TK et al. The InterPro Database, 2003 brings increased coverage and new features. Nucleic Acids Res 2003; 31:315-318.
21. Jordan IK, Makarova KS, Spouge JL et al. Lineage-specific gene expansions in bacterial and archaeal genomes. Genome Res 2001; 11:555-565.
22. Hughes AL. The evolution of functionally novel proteins after gene duplication. Proc R Soc Lond B Biol Sci 1994; 256:119-124.

23. Initiative AG. Analysis of the genome sequence of the flowering plant Arabidopsis thaliana. Nature 2000; 408:796-815.
24. Karp PD, Riley M, Saier M et al. The ecoCyc database. Nucleic Acids Res 2002; 30:56-58.
25. Gavin AC, Bosche M, Krause R et al. Functional organization of the yeast proteome by systematic analysis of protein complexes. Nature 2002; 415:141-147.
26. Ihmels J, Friedlander G, Bergmann S et al. Revealing modular organization in the yeast transcriptional network. Nat Genet 2002; 31:370-377.
27. Pilpel Y, Sudarsanam P, Church GM. Identifying regulatory networks by combinatorial analysis of promoter elements. Nat Genet 2001; 29:153-159.
28. Force A, Lynch M, Pickett FB et al. Preservation of duplicate genes by complementary, degenerative mutations. Genetics 1999; 151:1531-1545.
29. Lynch M, Force A. The probability of duplicate gene preservation by subfunctionalization. Genetics 2000; 154:459-473.
30. Stoltzfus A. On the possibility of constructive neutral evolution. J Mol Evol 1999; 49:169-181.
31. Hughes AL. Adaptive evolution of genes and genomes. New York: Oxford University Press, 1999.
32. Hughes AL. Adaptive evolution after gene duplication. Trends Genet 2002; 18:433-434.
33. Piatigorsky J. Crystallin genes: Specialization by changes in gene regulation may precede gene duplication. J Struct Funct Genomics 2003; 3:131-137.
34. Kondrashov FA, Rogozin IB, Wolf YI et al. Selection in the evolution of gene duplications. Genome Biol 2002; 3:RESEARCH0008.
35. Lynch M, Conery JS. The evolutionary fate and consequences of duplicate genes. Science 2000; 290:1151-1155.
36. Waterston RH, Lindblad-Toh K, Birney E et al. Initial sequencing and comparative analysis of the mouse genome. Nature 2002; 420:520-562
37. Yanai I, Graur D, Ophir R. Incongruent expression profiles between human and mouse orthologous genes suggest widespread neutral evolution of transcription control. OMICS 2004; 8:in press.
38. Kimura M. Evolutionary rate at the molecular level. Nature 1968; 217:624-626.
39. King JL, Jukes TH. NonDarwinian evolution. Science 1969; 164:788-798.
40. Clark AG. Invasion and maintenance of a gene duplication. Proc Natl Acad Sci USA 1994; 91:2950-2954.
41. Nowak MA, Boerlijst MC, Cooke J et al. Evolution of genetic redundancy. Nature 1997; 388:167-171.
42. Ohta T. Evolution of gene families. Gene 2000; 259:45-52
43. Walsh JB. How often do duplicated genes evolve new functions? Genetics 1995; 139:421-428.
44. Conant GC, Wagner A. Convergent evolution of gene circuits. Nat Genet 2003; 34:264-266.
45. Milo R, Shen-Orr S, Itzkovitz S et al. Network motifs: Simple building blocks of complex networks. Science 2002; 298:824-827.
46. Wagner A. Does selection mold molecular networks? Sci STKE 2003a; 2003:PE41.
47. Dokholyan NV, Shakhnovich B, Shakhnovich EI. Expanding protein universe and its origin from the biological Big Bang. Proc Natl Acad Sci USA 2002; 99:14132-14136
48. Unger R, Uliel S, Havlin S. Scaling law in sizes of protein sequence families: From super-families to orphan genes. Proteins 2003; 51:569-576.

CHAPTER 14

Scaling Laws in the Functional Content of Genomes:

Fundamental Constants of Evolution?

Erik van Nimwegen*

Power Laws in Genomic Quantities

A few years ago it was noticed that the distributions of gene-family sizes in fully-sequenced genomes follow power-law distributions.[1,2] Since then different authors have shown that there is in fact a large array of genomic features that show power law distributions. Almost all of these concern the distributions of genomic features within a single genome. For instance, it is shown in reference 3 that the number of genomic occurrences of DNA words, protein folds, superfamilies, and families all follow power-law distributions. Power-law distributions are also found in the structure of the 'protein universe';[4] the number of protein families per fold is power-law distributed, and so is the number of different assigned biological functions per fold.[3] Power-laws also appear in the structure of cellular interaction and regulatory networks. For example, the number of genes that a given gene interacts with is power-law distributed. This holds both when one defines 'interaction' between genes on the level of the proteins that they encode[5-7] or if one defines it at the level of coregulation of the expression of the genes.[8] The experimental data on transcription regulatory networks is rather incomplete but they also suggest that the number of genes regulated per transcription factor might have power-law tails.[9,10] Finally, power-laws also appear in cellular metabolic networks; the number of substrates that any given substrate interacts with is power-law distributed.[11,12]

Comparing Genomic Features across Genomes

Note that almost all the power-law distributions just mentioned refer to statistics that are taken over a **single** genome or cellular network. The statistics of genomic features **across** genomes has been much less (if at all) investigated. To a large extent this may be because until recently there simply weren't enough fully-sequenced genomes to obtain meaningful statistics across genomes. However, this situation is changing rapidly.

There are currently about 150 fully-sequenced microbial genomes in genbank and this number appears to grow exponentially as I have shown in reference 13.

Figure 1 shows an updated plot of the number of fully-sequenced microbial genomes as a function of time (see Methods section). The current number of available microbial genomes is only large enough to allow for meaningful cross-genome comparisons of the most

*Erik van Nimwegen—Division of Bioinformatics, Biozentrum, University of Basel, Klingelbergstrasse 50-70, CH-4056 Basel, Switzerland. Email: erik.vannimwegen@unibas.ch

Power Laws, Scale-Free Networks and Genome Biology, edited by Eugene V. Koonin, Yuri I. Wolf and Georgy P. Karev.

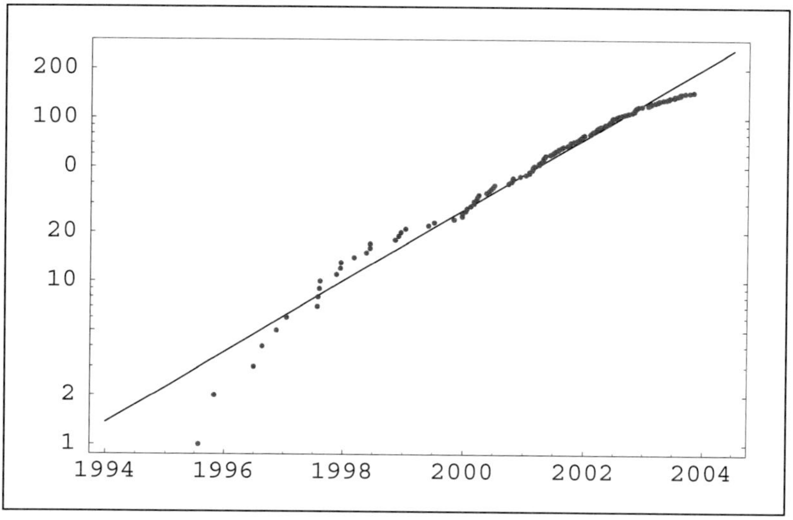

Figure 1. The number of fully-sequenced microbial genomes submitted to the genbank database as a function of time (in years). The vertical axis is shown on a logarithmic scale. The black line is the least-square fit to an exponential. $n = 2^{(t-1993.4)/1.38}$

basic statistics of gene-content and organization and this is what I will focus on in this chapter. However, the exponential fit in Figure 1 predicts that the number of sequenced genomes doubles roughly every 17 months. This implies that by 2010 we may have as many as 3000 fully-sequenced microbial genomes available. It is therefore clear that much more detailed comparative genomic analyses than the ones presented in this chapter will become possible over the next decade.

In reference 13 I compared the number of genes in high-level functional categories across all sequenced genomes and showed that they follow power-laws as a function of the total number of genes in the genome. In this chapter I will recapitulate these results and augment them in several ways. In particular, I have extended the analysis to all functional categories that are represented by at least 1 gene in each bacterial genome and have recalculated the observed exponents based on the latest genomic data. Second, I will go into more detail regarding the implications of the observed scaling laws for the general organization of gene-content across genomes and discuss an evolutionary model that relates the observed scaling behavior in gene content to fundamental constants of the evolutionary process. Finally, I will discuss theoretical explanations for the category of transcription regulatory genes which shows approximately quadratic scaling with the total number of genes in the genome.

Scaling in Functional Gene-Content Statistics

To count and compare the number of genes in different functional categories for all sequenced genomes one needs to first define a set of functional categories and then annotate all genomes in terms of these functional categories. I used the biological process hierarchy of the Gene Ontology[14] to define functional categories, and Interpro annotations of fully-sequenced genomes to associate genes with GO categories. The details of the annotation procedure are described in the Methods section. The result is a count of the number of genes associated with each of the GO categories in the biological process hierarchy for each of the sequenced genomes.

The set of genomes in this study consists of 116 bacteria, 15 archaea, and 10 eukaryota. In this chapter I will focus solely on the bacterial data since this is the only kingdom for which

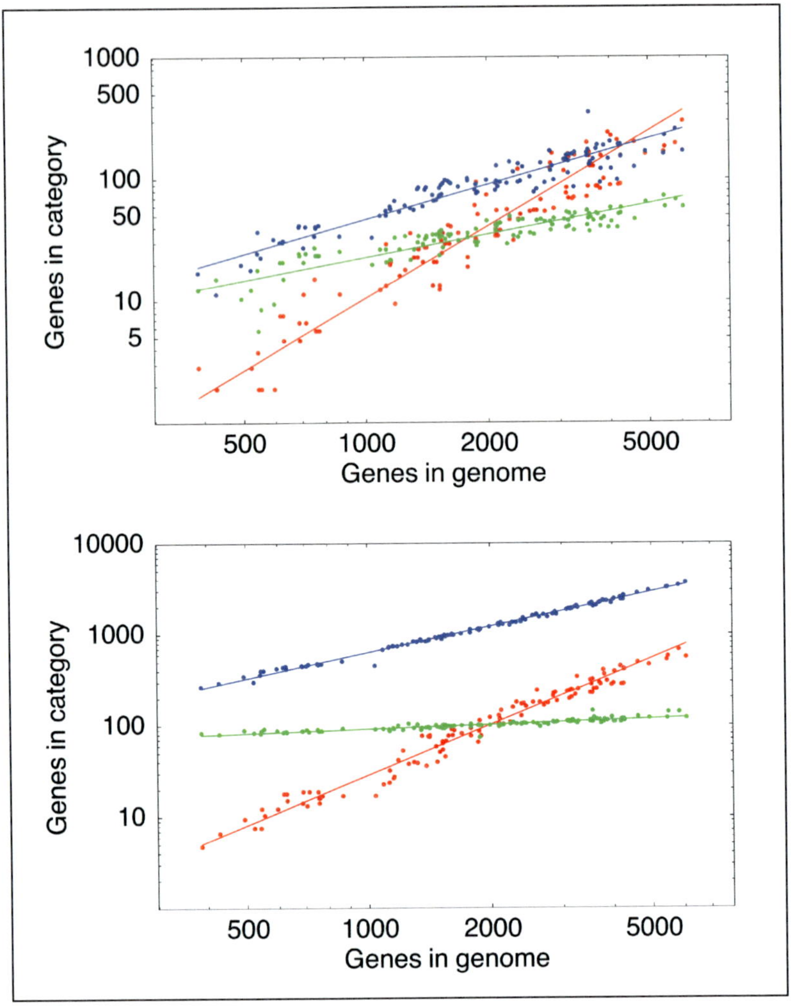

Figure 2. The number of genes involved in (uper panel) signal transduction (red), carbohydrate metabolism (blue), and DNA repair (green), and (lower panel) any biological process (blue), transcription regulation (red), and protein biosynthesis (green) as a function of the total number of genes in the genome that have any functional annotation at all. Each colored dot represents the counts for a single genome. Both axis are shown on a logarithmic scale. The straight lines show power-law fits: (upper panel) $n = 0.000015g^{1.95}$ (red), $n = 0.063g^{0.96}$ (blue), $n = 0.37g^{0.605}$ (green), and (lower panel) $n = 0.000095g^{1.83}$ (red), $n = 0.92g^{0.95}$ (blue), and $n = 30.8g^{0.16}$ (green).

there is sufficient data to obtain meaningful statistics. The reader is referred to reference 13 for a discussion of the observed scaling laws in archaea and eukaryota.

There are 154 GO categories in the biological process hierarchy that have at least 1 associated gene in each of the 116 bacterial genomes. I will refer to these categories as the 'ubiquitous' categories. The results for a selection of 6 of these ubiquitous GO categories are shown in Figure 2. The figure shows the dependence of the number of genes in each category on the total number of genes with annotation in the genome.

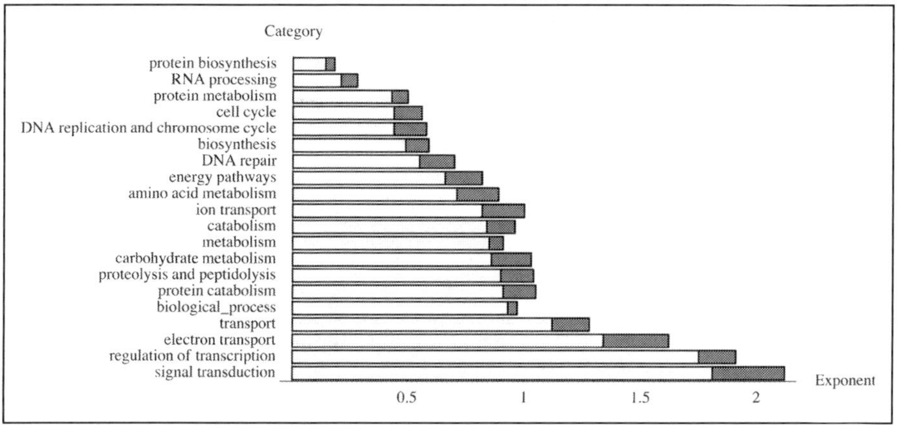

Figure 3. The 99% posterior probability intervals for the scaling exponents of a selection of functional categories. The functional categories are indicated on the left of each bar and the dark section of the bar indicates the 99% posterior probability interval for the exponent of that category. The categories are ordered from top to bottom by increasing lower-bound of the posterior probability interval.

The upper panel shows the categories "signal transduction" (red), "carbohydrate metabolism" (blue), and "DNA repair" (green), while the lower panel shows the categories "transcription regulation" (red), "biological process" (blue), and "protein biosynthesis" (green). Each dot represents the counts in a single bacterial genome and both axes in Figure 2 are shown on a logarithmic scale. Note that in reference 13 a similar plot was shown but with the horizontal axis representing the total number of genes rather than the total number of annotated genes. Since the total number of annotated genes is to a very good approximation proportional the total number of genes, the results are virtually identical whether one uses the total number of genes or the total number of annotated genes. I decided to use the total number of annotated genes on the horizontal axis in Figure 2 partly to illustrate this fact. In addition, the genome size in bacteria is also to a very good approximation proportional to the total number of genes in the genome. Thus, if we had used the genome size instead of the number of annotated genes on the horizontal axis Figure 2 would again have looked virtually identical.

The dots of each color in Figure 2 fall approximately on a straight line. Thus, the logarithms of the number of genes n_c in a category c and the total number of genes g (or the number of annotated genes or the genome size) are approximately linearly related:

$$\log(n_c) = \alpha_c \log(g) + b_c \tag{1}$$

In other words, the number of genes n_c in a category increases as a *power-law* in the total number of genes g:

$$n_c = \lambda_c g^{\alpha_c} \tag{2}$$

For the 6 functional categories shown, the exponents of the best power-law fits are indicated in the figure caption. The fits were obtained using the procedure described in the Methods section. The exponents range from $\alpha = 0.16$ for protein biosynthesis to $\alpha = 1.95$ for signal transduction.

To further show the range and variation of the observed exponents Figure 3 shows the inferred exponents and their 99% posterior probability intervals for a selection of 20 functional categories.

The exponents range from close to zero to roughly 2. Note that, for a category c with exponent α_c the relative **proportion** p_c of genes in the genome scales as $p_c = \lambda_c g^{\alpha_c - 1}$. That is, when $\alpha_c < 1$ the proportion of genes in the category will decrease with genome size, while for $\alpha_c > 1$ the proportion of genes in the category will increase with genome size. Thus, for a category c with exponent close to 2, the proportion p_c will increase almost linearly with genome size. The behavior of different categories thus ranges from categories where the number of genes is almost constant with genome size ($\alpha_c \approx 0$), to categories where the proportion of genes in the genome increases linearly with genome size ($\alpha_c \approx 2$).

The general picture that emerges from Figure 3 is that the proportion of genes in essential low-level functional categories such as protein biosynthesis and DNA replication decreases with genome size, whereas the proportion of genes that play regulatory roles such as genes involved in signal transduction and transcription regulation increases approximately linearly with genome size. In between these extremes is a number of categories, including different metabolic functions, for which the exponent is roughly 1, indicating that the genomic percentage of genes in these categories is roughly independent of genome size.

Upper Bound on Genome Size

The observed quadratic scaling of the number of regulatory genes with the total number of genes in the genome obviously cannot extend to arbitrarily large genome sizes. If we extend the red curves, corresponding to "transcription regulation" and "signal transduction", in Figure 2 to the right, we will eventually reach a point where the number of signal transducers and transcription regulators would be larger than the total number of genes in the genome and this is obviously impossible. Thus, if all bacterial genomes obey the relations indicated in our figure, there must be an upper bound on bacterial genome size. A naive upper bound is obtained by demanding that the number of genes in any category is less than the total number of genes in the genome, i.e., $n_c = \lambda_c g^{\alpha_c} \leq g$. If one substitutes the values of the fits for transcription regulation into this equation one obtains an upper bound of approximately $g \leq 70000$ genes. A tighter upper bound is obtained when one demands that n_c cannot increase by more than 1 gene when g is increased by 1 gene, i.e., $\lambda_c(g+1)^{\alpha_c} \leq \lambda_c g^{\alpha_c} + 1$.* One then obtains an upper bound of $g \leq 34000$ when the values for transcription regulation are substituted. In reference 15 an upper bound is derived by assuming that the number of genes involved in transcription regulation has to be less than half of the genome size, i.e. $n_c \leq g/2$. With our fit this leads to an upper bound of about $g \leq 30000$.

It is clear that all these upper bounds substantially overestimate the approximately 10000 genes of the largest observed bacterial genomes and that a more realistic theory is needed to plausibly explain the apparent size constraint on bacterial genomes. In this regard it is also interesting to note that in all the upper bounds just proposed, the proportion of genes that are transcription regulators is at least 50% at the maximal genome size, whereas the percentage is at most 11% in the currently sequenced bacterial genomes.

Consequences for the Topology of the Transcription Regulatory Network

The approximately quadratic scaling of the number of transcription regulatory genes also has some interesting consequences for the structure of the transcription regulatory network as a function of genome size. The class of transcription regulatory genes consists for the most part of DNA-binding transcription factors that regulate transcription through the binding to specific regulatory motifs in intergenic regions. We can imagine the transcription

* Note that in principle nothing keeps a genome from increasing n_c by more than 1 gene when g is increased by 1 gene. That is, some genes in other categories than c may be removed and replaced by genes in category c.

regulatory network by a set of arrows pointing from each of the regulators to each of the genes that they regulate. The total number of arrows in this network for a genome of size g is the product of the number of genes g times the average number of incoming arrows per gene $\langle r(g)\rangle$, i.e. $\langle r(g)\rangle$ represents the average number of different regulators regulating each gene in a genome of size g. Note that we can also write the total number of arrows as the number of transcription factors $n_{tr}(g)$in a genome of size g times the average number of genes $\langle n(g)\rangle$ that each regulator regulates. We thus have

$$n_{tr}(g)\langle n(g)\rangle = \langle r(g)\rangle g \Leftrightarrow \frac{\langle r(g)\rangle}{\langle n(g)\rangle} \propto g \tag{3}$$

where the equation on the right follows from the quadratic scaling of the number of regulators:$n_{tr}(g)\propto g^2$.To elucidate what the equation on the right implies, let us consider what follows if we assume that either $\langle n(g)\rangle$ = constant or $\langle r(g)\rangle$ = constant. In the first case the average number of genes regulated per transcription factor, i.e., the regulon size, is independent of genome size. In that case the average number of different transcription factors $\langle r(g)\rangle$ regulating each gene must be increasing linearly with genome size, i.e.,$\langle r(g)\rangle\propto g$. If on the other hand $\langle r(g)\rangle$ were constant with genome size, then the average regulon size $\langle n(g)\rangle$ should be decreasing with genome size, i.e, $\langle n(g)\rangle\propto 1/g$. Between these two extremes there is a continuum of solutions where $\langle r(g)\rangle$ increases more slowly, and $\langle n(g)\rangle$ decreases more slowly, but such that still $\langle r(g)\rangle/\langle n(g)\rangle\propto g$.

There is currently very little data to decide if real regulatory networks are closer to the limit where $\langle r(g)\rangle$ increases linearly, or closer to the limit where $\langle n(g)\rangle\propto 1/g$. One piece of indirect evidence is the dependence of the number of operons and the amount of intergenic region on genome size. If the average number of transcription factors $\langle r(g)\rangle$ regulating each gene were to increase with genome size, then one might expect that, as genome size increases, the average operon size should decrease and that the amount of intergenic region per gene should increase. It is of course a nontrivial task to identify the number of operons from genome sequence alone. However, as a proxy we may consider runs of consecutive genes that are located on the same strand of the DNA (see ref. 16 for a method of estimating operon number using this statistic). Since all genes in an operon necessarily have to be transcribed in the same direction, a decrease in operon number would likely be reflected by a decrease in the average length of runs of consecutive genes on the same strand. Figure 4 (upper panel) shows this average length of iso-strand genes as a function of genome size for all bacterial genomes that are currently in the NCBI database of fully-sequenced genomes.

The figure suggests a slight decrease of the length of these runs, consistent with what was reported in reference 16, but there is a large amount of variation and the trend is far from convincing, i.e,r^2=0.23 under simple regression. Note also that the drop in operon size is at most a factor of two between the largest and smallest genomes, while total gene number increases almost a factor of 20. The lower panel shows the average number of intergenic bases per gene as a function of the total number of genes in the genome. In this case a correlation between genome size and the amount of intergenic region is completely absent, i.e, r^2 = 0.0005.Thus these two statistics provide little evidence that $\langle r(g)\rangle$ increases substantially with genome size. However, one cannot exclude the possibility that in small genomes a large proportion of genes is not transcriptionally regulated at all, and that as genome size increases this proportion drops dramatically. This would still lead to a substantial increase of $\langle r(g)\rangle$ with genome size. It does seem plausible, however, that larger genomes may have a larger number of 'specialized' regulons that typically regulate a smaller number of genes compared to the more general regulators that one expects to find in all organisms, and that $\langle n(g)\rangle$ thus decreases with genome size.

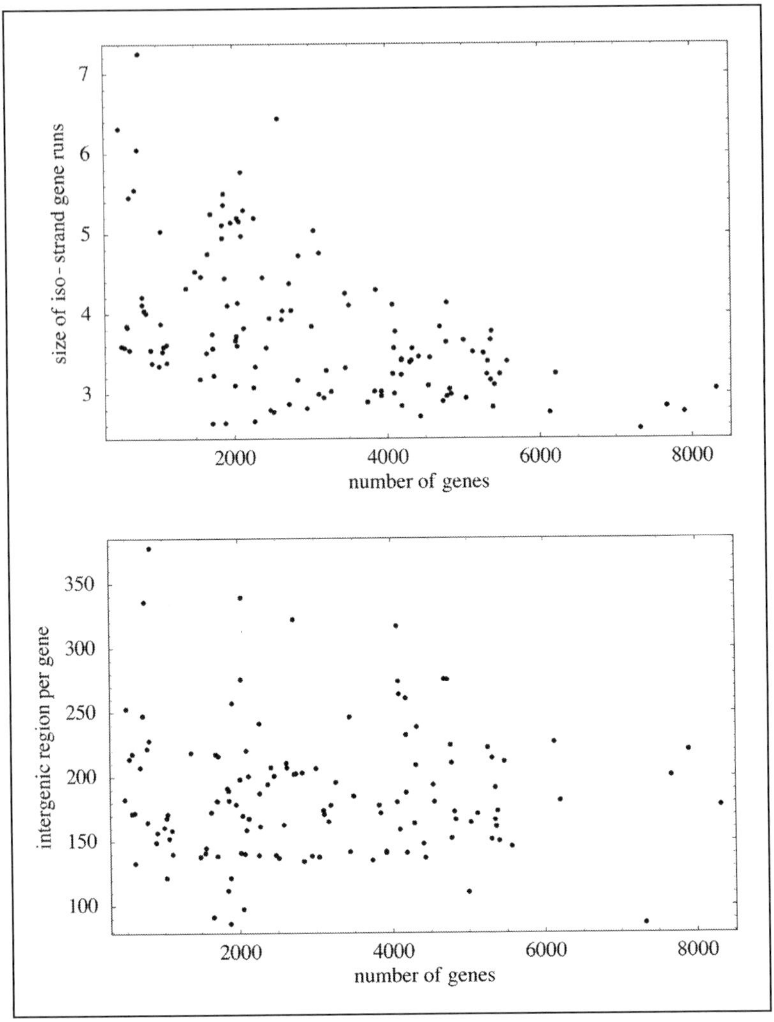

Figure 4. The average length of runs of genes that are transcribed from the same strand (upper panel) and the average number of intergenic bases per gene (lower panel) as a function of the total number of genes in the genome. Each dot corresponds to a bacterial genome from the NCBI database.

Quality of the Fits

The power-laws observed in Figure 2 are observed for the large majority of the 154 ubiquitous functional categories. I assessed the quality of the power-law fits by a measure F that measures the fraction of the variance in the data that is explained by the power-law fit (see Methods). Figure 5 shows the cumulative distribution of F for all 154 ubiquitous functional categories.

As can be seen from Figure 5 about two thirds of the categories have more than 90% of the variance explained by the fit. More than 95% of the categories have more than 80% of the variance explained by the fit. We thus see that most ubiquitous functional categories follow scaling laws like the ones shown in Figure 2.

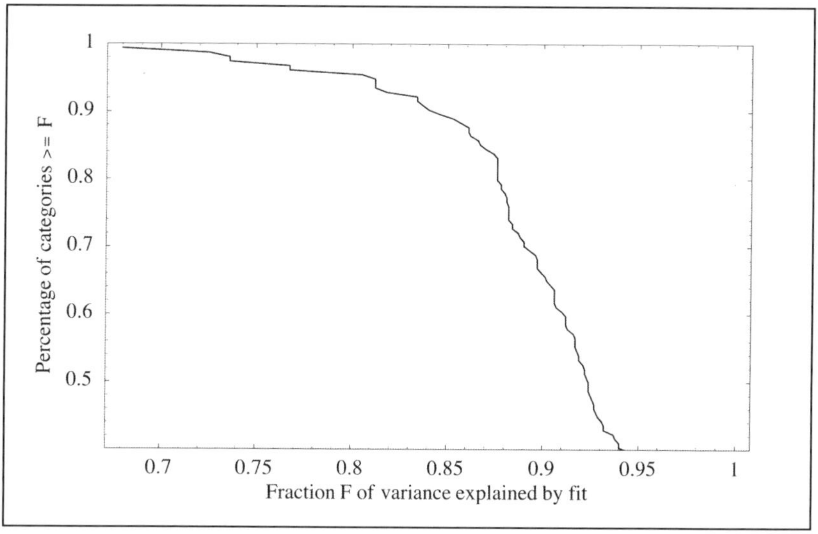

Figure 5. Cumulative distribution of the quality of the power-law fits for the 154 ubiquitous functional categories. The horizontal axis shows the fraction F of the variance in the data explained by the fit, and the vertical axis shows the percentage of categories that have at least a fraction F of their variance explained by the fit.

Principle Component Analysis

Instead of considering the scaling behavior of one functional category at a time, we can consider all functional categories at once. We can consider a 154-dimensional 'function space' in which each axis represents an ubiquitous functional category, and represent the 'functional gene content' of a genome by a point in this 154-dimensional space. That is, if we number the categories 1 through 154, and n_c is the number of genes in category c, then $\vec{n} = (n_1, n_2, \ldots, n_{154})$ represents the functional gene content of the genome as a point in the function space. The set of all sequenced genomes thus forms a scatter in this function space.

We can now ask what the shape is of this cloud in function space. To do this, it is convenient to again consider all axes on logarithmic scales. That is, we consider the scatter of points $\vec{x}$ with $x_c = \log(n_c)$.The results in the previous section showed that, to a good approximation, almost all functional categories obey the linear equations

$$x_c = \alpha_c \log(g) + \beta_c \tag{4}$$

where g is the total number of genes in the genome. If this equation were to hold exactly for all categories, then the numbers x_c and $x_{c'}$, for any two categories c and c', would also be linearly related:

$$x_c = \frac{\alpha_c}{\alpha_{c'}} x_{c'} + \beta_c - \frac{\alpha_c}{\alpha_{c'}} \beta_{c'} \tag{5}$$

Thus, the statement that equation (4) holds for all categories is equivalent to the statement that the scatter of points in function space falls on a straight line. Of course, since equation (4) holds only approximately for each category, the scatter of points falls only approximately on a line. To illustrate this Figure 6 shows three projections of the scatter of points onto three-dimensional subspaces each representing three functional categories.

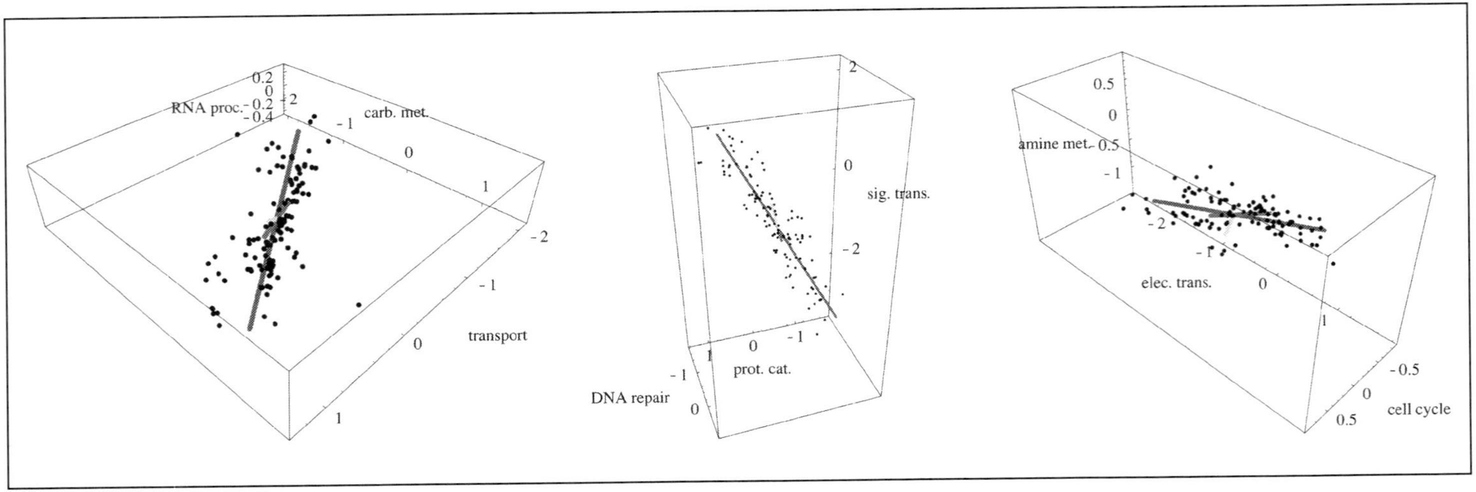

Figure 6. Three 'slices' through the scatter of genomes in function space. Each point corresponds to a genome, and its components along the three axes correspond to the logarithms of the gene numbers in each of the three categories that the axes represent. The scatter is shown with respect to the categories 'carbohydrate metabolism', 'transport', and 'RNA processing' (left panel); 'protein catabolism', 'DNA repair', and 'signal transduction' (middle panel); and 'cell cycle', 'electron transport', and 'amine metabolism' (right panel). The first three principle component axes are shown as the lines in the figures (see below).

That is, each of these three figures shows the scatter of points with respect to only three of the 154 axes. The functional categories that were used for the projections are indicated in each of the plots and in the caption. As can be seen, the scatter of points indeed approximately falls on a straight line in each of these projections.

The extent to which the scatter of points falls on a line can be quantified most directly using principle component analysis.[17] In principle component analysis one aims to represent a scatter of points in a high-dimensional space by a scatter in a much lower dimensional space. To this end it finds an ordered set of orthogonal coordinate axes in the n-dimensional space that have the property that, for any $m \leq n$, the sum of the squared distances of the data points to their projections on the first m axes of the coordinate system is minimized (i.e., no set of m axes has a lower sum of squared distances). That is, the first principle axis is chosen such that the average squared-distance of the data points to this axis is minimized. The second principle axis is chosen such that the average squared-distance of the data points to the surface spanned by the first and second principle axis is minimized. And so on for the further principle components.

The ability of this coordinate system to represent the scatter of points is again measured by the fraction of the variance in the data that is captured by consecutive axes. For the scatter of 116 bacterial genomes in the function space of 154 ubiquitous categories, the first principle axis captures 22% of all the variance in the data. The second principle axis captures 6% of the variance, the third 5% of the variance, etcetera. Thus, the amount of variance explained drops sharply between the first and second principle axis. After that, the amount of variance explained by the data decreases smoothly, dropping between 5 and 10 percent between consecutive axes. As many as 55 axes are needed to cover 95% of the variance in the data. These statistics suggest that only the first principle axis captures an essential characteristic of gene-content in bacterial genomes. Once this largest component of the variance is taken into account, one needs almost as many dimensions as there are genomes to explain the remaining variance in gene-content.

Table 1 summarizes the statistics of the first three principle axes. Each of the axes is a vector $\vec{a}$ whose direction in function space is reflected by the relative sizes of the components a_c. In the leftmost column of Table 1 I have listed, for each axis, the 4 categories with the highest components a_c and the 4 categories with the lowest components a_c (redundant categories were omitted). The values of these components a_c are shown in the second column. The projection of all genomes onto one of the principle axes gives another vector $\vec{v}$ where v_g is the component of genome g in the direction of the principle axis. The third column in Table 1 shows, for each principle axis, the 4 genomes with the highest components and the 4 genomes with the lowest components v_g. The components themselves are shown in column 4. These first three principle axes are also indicated as the red, blue, and green lines in Figure 6.

As can be easily derived from equation (4), the components a_c of the first principle axis correspond precisely to the inferred exponents of Figure 3. Thus, the entire collection of scaling laws is summarized by this single vector. In summary, a large fraction of the variation in functional gene-content among all bacterial genomes can be summarized by a single vector $\vec{a}$ which encodes how the numbers of genes in different functional categories increase and decrease as the total size of the genome varies. That is, as the genome size increases or decreases the numbers of genes in each functional category c increase or decrease as g^{α_c}. The vector $\vec{a}$ thus reflects a basic *functional architecture* of gene-content that holds across all bacterial genomes.

The meaning of the second and third principle axis is less clear, and given that they only capture a relatively small amount of the variance it is not clear that they are very meaningful at all. For both these axes the genomes at the extremes tend to be small parasitic organisms. This suggests that these axes may reflect different types of parasitic lifestyles.

Table 1. Summary of the first three principle axes

Top and Bottom Categories	Comp.	Top and Bottom Genomes	Comp.
		First Principle Axis	
cell communication	1.91	Streptomyces coelicolor	0.15
signal transduction	1.91	Streptomyces avermitilis	0.14
regulation of transcription	1.88	Bradyrhizobium japonicum	0.14
electron transport	1.46	Rhizobium loti	0.14
hydrogen transport	0.25	Buchnera aphidicola	-0.19
protein biosynthesis	0.17	Mycoplasma pneumoniae	-0.22
ATP metabolism	0.10	Ureaplasma parvum	-0.24
rRNA modification	0.06	Mycoplasma genitalium	-0.24
		Second Principle Axis	
porphyrin metabolism	0.28	Mycobacterium leprae	0.16
fatty acid metabolism	0.20	Prochlorococcus marinus	0.16
carboxylic acid biosynthesis	0.19	Wigglesworthia glossinidia brevipalpis	0.16
heterocycle metabolism	0.14	Candidatus Blochmannia floridanus	0.15
DNA alkylation	-0.16	Mycoplasma penetrans	-0.22
epigenetic regulation of gene expression	-0.17	Borrelia burgdorferi	-0.22
nucleoside metabolism	-0.17	Ureaplasma parvum	-0.22
aspartyl-tRNA aminoacylation	-0.17	Mycoplasma pulmonis	-0.3
		Third Principle Axis	
protein secretion	0.31	Borrelia burgdorferi	0.46
cell communication	0.15	Chlamydia trachomatis	0.26
signal transduction	0.13	Treponema pallidum	0.23
DNA dependent DNA replication	0.09	Chlamydia muridarum	0.23
amino acid biosynthesis	-0.16	Streptococcus pneumoniae	-0.13
ribonucleotide metabolism	-0.17	Prochlorococcus marinus	0.14
purine nucleotide metabolism	-0.18	Wigglesworthia glossinidia brevipalpis	-0.15
ATP metabolism	-0.18	Candidatus Blochmannia floridanus	-0.19

Each principle axis is a vector $\vec{a}$ in function space with component a_c the component of the vector in direction of category c. For each principle axis the 4 categories with the highest, and 4 categories with the lowest a_c are shown (omitting redundant categories) in the leftmost column. The second column shows the a_c. The location of each genome in function space can be expressed as a linear combination of principle axes. For each principle axes shown above the 4 genomes with the highest and 4 genomes with the lowest components along the axis are shown in column 3, and the values of the components of these genomes are shown in column 4.

Evolutionary Interpretation

What is the origin of the scaling laws discussed in the previous sections? In this section I will show that the observed scaling laws in fact suggest that there are fundamental constants in the evolutionary dynamics of genomes.

Consider a particular genome with numbers of genes n_c in different functional categories c. Consider next the evolutionary history of this genome. With this I mean that the current genome can be followed back in time through the life of the cell the genome was taken from, through the cell division that produced it, through the life of its ancestral cell, and its ancestors, and so on. In this way the history of the genome can be traced back all the way to an ancestral prokaryotic cell from which all currently existing bacteria stem. During this evolutionary history the numbers n_c have of coursed increased and decreased in ways that are unknown to us. That is, there are (unknown) functions of time $n_c(t)$ that describe the evolution of the numbers of genes in each category c in the genome under study. Similarly, there will be a function $g(t)$ that describes the evolutionary history of the total number of genes in the genome.

One can now ask what constraints the functions $g(t)$ and $n_c(t)$ should obey such that the observed scaling laws hold. To this end it is convenient to write the dynamics of $n_c(t)$ and $g(t)$ in terms of effective duplication and deletion rates. That is, we write

$$\frac{dn_c(t)}{dt} = \beta_c(t)n_c(t) - \delta_c(t)n_c(t) \equiv \rho_c(t)n_c(t) \tag{6}$$

and

$$\frac{dg(t)}{dt} = \beta(t)g(t) - \delta(t)g(t) \equiv \rho(t)g(t) \tag{7}$$

In these equations, $\beta_c(t)$ is the (time-dependent) average duplication rate of genes in category c, $\beta(t)$ is the average duplication rate of all genes, $\delta_c(t)$ is the average deletion rate of genes in category c, and $\delta(t)$ is the average deletion rate of all genes. I have also defined the differences of duplication rates and deletion rates as $\rho_c(t)$ and $\rho(t)$. Notice that, since in the above equations $\rho_c(t)$ and $\rho(t)$ can be arbitrary functions of time, any time-dependent function can still be obtained as the solution of the above equations. Formally solving these equations we obtain

$$n_c(t) = n_c(0)\exp\left(\int_0^t \rho_c(\tau)d\tau\right) = n_c(0)\exp(\langle\rho_c\rangle t) \tag{8}$$

and

$$g(t) = g(0)\exp\left(\int_0^t \rho(\tau)d\tau\right) = g(0)\exp(\langle\rho\rangle t) \tag{9}$$

where I have written the integrals as the **time averages** of ρ_c and ρ times the total time t. Note that these averages are a function of the evolutionary history of the genome under study. If we now express $n_c(t)$ in terms of $g(t)$ we obtain

$$n_c(t) = \frac{n_c(0)}{g(0)^{\langle\rho_c\rangle/\langle\rho\rangle}}\left[g(t)\right]^{\langle\rho_c\rangle/\langle\rho\rangle} \tag{10}$$

Note that, although this equation may appear to already imply the general scaling relations that were found, in fact it is completely general and holds for **any** set of evolutionary histories $n_c(t)$ and $g(t)$. The equation is nothing more than a way of rewriting the relation between $n_c(t)$ and $g(t)$ in terms of the averages $\langle\rho_c\rangle$ and $\langle\rho\rangle$ of this genome's history.

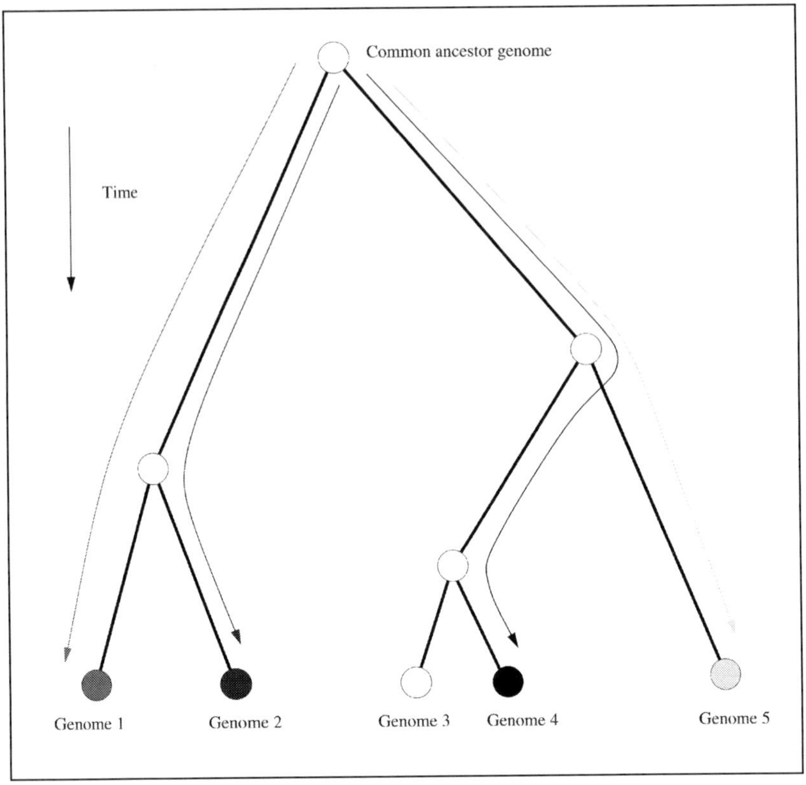

Figure 7. Example of the evolutionary histories of 5 genomes. Each leaf (ball) represents a genome. Each white ball corresponds to a common ancestral genome of two or more of the genomes. The root of the tree is the common ancestor genome of all 5 genomes. Each arrow represents the evolutionary history of a genome.

The observed scaling relations only follow from (10) if the **same** equation holds for all genomes. That is, if the variables $n_c(0)$, $g(0)$ and $\langle\rho_c\rangle/\langle\rho\rangle$ are the same for all evolutionary histories. This requirement is illustrated in Figure 7.

The figure shows the evolutionary histories of a set of genomes in a phylogenetic tree. Each leaf of the tree represents one of the bacterial genomes, and at the root is the common ancestor genome of all the genomes at the leafs. We thus have a separate equation (10) for each of the genomes at the leafs. The initial numbers $n_c(0)$and $g(0)$ in each of these equations are just the numbers of genes in category c and in the whole genome of the common ancestor, and it is therefore clear that the variables $n_c(0)$ and $g(0)$ are indeed the same for all the equations.*

Much more interesting is the requirement that the ratios $\langle\rho_c\rangle/\langle\rho\rangle$ are the same for all evolutionary histories. These different evolutionary histories are indicated as colored lines in Figure 7. The requirement that $\langle\rho_c\rangle/\langle\rho\rangle$ be equal for all evolutionary histories thus demands that if one takes the average of the duplication minus deletion rate of genes in category c over

* One caveat is that we implicitly assume that $n_c(t) > 0$ for all t. A slightly more complex treatment is needed for categories that were not present in the ancestor, or that disappeared and reappeared during the evolutionary history of some genomes.

the entire evolutionary history of a genome, and divides this by the average duplication minus deletion rate of all genes in the genome, then this ratio always comes out the same, independent of what evolutionary history is averaged over. That is, the ratios $\langle\rho_c\rangle/\langle\rho\rangle$ are universal **evolutionary constants** and correspond precisely to the exponents $\alpha_c = \langle\rho_c\rangle/\langle\rho\rangle$ of the observed scaling laws.

So what determines $\langle\rho_c\rangle$ and $\langle\rho\rangle$? It seems reasonable to assume that the rate at which genes are duplicated and deleted mostly reflects the effects of selection. That is, as a first approximation we may assume that the rate at which duplications and deletions are introduced during evolution are approximately the same for all genes but that different genes have different probabilities of being selectively beneficial when duplicated or deleted. The rate $\rho_c(t)$ is then given by some overall rate γ at which duplications and deletions are introduced* times the difference between the fraction of genes $f_c^+(t)$ in the category that would benefit the organism when duplicated and the fraction $f_c^-(t)$ of genes in the category that would benefit the organism when deleted:

$$\rho_c(t) = \gamma\left(f_c^+(t) - f_c^-(t)\right) \tag{11}$$

We then have

$$\alpha_c = \frac{\langle\rho_c\rangle}{\langle\rho\rangle} = \frac{\left\langle f_c^+ - f_c^-\right\rangle}{\left\langle f^+ - f^-\right\rangle} \tag{12}$$

The fractions $f_c^+(t)$ and $f_c^-(t)$ of course depend on the selective pressures of the environment at time t. Thus, as one follows the genome through its evolutionary history, the demand for genes of a certain function fluctuates up and down, and this will be reflected in the fluctuations of $f_c^+(t) - f_c^-(t)$. It seems intuitively clear that the size of the fluctuations is going to depend strongly on the functional class. That is, one would expect that the demand for genes that provide an essential and basic function fluctuates relatively little. For instance, one would expect that even as the selective environment changes it is rare that existing protein biosynthesis genes become dispensable or that duplicates of these genes become desirable. On the other hand, the desirability of transcription regulatory genes and signal transducers is going to depend crucially on the selective environment in which the organism finds itself. It is thus not implausible that, if one averages over sufficiently long evolutionary times that the ratio $\langle f_c^+ - f_c^-\rangle/\langle f^+ - f^-\rangle$ always reaches the same limit, and that this limit is large for highly environment-dependent categories such as transcription regulation, and small for environment-insensitive categories such as protein biosynthesis and replication. The main open question that remains is the origin of the precise numerical values of the ratios $\langle f_c^+ - f_c^-\rangle/\langle f^+ - f^-\rangle$ for different functional categories.

The Exponent for Transcription Regulatory Genes

The exponent α_c for the category of genes involved in transcription regulation is close to 2, with the 99% posterior probability interval running from 1.72 to 1.95. Thus, even though the current data suggests that the exponent is slightly less than 2, it is tempting to think that the scaling might be simply quadratic. This is especially tempting given that this allows one to speculate more easily about the origin of this exponent. That is, it is easier to theorize about a quadratic scaling law then about a scaling law with exponent 1.83. Some

* In reality the rate at which duplications are introduced is unlikely to equal the rate at which delections are introduced. We ignore this complication for notational simplicity. The theoretical development with this complication would be analogous.

theoretical explanations for the quadratic scaling of transcription regulators have recently been put forward.[15] In this section I want to discuss these proposals and contrast them with my own suggestions in this regard in reference 13.

One of the first things that of course come to mind when attempting to explain a scaling of the form $n_c \propto g^2$ is that in a genome with g genes, the number of **pairs** of genes scales precisely as g^2.That is, there are $g(g - 1)/2$ possible interactions in a genome with g genes. Therefore, if one could find and argument that suggests that the number of transcription regulators should be proportional to the number of pairs of genes, this would provide an possible explanation. In a recent paper[15] Croft and coworkers put forward two different models for the observed quadratic scaling of the number of transcription regulatory genes that both in essence argue that the number of regulators should be proportional to the number of potential interactions between genes, i.e., the number of pairs of genes in the genome.

Although it is attractive to seek such a simple combinatorial explanation I believe that a simple survey of what is known about the regulatory role of transcription factors in bacteria shows that such models are in fact highly implausible. If each regulator were to somehow 'correspond' to one or more pairwise interactions of genes then one would expect that the role of most regulatory genes would be to ensure that certain pairs of genes are expressed together or to ensure that certain pairs of genes are not expressed together.

This is, however, not what is observed. To mention just a few known E. coli regulons: the factor crp responds to the concentration of cAMP in the cell, and its main role is to make the activation of many catabolic pathways conditional on the presence or absence of cAMP. The factor lexA responds to single stranded DNA and activates a number of genes that are involved in the repair of DNA damage. PurR responds to the concentrations of hypoxanthine and guanine and represses a set of genes involved in de novo purine biosynthesis. FadR senses the presence of long chain fatty acyl-coA compounds and in response regulates genes that transport and synthesize fatty acids. Finally, tyrR responds to the levels of phenylalanine, tyrosine, and tryptophan and regulates genes that synthesize and transport aromatic amino-acids.

In all these cases, the role of the regulator is to sense a particular signal and to respond to this signal by activating or repressing a set of genes that implement a specific biological function which is related to the signal. In none of these example does it appear that the role of the regulator is to regulate the interactions between pairs of genes. It thus seems that the number of different transcription regulators that the cell has is much more a reflection of the number of different cellular responses to different environments that the cell is capable of.

In reference 13 I provided a qualitative argument for the approximately quadratic scaling of the number of transcription regulatory genes. According to equation (12) the average difference $\langle f_c^+ - f_c^- \rangle$ between the fractions of transcription regulatory genes that would increase fitness when duplicated and those that would increase fitness when deleted is almost twice as large as the average difference $\langle f^+ - f^- \rangle$ between the fractions of all genes that would increase fitness when duplicated and those that would increase fitness when deleted. This requirement will of course be satisfied if both $\langle f_c^+ \rangle \approx 2\langle f^+ \rangle$ and $\langle f_c^- \rangle \approx 2\langle f^- \rangle$. That is, transcription regulators are twice as likely to lead to fitness increase when duplicated as average genes, and transcription regulators are also twice as likely to lead to fitness increase when deleted.

I want to suggest that the origin of the factor 2 in these rates is the switch-like function of transcription factors. Imagine a gene that has just emerged through a gene duplication. Originally the duplicate will be the same as its parent. At that point, the main change caused by this duplication that may affect fitness is a change in the dosage of the gene, i.e., from one to two copies. One would expect that the probability for this dosage change to have strong deleterious effects is approximately equal for transcription regulators as for genes in general. If the duplicated gene is to get fixed in the genome on a longer time scale, then a process of mutation and selection should modify the duplicated gene into a gene that increases the fitness of the organism. I suggest that the probability for this process to be successful is twice as high in transcription regulators as in nonregulatory genes. In nonregulatory genes the only avenue for beneficial change is a change in the molecular function of the gene, e.g., a metabolic gene may evolve to catalyze a new chemical reaction. Transcription regulators, however, may evolve to respond to a signal which the cell was previously insensitive to. They may evolve to affect the state of the cell both when this signal is present and when the signal is absent. Since this gives twice as many opportunities for a beneficial change, one may expect that there is twice as much probability of success.

A similar argument holds for the rate of deletion. When deleting a nonregulatory gene, the cell simply changes to a state without the gene. When a transcription regulator is removed, one effectively removes a 'switch' from the genome: the cell becomes insensitive to the signal that the regulator responded to. But there are again two 'ways' of implementing this insensitivity. The genes regulated in response to the signal may be turned constitutively on, or constitutively off. Thus there are two independent ways of removing a transcription factor, and one would thus expect the effective rate of transcription factor deletion to be twice the rate of deletion of general genes.

These arguments are of course highly speculative and may well turn out to be incorrect. However, they at least seem consistent with what we know about transcription regulation in bacteria and the process of evolution through gene duplication and deletion. Note also that very similar arguments as the ones just presented for transcription regulatory genes can be put forward for the category of signal-transducing genes. These are indeed also observed to scale approximately quadratically with genome size. Thus, the argument just presented explains the scaling exponent for both the transcription regulation category and the signal transduction category. However, I suspect that it will be impossible to come to any solid conclusions regarding the cause of the approximately quadratic scaling for transcription regulators and signal transducers until we have better data on the genome-wide topology of the transcription regulatory and signal transduction networks in bacteria of differing sizes.

Finally, I note that most of the functional categories have exponents that do not appear to equal small integers or even simple rational numbers. It is thus clear that simple arguments such as the ones discussed above will not be capable of explaining these exponents. What determines the average $\langle f_c^+ - f_c^- \rangle / \langle f^+ - f^- \rangle$ for these categories is a fascinating question that at this point is completely open.

Methods

The Number of Genomes as a Function of Time

The data for Figure 1 were obtained by extracting the submission dates of all the microbial genomes in genbank at 'ftp://ftp.ncbi.nih.gov/genbank/genomes/Bacteria' from the 'gbs' file for each genome. When there were multiple submission dates the earliest date was taken. The logarithm of the number of genomes as a function of time was then fitted to a straight line using standard least-square regression. This least-square fit gives the number of genomes $n(t)$ as a function of time (in years) as $n(t) = 2^{(t-1993.4)/1.38}$. The exponential provides a reasonable fit, i.e. $r^2 = 0.98$ even though the data clearly suggest a decrease in the rate at which new genomes appear for the last 1 to 1.5 years.

Gathering Functional Gene-Content Statistics

The numbers of genes in different functional categories for each sequenced genome were obtained in the following way. Interpro annotations[18] of fully-sequenced genomes were obtained from the European Bioinformatics Institute.[19] Functional categories were taken from the gene ontology biological process hierarchy.[14] A mapping from Interpro to GO-categories was also obtained from the gene ontology website. Using this mapping I gathered, for each GO-category, all Interpro entries that map to it or to one of its descendants in the biological process hierarchy. I then counted, for each gene in each fully-sequenced genome and each GO-category, the number of Interpro hits h that the gene has to Interpro categories associated with the GO-category. A gene with many independent Interpro hits that are associated with the same GO-category is of course more likely to be a member of the GO-category than a gene with only a single hit. To quantify this, I chose the probability for a gene to be a member of a GO-category to which it has h hits to be $1 - \exp(-\lambda h)$, with $\lambda = 3$. The results presented in this chapter are largely insensitive to changes in λ (see the discussion in ref. 13).

In this way the number of genes associated with each GO-category in each genome was counted. I then selected all GO categories that have a nonzero number of counts in all bacterial genomes. There are 154 such ubiquitous GO-categories. I also counted the total number of genes that have any annotation at all for each genome. These are defined as genes that have at least one Interpro hit. Further discussion of this annotation procedure and its robustness can be found in reference 13.

Power-Law Fitting

In order to fit the data to power-law distributions I used a Bayesian procedure that is described in reference 13. The main advantage of this fitting procedure with respect to simple regression is that the results are explicitly rotationally invariant. That is, the best line that the fitting procedure produces doesn't depend on the orientation of the coordinate axes with respect to the scatter of data points.

The result is that the posterior distribution $P(\alpha \mid D)d\alpha$ for the slope α of the line, given the data D, is given by

$$P(\alpha \mid D)d\alpha = C\frac{\left(\alpha^2+1\right)^{(n-3)/2}d\alpha}{\left(\alpha^2 s_{xx} - 2\alpha s_{yx} + s_{yy}\right)^{(n-1)/2}} \tag{13}$$

where n is the number of genomes in the data, s_{xx} is the variance in x-values (logarithms of the total gene numbers), s_{yy} the variance in y-values (logarithms of the number of genes in the category), s_{yx} is the covariance, and C is a normalizing constant. For each of the 154 ubiquitous categories we then calculated the 99% posterior probability interval for the slope from equation (13).

Quality of the Power-Law Fits

To calculate the quality of the power-law fits, I first log-transform the data points (g_i,n_i) to $(x_i,y_i) = (\log(g_i),\log(n_i))$. The best power-law fit is a straight line $y = \alpha x+\beta$ in the (x,y) plane. For each data point (x_i,y_i) I then find the distance $d_i(l)$ to this line, and the distance $d_i(c)$ to the center of the scatter of points. That is, with $\langle x\rangle$ the average of the x-values, and $\langle y\rangle$ the average of the y-values, the distance $d_i(c)$ is given by

$$d_i(c) = \sqrt{\left(\left(x_i - \langle x\rangle\right)^2 + \left(y_i - \langle y\rangle\right)^2\right)} \tag{14}$$

and the distance to the line is given by

$$d_i(l) = \sqrt{\frac{\left(y_i - \alpha x_i - \beta\right)^2}{\alpha^2+1}} \tag{15}$$

I then define the quality of the fit F as the fraction of the variance that is explained by the fit.

$$F = 1 - \frac{\sum_i [d_i(l)]^2}{\sum_i [d_i(c)]^2} \tag{16}$$

References

1. Huynen M, van Nimwegen E. The frequency distribution of gene family sizes in complete genomes. Mol Biol Evol 1998; 15:583-589.
2. Gerstein M. A structural census of genomes: Comparing bacterial, eukaryotic, and archael genomes in terms of protein structure. J Mol Biol 1997; 274:562-576.
3. Luscombe NM, Qian J, Zhang Z et al. The dominance of the population by a selected few: Power-law behavior applies to a wide variety of genomic properties. Genome Biol 2002; 3(8).
4. Koonin EV, Wolf YI, Karev GP. The structure of the protein universe and genome evolution. Nature 2002; 420:218-222.
5. Uetz P, Giot L, Cagney G et al. A comprehensive analysis of protein-protein interactions in saccharomyces cerevisiae. Nature 2000; 403:623-627.
6. Ito T, Chiba T, Ozawa R et al. A comprehensive two-hybrid analysis to explore the yeast protein interactions. PNAS USA 2001; 98:4569-4574.
7. Maslov S, Sneppen K. Specificity and stability in topology of protein networks. Science 2002; 296:910-913.
8. Stuart JM, Segal E, Koller D et al. A gene-coexpression network for global discovery of conserved genetic modules. Science 2003; 302:249-255.
9. Guelzim N, Bottani S, Bourgine P et al. Topological and causal structure of the yeast transcriptional regulatory network. Nat Genet 2002; 31:60-63.
10. Milo R, Shen-Orr S, Itzkovitz S et al. Network motifs: Simple building blocks of complex networks. Science 2002; 298:824-827.
11. Jeong H, Tombor B, Albert R et al. The large-scale organization of metabolic networks. Nature 2000; 407:651-654.
12. Wagner A, Fell D. The small world inside large metabolic networks. Proc Roy Soc London Series B 2001; 268:1803-1810.
13. van Nimwegen E. Scaling laws in the functional content of genomes. Trends in Genet. 2003; 19:479-484.
14. The gene ontology consortium. Gene ontology: Tool for the unification of biology. Nat Genet 2000; 25:25-29.
15. Croft LJ, Lercher MJ, Gagen MJ et al. Is prokaryotic complexity limited by accelerated growth in regulatory overhead. Genome Biology 2003; 5:P2.
16. CherryJL. Genome size and operon content. J theor Biol 2003; 221:401-410.
17. Joliffe IT. Principle component analysis. New York: Springer-Verlag, 1986.
18. Apweiler R, Attwood TK, Bairoch A et al. The interpro database, an integrated documentation resource for protein families, domains and functional sites. Nucl Acids Res 2001; 29:37-40.
19. Apweiler R, Biswas M, Fleischmann W et al. Proteome analysis database: Online application of interpro and clustr for the functional classification of proteins in whole genomes. Nucl Acids Res 2001; 29:44-48.

Index

G

H

I

L

M

N

O

P

R

S

T

U